AF358483

Novel Drug Delivery Systems: Design, Evaluation and Application

Novel Drug Delivery Systems: Design, Evaluation and Application

Guest Editors

Yongtai Zhang
Zhu Jin

Basel • Beijing • Wuhan • Barcelona • Belgrade • Novi Sad • Cluj • Manchester

Guest Editors

Yongtai Zhang
Shanghai University of
Traditional Chinese Medicine
Shanghai
China

Zhu Jin
Shanghai Jiao Tong University
Shanghai
China

Editorial Office
MDPI AG
Grosspeteranlage 5
4052 Basel, Switzerland

This is a reprint of the Special Issue, published open access by the journal *Biomedicines* (ISSN 2227-9059), freely accessible at: https://www.mdpi.com/journal/biomedicines/special_issues/C06A490LAM.

For citation purposes, cite each article independently as indicated on the article page online and as indicated below:

Lastname, A.A.; Lastname, B.B. Article Title. *Journal Name* **Year**, *Volume Number*, Page Range.

ISBN 978-3-7258-2629-2 (Hbk)
ISBN 978-3-7258-2630-8 (PDF)
https://doi.org/10.3390/books978-3-7258-2630-8

Contents

About the Editors

Yongtai Zhang

Dr. Yongtai Zhang is currently a professor at Shanghai University of Traditional Chinese Medicine (SUTCM), where he received his PhD in Traditional Chinese Medicine (TCM) in 2012, and has worked as a visiting scholar at the University of Michigan in the United States. His main research interests are in novel drug delivery systems for TCM, especially in transdermal drug delivery technologies and formulations. Dr. Zhang has published more than 120 papers and more than 10 monographs and has been listed on the Stanford University Top 2% Scientists List in 2023 and 2024.

Zhu Jin

Zhu Jin is an Assistant Researcher at the School of Pharmacy of Shanghai Jiao Tong University, where he obtained his Ph.D. in April 2020. His research focuses on the combination of innovative drug delivery systems and non-vascular stents for treating non-vascular tumors and stenosis. He is particularly interested in developing combination therapies that integrate advanced additive manufacturing technology with a "lesion-tailored" design approach, aiming to create sustained-release systems and "drug-device combination" implants for efficient and precise drug delivery. Additionally, Jin serves as a committee member of the Young Committee of Pharmaceutics at the Shanghai Pharmaceutical Society and acts as a guest editor for several journals, including *Frontiers in Chemistry*, *Biomedicines*, and *Journal of Functional Biomaterials*.

Article

Drug-Delivery Silver Nanoparticles: A New Perspective for Phenindione as an Anticoagulant

Stoyanka Nikolova [1,*], Miglena Milusheva [1,2], Vera Gledacheva [3], Mehran Feizi-Dehnayebi [4], Lidia Kaynarova [5], Deyana Georgieva [5], Vassil Delchev [6], Iliyana Stefanova [3], Yulian Tumbarski [7], Rositsa Mihaylova [8], Emiliya Cherneva [9,10], Snezhana Stoencheva [11,12] and Mina Todorova [1]

[1] Department of Organic Chemistry, Faculty of Chemistry, University of Plovdiv, 4000 Plovdiv, Bulgaria; miglena.milusheva@uni-plovdiv.bg or miglena.milusheva@mu-plovdiv.bg (M.M.); minatodorova@uni-plovdiv.bg (M.T.)

[2] Department of Bioorganic Chemistry, Faculty of Pharmacy, Medical University of Plovdiv, 4002 Plovdiv, Bulgaria

[3] Department of Medical Physics and Biophysics, Faculty of Pharmacy, Medical University of Plovdiv, 4002 Plovdiv, Bulgaria; vera.gledacheva@mu-plovdiv.bg (V.G.); iliyana.stefanova@mu-plovdiv.bg (I.S.)

[4] Department of Chemistry, Faculty of Science, University of Sistan and Baluchestan, Zahedan P.O. Box 98135-674, Iran; m.feizi@sutech.ac.ir

[5] Department of Analytical Chemistry and Computer Chemistry, Faculty of Chemistry, University of Plovdiv, 4000 Plovdiv, Bulgaria; l.kainarova@uni-plovdiv.bg (L.K.); georgieva@uni-plovdiv.bg (D.G.)

[6] Department of Physical Chemistry, Faculty of Chemistry, University of Plovdiv, 4000 Plovdiv, Bulgaria

[7] Department of Microbiology, Technological Faculty, University of Food Technologies, 4002 Plovdiv, Bulgaria; tumbarski@abv.bg

[8] Laboratory of Experimental Chemotherapy, Department "Pharmacology, Pharmacotherapy and Toxicology", Faculty of Pharmacy, Medical University, 1431 Sofia, Bulgaria; rositsa.a.mihaylova@gmail.com

[9] Department of Chemistry, Faculty of Pharmacy, Medical University of Sofia, 2 Dunav Str., 1000 Sofia, Bulgaria; cherneva@pharmfac.mu-sofia.bg

[10] Organic Chemistry with Centre of Phytochemistry, Bulgarian Academy of Sciences, Acad. G. Bonchev Str., BI 9, 1113 Sofia, Bulgaria

[11] University Hospital "Sveti Georgi" EAD, 4002 Plovdiv, Bulgaria

[12] Department of Clinical Laboratory, Faculty of Pharmacy, Medical University of Plovdiv, 4002 Plovdiv, Bulgaria

* Correspondence: tanya@uni-plovdiv.bg

Citation: Nikolova, S.; Milusheva, M.; Gledacheva, V.; Feizi-Dehnayebi, M.; Kaynarova, L.; Georgieva, D.; Delchev, V.; Stefanova, I.; Tumbarski, Y.; Mihaylova, R.; et al. Drug-Delivery Silver Nanoparticles: A New Perspective for Phenindione as an Anticoagulant. *Biomedicines* **2023**, *11*, 2201. https://doi.org/10.3390/biomedicines11082201

Academic Editors: Yongtai Zhang and Zhu Jin

Received: 29 June 2023
Revised: 29 July 2023
Accepted: 31 July 2023
Published: 4 August 2023

Abstract: Anticoagulants prevent the blood from developing the coagulation process, which is the primary cause of death in thromboembolic illnesses. Phenindione (PID) is a well-known anticoagulant that is rarely employed because it totally prevents coagulation, which can be a life-threatening complication. The goal of the current study is to synthesize drug-loaded Ag NPs to slow down the coagulation process. Methods: A rapid synthesis and stabilization of silver nanoparticles as drug-delivery systems for phenindione (PID) were applied for the first time. Results: Several methods are used to determine the size of the resulting Ag NPs. Additionally, the drug-release capabilities of Ag NPs were established. Density functional theory (DFT) calculations were performed for the first time to indicate the nature of the interaction between PID and nanostructures. DFT findings supported that galactose-loaded nanostructure could be a proper delivery system for phenindione. The drug-loaded Ag NPs were characterized in vitro for their antimicrobial, cytotoxic, and anticoagulant activities, and ex vivo for spasmolytic activity. The obtained data confirmed the drug-release experiments. Drug-loaded Ag NPs showed that prothrombin time (PT, sec) and activated partial thromboplastin time (APTT, sec) are approximately 1.5 times longer than the normal values, while PID itself stopped coagulation at all. This can make the PID-loaded Ag NPs better therapeutic anticoagulants. PID was compared to PID-loaded Ag NPs in antimicrobial, spasmolytic activity, and cytotoxicity. All the experiments confirmed the drug-release results.

Keywords: silver nanoparticles; galactose-assisted; phenindione (2-phenyl-1,3-indandione); density functional theory (DFT); anticoagulant; thrombosis; antimicrobial; cytotoxicity; spasmolytic

1. Introduction

The vascular system must be in a homeostatic state for coagulation and fibrinolytic processes to function properly. These processes are crucial to the protective physiological systems of the organism. In healthy individuals, microscopic blood clots are frequently required for the repair of vascular injuries. The fibrinolytic system effectively eliminates insoluble clots by breaking them down into soluble fragments through proteolytic digestion and repairing the damaged vessel's endothelial surface [1]. Anticoagulants prevent the blood from developing the coagulation process. Anticoagulant therapy is focused on preventing blood vessel clot formation, which is the primary cause of death in thromboembolic illnesses [2–4]. Orally ingestible medications are used in standard therapeutic protocols for the prophylaxis and treatment of venous thrombosis and to lower the risk of recurrent myocardial infarction [5,6]. In order to achieve more effective therapeutic results, combination therapy with anticoagulants and platelet aggregation inhibitors is routinely utilized in high-risk patients [7,8]. However, these medications have some undesirable side effects, such as aspirin-related gastrointestinal ulcers, allergy-related bleeding, or warfarin-induced skin necrosis [6,9]. Furthermore, the effectiveness of these medications is still not adequate. Therefore, medicinal chemists continue trying to find new drug candidates in this therapeutic area [10]. Among the known oral anticoagulants, coumarin, warfarin, and indanone derivatives are well distributed. Their therapeutic action depends on their ability to suppress the formation of several functional factors of blood coagulation in the liver [1,11].

Numerous medications have an impact on platelet and coagulation activity. They might work by amplifying the innate inhibition of coagulation, especially at the level of thrombin and factor Xa. Many oral anticoagulants work by inhibiting vitamin K, which prevents the formation of factors II, VII, and X, protein C, and other proteins that are necessary for blood clotting. The more popular oral anticoagulants are vitamin K-like coumarin derivatives, such as warfarin, dicoumarides, and phenindione (2-phenyl-1,3-indandione, PID). Clinically, warfarin is used to reduce the incidence of thromboembolism associated with prosthetic heart valves and atrial fibrillation. PID's effects are comparable to those of warfarin, but it is no longer often used due to the risk of serious side effects [1,11–14].

Indane and its analogues present a variety of biological applications by serving as pharmacophores and by providing scaffolds for the rational design of medications deliberated at diverse biological targets [15]. The indanes' bioactive profile stimulated the creation of therapeutically effective medications and therapeutically useful molecules. Indanes enhance the molecular interactions with target peptides and enzymes, enabling the efficient modulation of enzyme expression to yield the optimal pharmacological effect [16]. Indanes are effective scaffolds for the synthesis of anticancer and anti-inflammatory compounds [17–19]. The bioactive compounds based on these scaffolds allow for the targeting of composite metastatic pathways, providing an additional benefit for treating malignancies with multiple subtypes. Similarly, indanes and indanones form the core structure of several bioactive natural compounds and secondary metabolites with radical scavenging properties that serve as a blueprint for the structural optimization of bioactive indane-based synthetic compounds [20].

The indanone nucleus, its analogs, and derivatives offer intriguing biological uses and medicinally significant molecules that are therapeutically effective. Indanone-based medications provide an expeditious profile in the current drug development paradigm focused on anti-inflammatory and anti-Alzheimer therapies. Indane nucleus is a successful medication for Parkinson's disease treatment. The indanone structural motif is applied in the treatment of pulmonary embolism, mural thrombosis, and cardiomyopathy or serves as an anticoagulant [21]. Therefore, indanes are strong candidates for molecules in the modern drug development paradigm.

Silver nanoparticles (Ag NPs), on the other hand, possess anticoagulant and anti-inflammatory activity, which makes them ideal candidates for biological applications [22–30]. Finding an environmentally friendly method to make NPs is the main challenge in contemporary synthetic organic chemistry [31–33]. The noble metal cations can be converted to their nano-forms using biological procedures that utilize various biological resources as reducing and stabilizing agents [34–37]. Polymer-based NPs were synthesized due to the spectrum of biological interest [38]. Recently, the application of carbohydrates has become a popular area in NP synthesis [39–41]. Sugars as reducing agents for the green synthesis of metal NPs have been applied in the last ten years [42–46]. Many mono-, di-, and polysaccharides were used as reducing and capping agents for the preparation of noble metal NPs due to their sustainability, abundance, low cost, harmlessness, renewability, biodegradability, and compatibility with biological systems [47–51]. The capping agents have an effective role in NP's growth and control of their size but could be hazardous [52], causing many doubts about their use in biological applications [53].

Bearing in mind the structural moiety of indanones and Ag NPs and their biological activities, novel Ag NPs as drug-delivery systems for PID were obtained. The synthesized drug-loaded Ag NPs could be effective as a new therapeutic anticoagulant in both thrombosis treatment and medical device coating.

2. Materials and Methods

All solvents and reagents were purchased from Merck (Merck Bulgaria EAD, Sofia, Bulgaria). The melting point was determined on a Boetius hot stage apparatus and is uncorrected. Elemental analyses were performed with a TruspecMicro (LECO, Mönchengladbach, Germany). PID was characterized using IR, ^{1}H-NMR, and ^{13}C-NMR. IR spectra were determined on a VERTEX 70 FT-IR spectrometer (Bruker Optics, Ettlingen, Germany). ^{1}H-NMR and ^{13}C-NMR spectra were recorded on a Bruker Avance III HD 500 spectrometer (Bruker, Billerica, MA, USA) at 500 MHz (^{1}H-NMR) and 125 MHz (^{13}C-NMR), respectively. Chemical shifts are given in relative ppm and were referenced to tetramethylsilane (TMS) ($\delta = 0.00$ ppm) as an internal standard; the coupling constants are indicated in Hz. The NMR spectra were recorded at room temperature (ac. 295 K). The purity was determined by TLC using several solvent systems of different polarities. TLC was carried out on precoated 0.2 mm Fluka silica gel 60 plates (Merck Bulgaria EAD), using chloroform:n-hexane:methanol: acetone = 4:4:1:1 as a chromatographic system.

2.1. Synthetic Methods

2.1.1. Synthesis of 2-Phenyl-1,3-Indandione

2-phenyl-1,3-indandione is synthesized using a known procedure, via condensation of phenylacetic acid **2** with phthalic anhydride **1** to phenylmethylenphthalide **3**, and further rearrangement in the presence of sodium ethoxide to PID **4** [54] (Scheme 1).

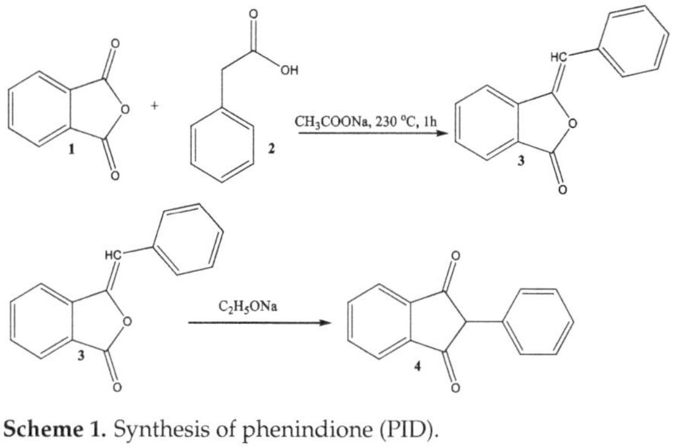

Scheme 1. Synthesis of phenindione (PID).

2-phenyl-1H-indene-1,3(2H)-dione (**4**): ^{1}H-NMR: 4.29 (s, 1H, CH), 7.21–7.22 (m, 2H, Ar), 7.31–7.39 (m, 3H, Ar), 7.91–7.98 (dd, J = 10, 5, 2H, Ar), 8.06–8.10 (dd, J = 10, 5, 2H, Ar); ^{13}C-NMR: 198.3 (CO), 142.7, 136.0, 133.2, 129.0, 128.8, 123.8, 59.8 (CH). IR [KBr]: ν, cm^{-1}

3088, 3078, 3025, 3006 (ν C-H (Ar-H)), 2888 (ν (CH), methine group), 1746, 1705 (ν (C=O)), 1582, 1498, 1473, 1453 ν(CC) Ar, 1344 (δ(CH), methine group), 768 (γ C-H(Ar-H)); Anal. calcd. for $C_{15}H_{10}O_2$: C 81.07; H 4.54; Found: C 81.10; H 4.57 (Supplementary Materials Figures S1–S3).

2.1.2. Synthesis of Galactose-Assisted PID-Loaded Ag NPs

A 0.01 M $AgNO_3$ solution was added to 1.25 g (0.007 mol) of galactose that had been dissolved in 25 mL of water and refluxed for 2 min. Different concentrations of PID were added to the solutions to examine how the ratio of Ag NPs to the drug molecule would impact drug release and biological activity. Galactose and $AgNO_3$ concentrations were constant during the course of the trials, whereas medication concentrations varied based on the ratio. Drug molecules were arranged in the following ratios to Ag NPs: 1:1, 1:5, 1:10, 1:20, and 1:50 (concentration range from 4×10^{-4} to 2×10^{-2} mol/L). In about 5 min, the color of the solution turned pale yellow, indicating the formation of Ag NPs.

2.2. Characterization of the Ag NPs. Analytical Techniques

The solution was utilized for UV-Vis, TEM, single particle ICP-MS (sp-ICP-MS), dynamic light scattering (DLS), and zeta potential. The suspension was centrifuged at 5000 rpm for 15 min, filtered (0.22 m, Chromafil®, Macherey-Nagel, Düren, Germany), and the precipitate was used for Fourier transform infrared spectra (FTIR) and X-ray diffraction (XRD) analyses.

2.2.1. UV-Vis Spectra

UV–Vis spectra were recorded in the range between 320 and 800 nm using a Cary-60 UV-Vis spectrophotometer (Agilent Technologies, Santa Clara, CA, USA). The produced nanoparticle dispersions' absorbance was measured in a quartz cuvette with a 1 cm optical path.

2.2.2. FTIR Spectra

IR spectra were determined on an FT-IR spectrometer VERTEX 70. The spectra were collected in the range from 600 cm^{-1} to 4000 cm^{-1} with a resolution of 4 nm and 20 scans. The instrument is equipped with a diamond-attenuated total reflection (ATR) accessory. The IR spectra were analyzed with the OPUS-Spectroscopy Software, Bruker (Version 7.0, Bruker, Ettlingen, Germany).

2.2.3. spICP-MS

A 7700 Agilent ICP-MS spectrometer equipped with a MicroMist™ nebulizer and Peltier-cooled double-pass spray chamber was used for the characterization of silver nanoparticles at 107 amu. The ICP-MS operating parameters are as follows: RF power—1.55 kW; sample flow rate—0.322 mL min^{-1}; carrier Ar gas flow rate—1.2 L min^{-1}; acquisition time 60 s; dwell time 5 ms; and transport efficiency 0.032. The transport efficiency was determined by using the particle-size method [55]. Ultrapure water (UPW) was used throughout the experiments (PURELAB Chorus 2+ (ELGA Veolia, High Wycombe, UK) water purification system). For the sonication of silver colloids, an ultrasonic bath (Kerry US, Burlingame, CA, USA) was used. Reference materials (RM) of citrate-stabilized silver dispersions Ag NPs (Sigma-Aldrich, Saint Louis, MO, USA) with mean size 40 ± 4 nm and total mass concentration of silver 0.02 mg mL^{-1} were used in this study for transport efficiency determination and calibration.

A standard solution of Ag 9.974 ± 0.041 mg L^{-1} in 2% HNO_3, (CPAchem Ltd., Bogomilovo, Bulgaria) was used for the preparation of ionic standards.

2.2.4. TEM

The morphology and size of drug-loaded Ag NPs were also observed using TEM. A drop of the nanosuspension was placed on a 200 mesh formvar-coated copper grid and allowed to dry for 24 h. Images were obtained using Talos F200C G2 Transmission Electron

Microscope (Talos 1.15.3, Thermo Fisher Scientific, Waltham, MA, USA) operating at 200 kV and analyzed using Velox Imaging Software (Velox 2.15.0.45, Waltham, MA, USA).

2.2.5. DLS and Zeta Potential

DLS measurements were carried out on a Brookhaven BI-200 goniometer with vertically polarized incident light at a wavelength of $l = 632.8$ nm supplied by a He–Ne laser operating at 35 mW and equipped with a Brookhaven BI-9000 AT digital autocorrelator. The scattered light was measured for dilute aqueous dispersions in the concentration range 0.056–0.963 mg mL^{-1} at 25, 37, and 65 °C. Measurements were made at angles ζ in the range of 50–130°. The system allows measurements of ζ-potential in the range from -200 mV to $+200$ mV. All analyses were performed in triplicate at 25 °C.

2.2.6. X-ray Diffraction (XRD)

The level of crystallinity of the synthesized nanoparticles was studied by using X-ray powder diffractometry. Using a SIEMENS D500 X-ray powder diffractometer (KS Analytical Systems, Aubrey, TX, USA), the diffraction patterns of Ag NPs (blank) and drug-loaded Ag NPs were recorded at a 2θ range from 10° to 80°. All measurements were carried out at a 35 kV voltage and a 25 mA current. Using a Cu-anticathode (K1), monochromatic X-rays (1.5406) were produced.

2.3. In Vitro Drug Release

Using the dialysis bag approach, in vitro drug release was evaluated. A dialysis membrane was hydrated in distilled water for 24 h (MWCO 12 kDa, Sigma-Aldrich, St. Louis, MO, USA). Drug-loaded Ag NPs (equivalent to the amount of PID in ratios of 1:1, 1:5, 1:10, 1:20, and 1:50) were dispersed in 10 mL of phosphate-buffered saline (PBS) and then transferred to the dialysis bag, closed with a plastic clamp. Each bag was placed into a beaker with 40 mL of PBS (dialysis medium, pH 7.4). Aliquots with 2 mL of each dialysis medium were taken for measurements, and then the fresh medium was added at specified intervals. The drug release experiment lasted 24 h. The mean results of triplicate measurements and standard deviations were reported. The solution in the Falcon tube was shaken before each evaluation of UV–visible absorbance. For data analysis, PID's maximum absorption band values ($\lambda = 274, 230$, and 204 nm) were employed. Additionally, drug-only controls (one drug control for each ratio) were made.

2.4. Computational Perspective

In the current study, the 3D structure of PID and the surface of nanoparticle containing three galactose molecules were built and optimized by Gauss View 6.0 and Gaussian 09W package [56] for delivering the PID as an anticoagulant. The density functional theory (DFT) approach was carried out utilizing the B3LYP method with the 6–311 g++(d,p) basic set for all atoms. All these calculations were conducted at the ground state in the gas phase. The harmonic frequencies of all the structures were checked for positivity to determine that they were at their true minimum. The aim was to obtain optimized geometries, molecular electrostatic potential (MEP) surfaces, HOMO (highest occupied molecular orbital)-LUMO (lowest unoccupied molecular orbital) analysis and determine the type of interaction between PID and nanoparticle-containing three galactoses. Using the following formula, the E_{ads} (adsorption energy) of PID on the surface of nanoparticle containing galactose was estimated [57]:

$$E_{ads} = (E_{PID \ on \ nanostructure}) - (E_{nanostructure} + E_{PID})$$

Herein, $E_{PID \ on \ nanostructure}$, $E_{nanostructure}$, and E_{PID} are denoted as the total energy of PID on the nanostructure, the primal energy of free nanostructure, and the primal energy of free PID, respectively.

2.5. Microbiological Tests

Tested Microorganisms

Twenty tested microorganisms, including six Gram-positive bacteria (*Bacillus subtilis* ATCC 6633, *Bacillus amyloliquefaciens* 4BCL-YT, *Staphylococcus aureus* ATCC 25923, *Listeria monocytogenes* NBIMCC 8632, *Enterococcus faecalis* ATCC 19433, and *Micrococcus luteus* 2YC-YT), six Gram-negative bacteria (*Salmonella enteritidis* ATCC 13076, *Klebsiella* sp.—clinical isolate, *Escherichia coli* ATCC 25922, *Proteus vulgaris* ATCC 6380, *Pseudomonas aeruginosa* ATCC 9027, *Salmonella typhimurium* NBIMCC 1672), two yeasts (*Candida albicans* NBIMCC 74, *Saccharomyces cerevisiae* ATCC 9763) and six fungi (*Aspergillus niger* ATCC 1015, *Aspergillus flavus*, *Penicillium* sp., *Rhizopus* sp., *Mucor* sp.—plant isolates, *Fusarium moniliforme* ATCC 38932) from the collection of the Department of Microbiology at the University of Food Technologies—Plovdiv, Bulgaria, were selected for the antimicrobial activity test.

Culture media

Luria-Bertani agar medium supplemented with glucose (LBG agar)

LBG agar was prepared according to the manufacturer's (Laboratorios Conda S.A., Madrid, Spain) prescription: 50 g of LBG-solid substance mixture (containing 10 g tryptone, 5 g yeast extract, 10 g NaCl, 10 g glucose, and 15 g agar) was dissolved in 1 L of deionized water (pH 7.5), and the then medium was autoclaved at 121 °C for 20 min.

Malt extract agar (MEA)

MEA was prepared by the manufacturer's (HiMedia®, Mumbai, India) prescription: 50 g of the MEA-solid substance mixture (containing 30 g malt extract, 5 g mycological peptone, and 15 g agar) were dissolved in 1 L of deionized water (pH 5.4), and then the medium was autoclaved at 115 °C for 15 min.

Antimicrobial activity assay

The antimicrobial activity of the samples was determined by using the agar-well diffusion method. The tested bacteria *B. subtilis* and *B. amyloliquefaciens* were cultured on LBG agar at 30 °C. The tested bacteria *S. aureus*, *L. monocytogenes*, *E. faecalis*, *S. enteritidis*, *Klebsiella* sp., *E. coli*, *P. vulgaris*, and *P. aeruginosa* were cultured on LBG agar at 37 °C for 24 h. The yeast *C. albicans* was cultured on MEA at 37 °C, while *S. cerevisiae* was cultured on MEA at 30 °C for 24 h. The fungi *A. niger*, *A. flavus*, *Penicillium* sp., *Rhizopus* sp., *Mucor* sp., and *F. moniliforme* were grown on MEA at 30 °C for 7 days or until sporulation.

The inocula of the tested bacteria/yeasts were prepared by homogenization of a small amount of biomass in 5 mL of sterile 0.5% NaCl. The inocula of tested fungi were prepared by the addition of 5 mL of sterile 0.5% NaCl into the tubes. After stirring by vortex V-1 plus (Biosan, Riga, Latvia), they were filtered and replaced in other tubes before use. The number of viable cells and fungal spores was determined using a bacterial counting chamber Thoma (Poly-Optik, Görlitz, Germany). Their final concentrations were adjusted to 10^8 cfu/mL for bacterial/yeast cells and 10^5 cfu/mL for fungal spores and then inoculated in preliminarily melted and tempered at 45–48 °C LBG/MEA agar media. Next, the inoculated media were transferred in a quantity of 18 mL in sterile Petri plates (d = 90 mm) (Gosselin™) and allowed to harden. Then six wells (d = 6 mm) per plate were cut, and triplicates of 60 µL of each extract were pipetted into the agar wells. The Petri plates were incubated at identical conditions.

The antimicrobial activity was determined by measuring the diameter of the inhibition zones around the wells on the 24th and 48th hour of incubation. Tested microorganisms with inhibition zones of 18 mm or more were considered sensitive; moderately sensitive were those in which the zones were from 12 to 18 mm; resistant were those in which the inhibition zones were up to 12 mm or completely missing [58].

2.6. Cytotoxic Activity

In order to evaluate the in vitro biocompatibility of the PID-loaded Ag NPs, a series of cell viability assays were performed against human malignant cell lines of hematological (bcr-abl positive LAMA-84 and K-562 chronic myeloid leukemia cells) and epithelial (T-

24 urothelial bladder carcinoma cells) origin, as well as normal murine fibroblast cells (CCL-1). All cell lines were purchased from the German Collection of Microorganisms and Cell Cultures (DSMZ GmbH, Braunschweig, Germany). Cell cultures were cultivated in a growth medium RPMI 1640 supplemented with 10% fetal bovine serum (FBS) and 5% L-glutamine and incubated under standard conditions of 37 °C and 5% humidified CO_2 atmosphere.

Cell viability assay

The experimental design involved several cytotoxicity assays that measured cell growth inhibition by the newly synthesized drug-loaded Ag NPs. Cell viability was evaluated using a standard MTT-based colorimetric assay. Exponential-phased cells were harvested and seeded (100 μL/well) in 96-well plates at the appropriate density (3×10^5) for the suspension cultures (LAMA-84 and K-562) and 1.5×10^5 for the adherent ones (CCL-1, T-24). Cells were treated and incubated with various concentrations of the experimental compounds in the concentration range of 400–6.25 μM. After an exposure time of 72 h, filter-sterilized MTT substrate solution (5 mg/mL in PBS) was added to each well of the culture plate. A further 1–4 h incubation allowed for the formation of purple, insoluble formazan precipitates. The latter was dissolved in isopropyl alcohol solution containing 5% formic acid prior to absorbance measurement at 550 nm using a microplate reader (Labexim LMR-1). Collected absorbance values were blanked against MTT and isopropanol solution and normalized to the mean value of untreated control (100% cell viability). Semi-logarithmic "dose-response" curves were constructed, and the half-inhibitory concentrations of the screened compounds against each tested cell line were calculated. Values of $p \leq 0.05$ were considered statistically significant.

2.7. Anticoagulant Activity

Platelet-poor plasma is used for screening coagulation tests and markers of activation of coagulation. It is obtained by centrifugation of the blood with sodium citrate at 3000–3500 U/min and $2000 \times g$ for 10–15 min. Prothrombin time (PT) activated partial thromboplastin time (APTT), and fibrinogen was immediately analyzed on an automated coagulation analyzer Sysmex SC 2000i (Siemens Corporation, Wakinohama-Kaigandori, Chuo-Ku, Kobe 651-0073, Japan). Fibrinogen was measured by von Clauss chronometric method [59].

2.8. Smooth Muscle Activity

2.8.1. Ex Vivo Experiments on Gastric Smooth Muscle Preparations (SMPs) from Rat Wistar

SMPs with dimensions 1.0–1.5 mm wide and 10–12 mm long were obtained from adult male Wistar rats weighing about 270 g. Strips were circularly dissected from corpus gastric muscle and mounted in a tissue bath and superfused with warmed (37 °C) Krebs solution. The number of SM preparations used for each data point is indicated by n.

The pH of the solution was measured before each experiment by a pH meter, HI5521 (Hanna Instruments, Woonsocket, RI, USA). Krebs bathing solution was continuously aerated with a mixture of 95% O_2 and 5% CO_2. Krebs contained the following (in mmol/L): NaCl 120; KCl 5.9; $CaCl_2$ 2.5; $MgCl_2$ 1.2; NaH_2PO_4 1.2; $NaHCO_3$ 15.4; glucose 11.5 at pH 7.4.

2.8.2. Method of Studying a Mechanical Activity of Isolated SM Preparation

The contractile activity (CA) of the SMPs and the changes in substance-evoked reactions were detected isometrically by using Tissue Organ Bath System 159,920 Radnoti (Dublin, Ireland). The initial mechanical stress of the preparations obtained by the stretch tension system corresponded to a tensile force of 10 mN. SM tissue vitality was tested by adding 1×10^{-6} mol/L Ach at baseline. The tissue was equilibrated for 60 min and washed every 15–20 min by replacing the Krebs solution. In the meanwhile, the compounds were prepared for the experiment. This requires a more concentrated solution than the actual concentration in the bath, so only a small volume (1/100) of the drug stock was needed to achieve the desired concentration. Then, an agonist was picked (a compound that causes active contraction) to which the tissue responds. The intactness of the contractile apparatus of SMPs during and at the end of experiments was checked by adding 1×10^{-6} mol/L Ach between each treatment with drugs.

2.9. Ethics Statement

Animals used in experiments were male Wistar rats. The experiments were approved by the Ethical Committee of the Bulgarian Food Agency with No 229/09.04.2019 and were carried out following the guidelines of the European Directive 2010/63/EU. The animals were provided by the Animal House of Medical University Plovdiv, Bulgaria.

2.10. Statistical Analysis

The Instat computer program for analysis of the variance was used. The mean and standard error of the mean (SEM) for each group were calculated. A two-way ANOVA for repeated measurements was used to compare different groups with the respective controls. A p-value of $p < 0.05$ was considered representative of a significant difference.

IBM SPSS Statistics v. 26 statistical package was used for statistical analyses.

3. Results and Discussion

Drug-loaded Ag NPs were synthesized using a previously described procedure under green one-pot reaction conditions: in water as a solvent, with no organic solvents, and at the boiling point of water [44]. Anhydrous $AgNO_3$ was used as a precursor for Ag NPs. The capping and reducing agent that can quickly reduce Ag^+ to the ground state at the boiling point of water (100 °C) was galactose [44]. Galactose-capped Ag NPs were chosen due to their lower toxicity [60,61], biocompatibility, and suitability for exploitation in medical applications. PID was used as a compound with known anticoagulant activity.

3.1. Physical Characterizations of Galactose-Assisted Drug-Loaded Ag NPs

Initially, the synthesis of drug-loaded Ag NPs was monitored by recording the UV–visible spectra in the region 190–800 nm (Figure 1). The absorption maximum in galactose solution was detected at 287 nm, while PID was observed at 274, 230, and 204 nm. The appearance of a surface plasmon resonance (SPR) at 418 nm showed the formation of the Ag NPs [41,62]. The SPR property of metallic NPs is one of the most important characteristics, which depends on the size and shape of synthesized metals [63].

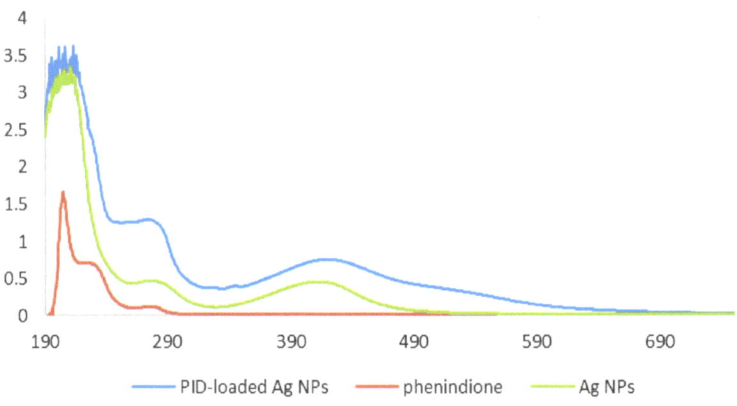

Figure 1. Absorption spectra of PID (red), AgNPs (light green), and PID-loaded Ag NPs (blue).

The position of the SPR in the visible part of the spectrum also serves as a reference for the shape of the nanoparticles [64–68]. The wavelength of the SPR peak is in the region 410–450 nm, which is indicative of the spherical shape of the obtained nanoparticles [41,69,70].

UV–visible spectroscopy is used also to investigate Ag NP colloidal aggregation [71]. In our previous study, we observed a symmetric plasmon band for the drug-loaded Ag NPs, which corresponds to their low degree of aggregation [41]. In the present study, the observed plasmon band is rather asymmetric. A possible explanation is the existence of two tautomeric forms for PID.

The synthesis of drug-loaded Ag NPs was also investigated by using FTIR and compared to any of the reactants (Figure 2). IR analysis depicted shifts in the characteristic peaks of galactose, Ag NPs, and PID-loaded Ag NPs indicating interactions amongst the molecules.

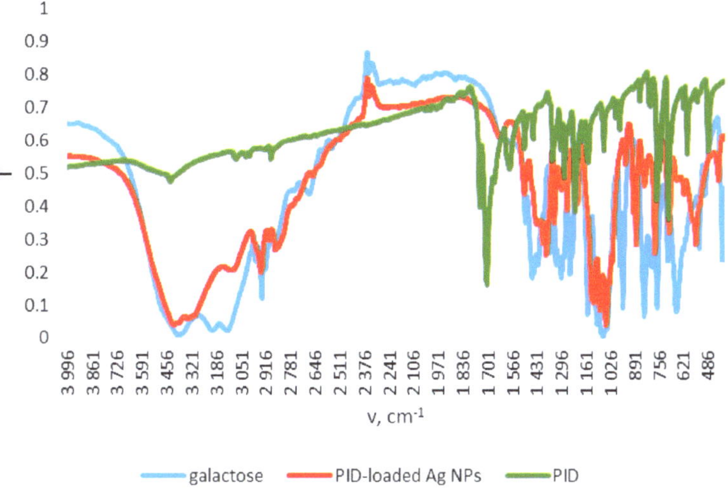

Figure 2. IR spectra of PID, PID-loaded Ag NPs, and galactose in the region of 4000–400 cm^{-1}.

IR analysis depicted the changes resulting from modified galactose-assisted PID-loaded Ag NPs carrier functionality due to incorporating silver nanoparticles. The participation of the carbohydrate in the synthesis of the nanoparticles is confirmed by the observed band shifts in the IR spectrum of drug-loaded Ag NPs compared to pure galactose [72]. Certain functional groups, such as C-O and glycosidic hydroxyl groups, have a significant contribution to the fixation of nanoparticles [73].

Wiercigroch et al. separated the IR spectra of galactose into five regions [74]. According to the authors, the first region involves deformation vibration (β) of the pyranose ring, In the pyranose ring of galactose, stretching vibrations v(C-O), (C-C), and in-plane bending (COH) were seen at 1151, 1104, and 1045 cm^{-1} [75], respectively, whereas shifts were seen for the PID-loaded Ag NPs at 1211, 1121, 1119, 1094, 1069, and 1051 cm^{-1}. The presence of an α-anomer is shown by an 837 cm^{-1} band in both galactose and PID-loaded Ag NPs IR spectra [76,77]. The bands corresponding to the vibrations of the alpha and beta anomeric forms of galactose as well as the vibrations of the pyranose ring are observed in the second region of the infrared spectrum. The bands that correspond to the hydroxyl group vibrations are in the fifth region. The bands for the C-H stretching vibrations of the methylene and methine groups are observed in the fourth region, while the bands for the deformation C-H vibrations of the methylene groups are located in the third area. A stretching vibration for the hydroxyl groups in the galactose spectrum appeared at 3386 cm^{-1}, 3205 cm^{-1}, and 3131 cm^{-1}, while shifts were observed at 3390 cm^{-1} and

3216 cm^{-1} for PID-loaded Ag NPs. The bands for the stretching C-H vibration ν(C-H) in the galactose molecule appeared at 2949 cm^{-1}, 2937 cm^{-1}, and 2916 cm^{-1}. Four bands were observed for drug-loaded Ag NPs at 2939, 2867, 2863, and 2849 cm^{-1}. The bands for the deformation vibrations of the CH$_2$ group in the IR spectrum of galactose are observed in the area 1457–1248 cm^{-1}, and a shift from 1496 cm^{-1} to 1238 cm^{-1} was observed for PID-loaded Ag NPs.

Ag NPs were synthesized in the presence of [Ag(NH$_3$)$_2$]$^+$ with the assistance of carbohydrates [41,78]. It is well known that biomolecules interact with the silver's upper face, where the initial bond to the metal was formed [79,80]. The size and rate of aggregation are decreased by silver nanoparticle coating [81]. The dynamic surface location of the silver nanoparticles allowed the measurement of particle size, shape, and accumulation rate. The reaction parameters, which include temperature, pH, reactant concentration, and duration, have an impact on the shape and size of the silver accumulating [82]. Silver nanoparticles with less than 10 nm diameter have the potential to enter the nuclear cavity and interact with genetic material. Cytotoxicity of nanoparticles relies on their structure; for example, plate-like shape Ag NPs are more dangerous than those with a wire or spherical shape [83–86].

To establish the size and shape of the Ag NPs, spICP-MS, DLS, and Zeta potential were used.

The nanocolloid suspensions were examined by using spICP-MS to determine the size and distribution by size of the silver core in synthesized PID-loaded Ag NPs. The investigated parameters of the particles, such as mean, mode, and median diameter, as well as size distribution histogram, were evaluated by using an ionic calibration strategy [55]. For this purpose, a set of Ag$^+$ ionic standard solutions in 0.5% HNO$_3$ with a concentration range of 45–1200 ng L^{-1} were prepared. The levels of calibration standards were selected so that the mass of the delivered silver for chosen dwell time corresponded to Ag NPs with diameters in the range of 20–50 nm.

Reference material of citrate-stabilized Ag NPs with certified size 40 ± 4 nm and one ionic standard solution of Ag$^+$ (1 ng L^{-1}) were used for determination of the transport efficiency and instrumental sensitivity. The transport efficiency was calculated by the particle-size method [55].

All measurements were made at the following parameters: dwell time 5 ms, acquisition time 60 s, sample flow rate 0.322 mL·min^{-1}, and transport efficiency 0.061. The size histogram of the analyzed NPs is presented in Figure 3.

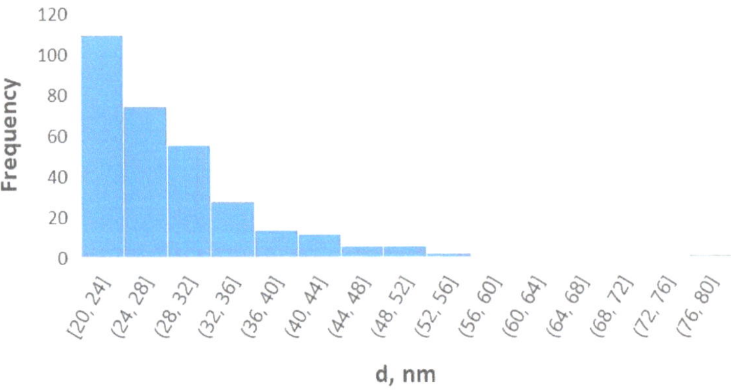

Figure 3. Particle-size distribution histogram of galactose-assisted Ag NPs (~300 NPs).

Based on the obtained histogram, it can be concluded that Ag NPs have an asymmetrical size distribution, with the highest NPs fraction between 20 and 24 nm, which is close to the method's detection limit—LOD$_{size}$ 17 nm. The estimated mean diameter, mode, and median are as follows: 28 nm, 22 nm, and 27 nm.

The results from the size distribution histograms (Figure 3) were confirmed by the TEM images (Figure 4a,b). The obtained data clearly show the size and form of the nanoparticles as well as the individual nanoparticles. The TEM image confirmed the synthesis of spherical particles of different sizes for drug-loaded Ag NPs (Figure 4). We assume that galactose, as well as the benzene ring in PID-loaded Ag NPs (Figure 4a,b), prevent the aggregation of the particles.

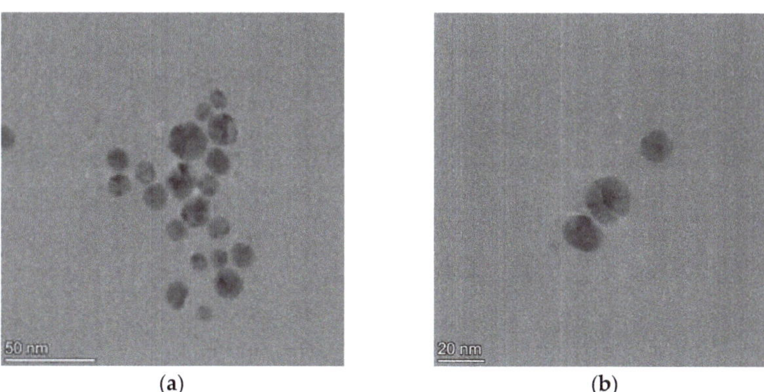

(a)   (b)

Figure 4. TEM images of galactose-assisted PID-loaded Ag NPs (**a**,**b**).

Based on the DLS, the median average size of the obtained particles was identified in the range of 20 to 40 nm for PID-loaded Ag NPs (Figure 5).

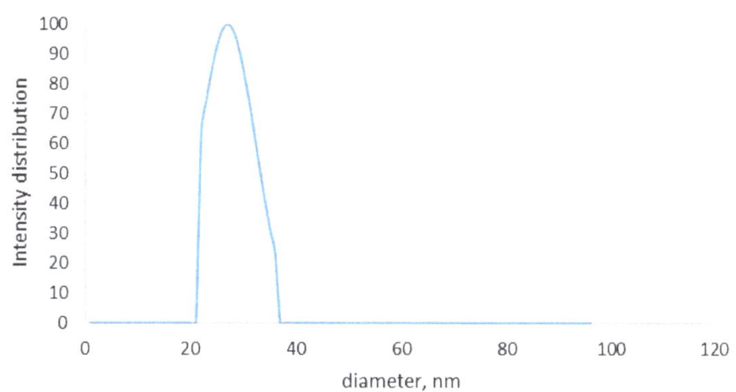

Figure 5. Dynamic light scattering histograms of PID-loaded Ag NPs.

For the complex production of silver, many groups, including hydroxyl, carboxyl, phenol, and carbonyl, are associated with oxygen and nitrogen by covalent bond linkage; as a result, they are likely absorbed on their surface [80]. Using NMR and DFT calculations, Sigalov [87] examined the PID for keto-enol tautomerism. According to the estimated data, a strong hydrogen bond contributes to the enol form's stability in the DMSO solution. It was suggested that the strong Ionic interaction with an anion in the enol forms promotes a rapid proton transfer between carbonyl oxygen atoms, which is the reason for the symmetry of their NMR spectra. It was shown that during keto-enol tautomerization, the 2-phenyl-1,3-indandione in its di-keto form interacts with an anion similarly. That was the reason to assume that the PID's enol-form can form hydrogen bonds to galactose OH groups while connecting to galactose coating.

The zeta potential of drug-loaded Ag NPs was -12.28 ± 0.28 mV. These results are close to previously described Ag NPs [41]. Carboxylic acids, as the oxidized products of sugars, are utilized to provide a negative surface charge density to counteract the Van der Waals forces responsible for particle coalescence. The carboxylic groups of generated galactonic acid could be a source of the negative charge of the zeta potential. To stabilize metal surfaces, self-assembling carboxylic acids ensure a dense covering [88].

Using the X-ray powder diffraction pattern of galactose, a Rietveld refinement analysis was performed. The proposed cell parameters in the monoclinic space group P 2/m were a = 9.75166 Å, b = 15.63694 Å, c = 7.88817 Å, β = 104.532° (cell volume: 1164.35 Å^3). The XRD analysis supports galactose's role in the synthesis of Ag NPs (Figure 6). XRD patterns of pure galactose (Figure 6a) and PID (Figure 6d) revealed multiple distinctive peaks in the 2θ area as a result of their crystalline structure. The X-ray diffraction pattern of PID-loaded Ag NPs showed the reflections for silver at 2θ = 37.80° and 43.84°. We were able to refine the X-ray diffraction pattern of PID, as well. The unit cell parameters found were a = 18.24746 Å, b = 5.24937 Å, c = 9.32495 Å, and β = 124.144°. The space group is P 2/m.

Compared to the typical galactose XRD pattern, the Ag NPs pattern was altered (Figure 6b). The reflections for silver were measured at 2θ = 37.80° and 43.84°. The 1 1 1 plane, which is associated with the spherical structure of silver [89], corresponds to the strong Bragg reflection at 38.16°.

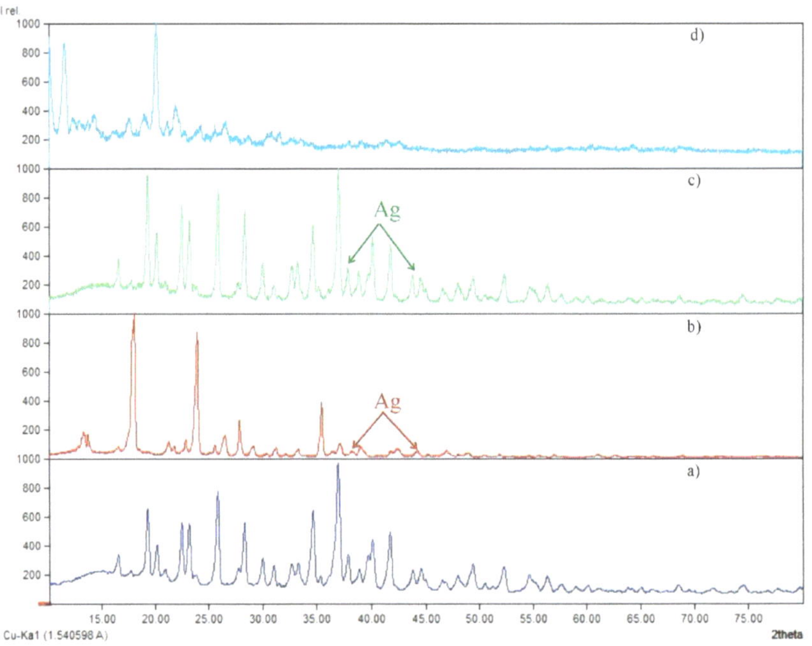

Figure 6. X-ray diffraction patterns of galactose (**a**), Ag NPs (**b**), PID-loaded Ag NPs (**c**), and PID (**d**).

3.2. Phenindione Adsorption onto Galactose-Loaded Nanostructure by DFT

At first, the PID molecule and nanostructure were optimized separately, then the PID drug was placed on the nanostructure, and optimization was performed. The length of the nanostructure was chosen to be 19.11 Å, and its structure consists of 20 carbons, 16 oxygen, and 36 hydrogens. The optimized geometry of PID, nanostructure, and PID-nanostructure is demonstrated in Figure 7. The PID molecule was first situated in various locations on the external surface of the nanostructure with diverse orientations. The ideal and most stable geometry of the PID on the nanostructure is shown in Figure 7c. As can be seen from this figure, the galactose molecules are held together through hydrogen bonds. In addition, the PID molecule interacts with one of the galactose molecules through a weak hydrogen bond. The estimated value of the adsorption energy (E_{ads}) for PID on nanostructure is obtained as -0.438 eV (Table 1). The PID's adsorption energy is negative indicating that the adsorption of this drug on nanostructure is exothermic. The interaction of the PID molecule with the nanostructure is weak due to the fact the E_{ads} value of PID on the nanostructure is small and the presence of a weak hydrogen bond between this molecule and the surface of the nanoparticle. As a result, galactose nanostructure can be utilized as a drug delivery system for the phenindione molecule.

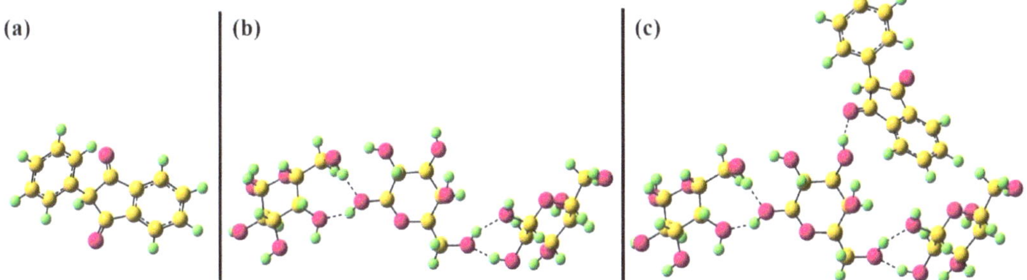

(a) **(b)** **(c)**

Figure 7. (**a**) The optimized structure of PID, (**b**) nanostructure, and (**c**) PID on nanostructure.

Table 1. The adsorption energy (E_{ads}) of PID on nanostructure, electronic energy (E_{elec}), HOMO-LUMO energy levels, and additional quantum parameters.

Parameters	PID	Ag NPs	PID-Loaded Ag NPs
E_{elec} (eV)	$-19{,}838.000$	$-56{,}172.470$	$-76{,}010.908$
E_{ads} (eV)	-	-	-0.438
E_{HOMO} (eV)	-6.86	-7.16	-6.83
E_{LUMO} (eV)	-2.73	-0.02	-2.71
ΔE_{gap} (eV)	4.13	7.14	4.12
η	2.06	3.57	2.06
σ	0.48	0.28	0.48
χ	4.79	3.59	4.77
Pi	-4.79	-3.59	-4.77
ω	5.57	1.80	5.52

The quantum descriptors including HOMO-LUMO energy gap ($\Delta E = E_{LU-MO} - E_{HOMO}$), absolute hardness ($\eta = (E_{LU-MO} - E_{HOMO})/2$), absolute softness ($\sigma = 1/\eta$), chemical potential ($P_i = -\chi$), global electrophilicity ($\omega = P_i^2/2\eta$) and absolute electronegativity ($\chi = -(E_{HOMO} + E_{LUMO})/2$) [90,91] were calculated and are summarized in Table 1. The stability and chemical reactivity of a surface or molecule can be investigated using the HOMO/LUMO energy gap. The low-energy gap indicates high chemical reactive nature and vice versa [92,93]. According to Table 1, ΔE value for PID on nanostructure is lower than free PID or nanostructure suggesting this complex possesses higher chemical reactivity than others. In addition, a high ω value illustrates that a compound is more capable of interacting with biological macromolecules such as DNA [94]. A high ω value for PID on nanostructure means that this complex has a high biological activity. Figure 8a demonstrates the HOMO-LUMO energy level for PID on the nanostructure. As clear, the electron density of HOMO is situated in the phenyl group of the PID molecule, and the electron cloud of LUMO is located in the indandione group of the PID molecule.

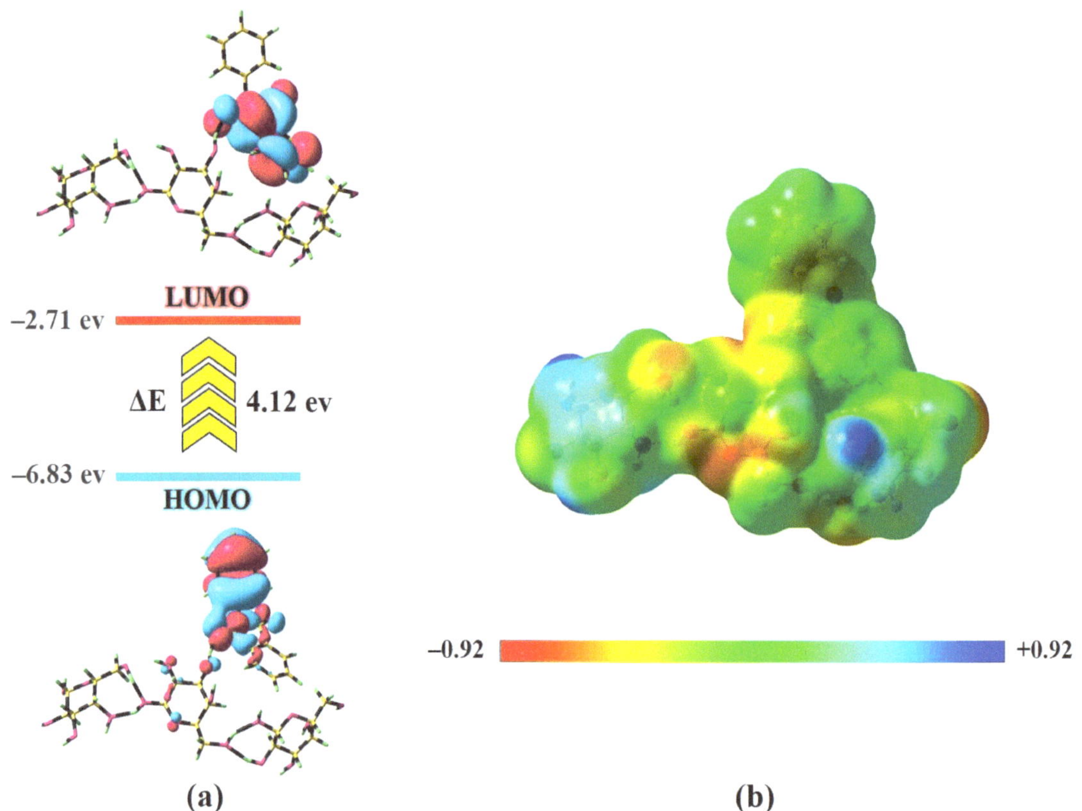

Figure 8. (**a**) HOMO–LUMO diagram of PID on nanostructure, and (**b**) MEP surface of PID on nanostructure.

MEP (molecular electrostatic potential) surface is estimated for PID on nanostructure using the DFT approach and represented in Figure 8b. This surface is beneficial for illustrating the distribution of charge within the system. The MEP utilizes various colors in a 3D scheme including red and blue color. The blue color denotes the electron-deficient zone (positive charge), and the red one demonstrates the electron-rich area (negative charge) [95]. Based on Figure 8b, it can be predicted that the interaction between PID and nanostructure leads to charge transfer from PID molecule to the nanostructure. The charge transfers between phenindione and nanostructure result in a weak interaction. Therefore, the result of MEP also confirms that the galactose nanostructure can act as an appropriate drug delivery system for PID.

3.3. Parametric Drug Release Optimization

To find out how the ratio of Ag NPs to the drug molecule would affect the drug release and biological activity, different amounts of PID were used in the preparation of the solutions. The most popular method of medication is oral. Oral medication, compared to intravenous treatments, is easier to take, less painful, and less expensive. A drug's oral bioavailability will be hampered if its water solubility is too low. Nanocarriers can be used to resolve such solubility issues [96]. The influence of Ag NPs to drug ratio was evaluated utilizing an in vitro approach with dialysis bags in order to get insights into the drug release trends of PID-loaded Ag NPs [97,98]. Due to a lack of universal, accepted practices, this is a typical technique to learn about innovative drug delivery systems [99]. To determine the in vivo–in vitro correlation of nanoparticle formulations, as well as to direct the development and quality control, drug release profiles from dialysis-based assays are used [100,101].

Profiles of PID-molecules only (controls) were compared to those seen in the presence of Ag NPs to determine the drug release properties of the PID-loaded Ag NPs. Different Ag NPs to drug ratios (1:1, 1:5, 1:10, 1:20, and 1:50) were utilized to examine how the ratio might impact the drug's release. One drug control per ratio was made using an ethanol solution of PID.

Under normal physiological conditions, the pH value of blood is 7.4. To simulate the physiological conditions, the drug release study was carried out in a neutral medium (PBS, pH 7.4). The solution of drug-containing Ag NPs was kept in a cellulose dialysis bag with a molecular weight cut-off of 12,400 Da. The pores in this cut-off were big enough to trap the Ag NPs inside the membrane while yet allowing the medicine to escape from the dialysate. Since the calibration curve for PID was linear, it was possible to conduct the quantification tests. Therefore, determining the drug release in the first 6 h was crucial for our investigation.

During the 24 h period, drug release was measured. We found that the absorbance intensity rose with time during the initial 5 to 10 h of the drug release experiment, indicating an increasing concentration of the PID in the dialysate (Figure 9).

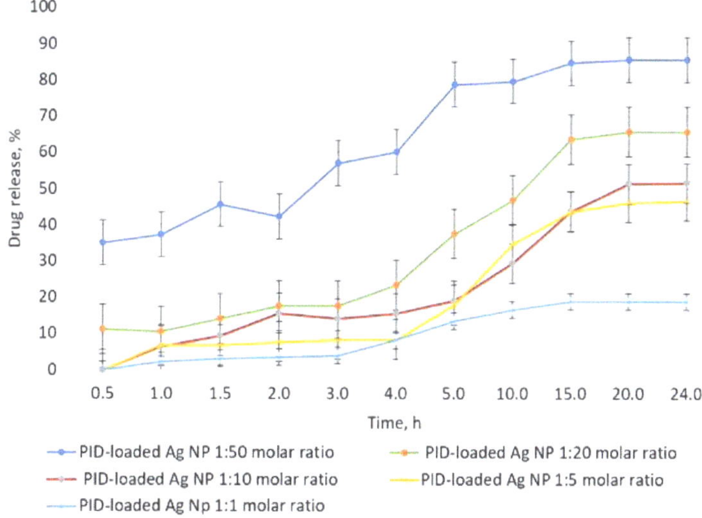

Figure 9. Drug release concentrations of PID-loaded Ag NPs in molar ratios of 1:1, 1:5, 1:10, 1:20, and 1:50.

The obtained data showed that the drug release pattern changed a little after the fifth hour, but the rate of increase in absorbance intensity remained constant until the 24th hour. Therefore, in the 24th hour, the maximal concentration of PID had been released in the dialysate, proving that the drug release was complete under the given conditions and that the concentration of the released drug was stable. Due to the existence of a galactose coating and the effectiveness of the drug loading for Ag NPs, the concentration for PID-only samples was greater.

However, at the sixth hour of the drug-release experiment, PID-loaded Ag NPs with a molar ratio of 1:50 demonstrated a very good drug release between 80 and 86%, whereas a molar ratio of 1:20 showed 60–68%, while a molar ratio of 1:1 showed roughly 19% drug release. The obtained results suggested the effectiveness of drug loading, as well as the possible application of Ag NPs as a drug delivery system for PID.

3.4. Antimicrobial Activity

One of the most pervasive problems in the world today is antibiotic resistance, and many potent medications have failed to control illnesses. According to the literature, Ag NPs possess antimicrobial activity [48,51,69,102,103]. Thus, we measured the antimicrobial activity of the PID-loaded Ag NPs compared to PID only. The antimicrobial activity was tested in vitro against human pathogenic bacteria and economically relevant phytopathogenic fungi. In our experiments, six Gram-positive bacteria, six Gram-negative bacteria, two yeasts, and six fungi were used. The inhibition zones of bacterial and fungal growth caused by the novel compounds are outlined in Table 2. The methanol used as a solvent for the samples did not show any antimicrobial effect.

Table 2. Antimicrobial activity of the tested PID vs. PID-loaded Ag NPs.

Inhibition Zones, mm		
Tested Microorganism	**PID**	**PID-Loaded Ag NPs**
Bacillus subtilis ATCC 6633	17	12
Bacillus amyloliquefaciens 4BCL-YT	25	13
Staphylococcus aureus ATCC 25923	27	27
Listeria monocytogenes NBIMCC 8632	22	15
Enterococcus faecalis ATCC 29212	23	17
Micrococcus luteus 2YC-YT	30	20
Salmonella enteritidis ATCC 13076	21	15
Salmonella typhimurium NBIMCC 1672	25	13
Klebsiella pneumoniae ATCC 13883	11	-
Escherichia coli ATCC 25922	18	13
Proteus vulgaris ATCC 6380	14	10
Pseudomonas aeruginosa ATCC 9027	18	12
Candida albicans NBIMCC 74	11	-
Saccharomyces cerevisiae ATCC 9763	9	-
Aspergillus niger ATCC 1015	11	13
Aspergillus flavus	9	-
Penicillium chrysogenum	13	10
Rhizopus sp.	11	8
Fusarium moniliforme ATCC 38932	11	14
Mucor sp.	11	-

Dose 0.6 mg/disk, dwell = 6 mm.

We observed that the PID, as well as PID-loaded Ag NPs, exerted very good activity against Gram-positive bacteria including *Micrococcus luteus* 2YC-YT, *Staphylococcus aureus* ATCC 25923, *Bacillus amyloliquefaciens* 4BCL-YT, Gram-negative bacteria including *Salmonella enteritidis* ATCC 13076, and modest activity against the other bacteria and fungi.

The obtained data showed that in general antimicrobial activity of PID is higher than drug-loaded Ag NPs. We can explain this fact with the time needed to release the drug from the Ag NPs. These observations were also confirmed by anticoagulant and spasmolytic activity measurements.

3.5. Cytotoxic Activity

A series of MTT experiments were conducted against normal murine fibroblast cells, malignant human cell lines of leukemic (LAMA-84, K-562), and urothelial bladder carcinoma (T-24) origin to accommodate the cytotoxicity of the target compounds. The results of the antiproliferative assays are presented in Table 3. According to the obtained data, cell growth of both normal and malignant in vitro cultures tends to be unaffected by free PID and the unloaded galactose-assisted Ag NPs (IC50 values invariably exceeding 200 µM), indicating favorable biocompatibility. However, a moderate enhancement of PID cytotoxicity was established in the PID-loaded Ag NPs formulation (particularly in the suspension-growing leukemic cell lines LAMA-84 and K-562), possibly due to improved cellular kinetics and intracellular delivery of the phenindione derivative. Moreover, a number of studies reported on the inhibitory effect of indandione analogues on various kinases and other enzymes (e.g., akt, braf, PI3K, SIK2, among others) [104–106], which may account for the antiproliferative activity of these compounds in the tested leukemia models, bearing in mind their BCR-ABL positive phenotype. Nevertheless, the growth-inhibitory potential of the drug-loaded Ag NP is twice as low in the normal fibroblast cellular system, displaying a certain selectivity towards the leukemic cells with elevated constitutive kinase activity.

Table 3. In vitro cytotoxicity of the tested compounds [μM ± SD] against cell lines of different origin.

Compound/Cell Line	CCL-1	LAMA-84	K-562	T-24
PID-loaded Ag NPs	65.4 ± 8.3	25.3 ± 1.2	27 ± 3.2	53.2 ± 4.7
Ag NPs	>200	>300	>300	>300
PID	>300	>200	>200	>300

CCL-1—murine fibroblast cell line; LAMA-84, K-562—BCR-ABL+ chronic myeloid leukemia; T-24—urothelial bladder carcinoma.

3.6. Anticoagulant Activity

All over the world, venous thromboembolism is the major cause of cardiovascular mortality, next in rank to myocardial infarction and stroke [107,108]. This clinical situation affects patients in various settings and age groups, including children [109,110]. The most prevalent manifestation of venous thromboembolism is deep vein thrombosis, the expression of which may manifest in its most severe form known as acute pulmonary thromboembolism [111,112]. The major treatment for both situations involves the administration of full anticoagulation, thus minimizing the recurrence of venous thromboembolism. One homeostatic mechanism that stops continued bleeding after tissue injury is coagulation. It happens when platelets, plasma clotting factors, and injured tissue interact. The fibrinolytic cascade, which controls fibrin and fibrinogen breakdown and minimizes excessive thrombus formation, counterbalances the coagulation cascade. The coagulation cascade, the fibrinolytic cascade, and the platelet function interact dynamically to produce normal clotting and hemostasis [113].

The capacity of anticoagulants to prevent the production of vitamin K-dependent clotting factors (II, VII, IX, and X) causes their anticoagulant effect. Warfarin, for example, shows its anticoagulant effect 8–12 h after oral or intravenous administration. [13]. Anticoagulants stop the blood coagulation process. Anticoagulant therapy is primarily focused on preventing the development of blood vessel clots, which are the main cause of death in thromboembolic disorders. In our experiments, for screening coagulation tests, a platelet-poor plasma was used. We observed that PID itself showed no coagulation. The measured PT, APTT, and fibrinogen, as well as the reference values, are shown in Table 4.

Table 4. Anticoagulant properties of PID-loaded Ag NPs.

PID-Loaded Ag NPs (Molar Ratio)	PT, s	APTT, s	Fibrinogen, g/L
1:1	14.1	35	0.91
1:5	13.9	34.9	1.34
1:10	13.7	35.3	1.31
1:20	14.1	35.7	1.25
1:50	no coagulation	43.1	1.31
reference	8–10.8	28.2–31.4	1.91–2.87

When PID was tested for its anticoagulant activity, it showed no coagulation. That means that the compound can prolong the coagulation of blood for a longer time causing bleeding in patients. Bleeding is a life-threatening complication.

Drug-loaded Ag NPs, on the other hand, showed that PT (sec) and APTT (sec) are approximately 1.5 times longer than the normal values. Only the molar ratio of 1:50 exerted no coagulation, as PID did itself for any of the molar ratios used. This could make the Ag NPs better anticoagulants than PID itself.

3.7. Ex Vivo Experiments on SM Activity

The intrinsic biological activity of substances could be studied both in vivo and ex vivo using isolated tissues that are still functionally active. The ex vivo approach, which is frequently used to assess the potential biological activity of recently synthesized experimental compounds and approved pharmaceutical drugs, is carried out on isolated tissues responsive to physiological stimuli [114]. Because they may still create active tension when separated from the body, SM cells were a natural choice for our study's platform for ex vivo contractility evaluation [115,116]. Many internal organs primarily have SM tissue. It is a complex superposition of bioelectrical (slow wave with characteristic frequency and amplitude) and contractile activity (CA) (tone, frequency, and amplitude of spontaneous or induced muscle contractions), which can be measured isometrically in isolated tissues [117–119] and related to the motor activity of the stomach.

The primary function of smooth muscle is contraction. On ex vivo conditions smooth muscles can preserve their elastic properties for up to 10 h and show a significant change in their tone, frequency, and amplitude of contraction under exogenously administered pharmacological agents [120,121].

The isolated tissue bath experiment is a conventional pharmacological technique for evaluating concentration-response correlations in a range of contractile tissues. Pharmacologists and physiologists continue to utilize this method, even though it has been around for more than a century, due to its versatility, simplicity, and reproducibility [122]. The isolated tissue bath continues to be a significant part of drug development and fundamental research because it enables the tissue to behave like a tissue. Since it permits the tissue to behave like a tissue, the isolated tissue bath continues to be a crucial component of drug development and fundamental research. The ability of the living tissue to function as a whole organ with physiologically realistic contraction or relaxation is the main advantage of this technology. It also has the advantage of bringing the effects of the drugs under research closer to how they would act in the body by allowing for the calculation of important pharmacological variables while retaining tissue function [122]. Previously [123–127], we applied the isolated tissue bath for ex vivo SM activity investigation of different compounds. Thus, we applied the same technique to the exogenous administration of 5×10^{-5} mol/L PID-loaded Ag NPs (n = 7), compared to Ag NPs (n = 6) and PID only (n = 6). We found that a single administration of Ag NPs showed no change in smooth muscle spontaneous contractile activity (SCA) for about 6 h (Figure 10). After the 6th hour, we found that the acetylcholine response slightly decreases.

According to González et al. [128] Ag NPs modify the contractile activity of ACh through nitrogen oxide production and possibly induce smooth muscle hyperreactivity. The obtained data (Table 5) showed that under the same experimental conditions in the tissue bath, PID showed a rapid significant relaxation response with a pronounced change in the tone and amplitude of spontaneous contractions.

The maximum tonic relaxation appears approximately after 10.47 ± 0.22 min and persists with the same force for 6 h. No change was observed in the ACh response before and after administration of the PID, which indicates that the main neurotransmitter pathway is not affected, unlike Ag NPs. This data confirmed the fact that indane and its analogues showed not only anticoagulant activity but can also affect smooth muscles [129]. When PID-loaded Ag NPs were applied, we found they affected the tone and the amplitude of contractile activity the same way as PID did (Table 5). For comparison, the maximum effect of PID-loaded Ag NPs (1:10; 1:20; 1:50) at concentration 5×10^{-5} M was a relaxation with a strength, which was approximately two times lower than the relaxation evoked by 5×10^{-5} M PID on the same tissues. After the sixth hour, the effect reaches its peak, which confirms our hypothesis that the nanoparticles release the PID gradually. PID-loaded Ag NPs (1:1 and 1:5) at concentration 5×10^{-5} M did not affect the SCA parameters of gastric

SMPs. Acetylcholine did not significantly change after the 6th hour, demonstrating that the Ag NPs have already released the maximum amount of PID. Thus, the obtained data correspond to the dialysis drug-release experiments. We assumed that the effects of all PID-loaded Ag NPs on the contractile processes are probably caused by a specific trigger action of the nanoparticles they exert in gastric SM tissue.

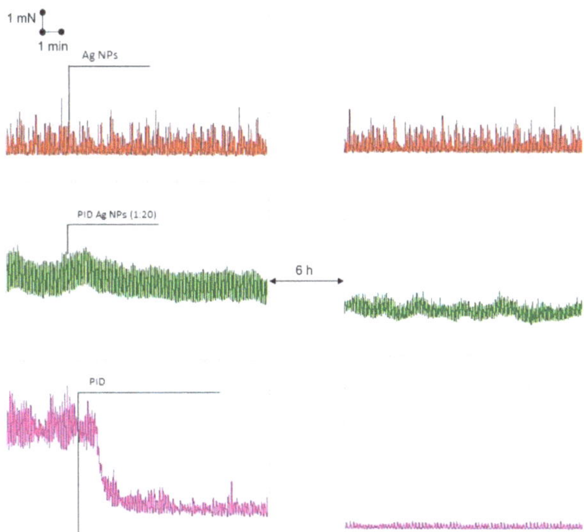

Figure 10. Representative tracings of gastric SM activity parameters before and after application with Ag NPs, PID, and the most pronounced effect of PID-loaded Ag NPs (1:20) at concentration 5×10^{-5} mol/L.

Table 5. Changes in the parameters of SCA of gastric SMPs under the influence of Ag NPs, compared to PID-loaded Ag NPs (in a molar ratio of 1:1, 1:5, 1:10, 1:20, 1:50), and PID itself.

Compound/Parameter	Tonic Activity (mN)	Strength of The Contractile Reaction (mN)	Frequency (Number of Contractions/min)
SCA (initial reaction)	1.81 ± 0.19	3.40 ± 0.19	4.97 ± 0.04
Ag NPs	1.69 ± 0.03	3.15 ± 0.16	4.93 ± 0.03
SCA (initial reaction)	1.89 ± 0.12	3.00 ± 0.18	4.80 ± 0.09
PID-loaded Ag NPs (1:1)	1.80 ± 0.07	2.88 ± 0.05	4.90 ± 0.07
SCA (initial reaction)	2.03 ± 0.15	3.64 ± 0.21	5.04 ± 0.05
PID-loaded Ag NPs (1:5)	2.07 ± 0.22	2.14 ± 0.11	5.09 ± 0.17
SCA (initial reaction)	1.99 ± 0.11	3.64 ± 0.21	5.03 ± 0.03
PID-loaded Ag NPs (1:10)	0.77 ± 0.09 *	2.14 ± 0.11 *	4.87 ± 0.11
SCA (initial reaction)	1.94 ± 0.07	2.44 ± 0.13	5.07 ± 0.10
PID-loaded Ag NPs (1:20)	0.91 ± 0.08 *	0.98 ± 0.07 *	4.90 ± 0.09
SCA (initial reaction)	1.67 ± 0.14	3.50 ± 1.21	4.88 ± 0.06
PID-loaded Ag NPs (1:50)	0.80 ± 0.14 *	1.73 ± 0.16 *	4.83 ± 0.08
SCA (initial reaction)	1.96 ± 0.09	2.42 ± 0.22	4.90 ± 0.04
PID	0.44 ± 0.07 *	0.13 ± 0.03 *	4.87 ± 0.03

The comparison is between spontaneous effects on CA in Krebs solution before and after the application of testing substances. All data were expressed as mean values ± standard error of the mean (mean ± SEM); * statistically significant differences ($p < 0.05$).

4. Conclusions

Rapid synthesis and stabilization of silver nanoparticles as drug-delivery systems for PID was applied for the first time. The size range for the produced Ag NPs is determined by using various approaches. Galactose was used as a capping and reducing agent. DFT calculations illustrate that the interaction of PID onto nanostructure is weak, suggesting that galactose-loaded nanostructure can be used as a drug delivery system for phenindione. MEP surface analysis confines that the interaction between phenindione and nanostructure leads to charge transfer from phenindione drug to nanostructure. Additionally, the drug-release capabilities of Ag NPs were established. The drug release showed very good release properties of the Ag NPs between 5 to 10 h of the experiments. The maximum concentration of PID was reached till the 20th h, proving that the drug release was complete under the given conditions and that the concentration of the released drug was stable. Due to the existence of a galactose coating and the effectiveness of the drug loading for Ag NPs, the drug release concentration for PID-only samples was greater. The drug-loaded Ag NPs were characterized in vitro for their antimicrobial and anticoagulant activity and ex vivo for spasmolytic activity compared to PID only. The obtained data showed that Ag NPs exerted very good antimicrobial activity against Gram-positive bacteria (*Micrococcus luteus* 2YC-YT, *Staphylococcus aureus* ATCC 25923, *Bacillus amyloliquefaciens* 4BCL-YT), and Gram-negative bacteria like *Salmonella enteritidis* ATCC 13076, but a bit lower than PID only showed itself due to the time needed to release the drug from the Ag NPs. These observations were also confirmed by anticoagulant and spasmolytic activity measurements. When PID was tested for its anticoagulant activity, it showed no coagulation. That means that the compound can prolong the coagulation of blood for a longer time, causing bleeding in patients, which can be a life-threatening complication. Drug-loaded Ag NPs, on the other hand, showed approximately 1.5 times higher PT and APTT than the normal values. This could make them better anticoagulants than PID itself. The ex vivo spasmolytic experiments showed that PID expressed a rapid and significant relaxation response with a pronounced change in the tone and amplitude of spontaneous contractions. For PID-loaded Ag NPs in the same experiments, the effect reaches its peak after the fifth hour, which confirmed our hypothesis that the nanoparticles release the PID gradually. PID-loaded Ag NPs (1:1 and 1:5) at concentration 5×10^{-5} M did not affect the SCA parameters of gastric SMPs. Acetylcholine did not significantly change after the 6th hour, demonstrating that the Ag NPs have already released the maximum amount of PID.

Supplementary Materials: The following supporting information can be downloaded at: https://www.mdpi.com/article/10.3390/biomedicines11082201/s1, Figure S1: ^{1}H-NMR spectrum of phenindione, Figure S2: ^{13}C-NMR spectrum of phenindione; Figure S3: FT-IR spectrum of phenindione.

Author Contributions: Conceptualization, M.T. and S.N.; methodology, M.T., S.N., L.K., D.G., M.F.-D., I.S., V.G., R.M. and Y.T.; investigation, M.T., M.M., S.N.—synthesis of Ag NPs; L.K., D.G., V.D., E.C. and M.T.—UV, FTIR, spICP-MS, XRD; M.F.-D.—DFT; I.S. and V.G.—spasmolytic activity; Y.T.—antimicrobial activity; S.S.—anticoagulant activity; R.M.—cytotoxicity; writing—original draft preparation, S.N., M.M., M.F.-D. and M.T.; writing—review and editing, S.N. and M.M.; visualization, S.N., M.M., M.T., M.F.-D. and V.G.; supervision, S.N.; project administration, S.N. and M.T. All authors have read and agreed to the published version of the manuscript.

Funding: This research and the APC were funded by the Bulgarian Ministry of Education and Science under the National Program "Young Scientists and Postdoctoral Students—2". National Program "Young Scientists and Postdoctoral Students—2".

Institutional Review Board Statement: The procedures used in this study agreed with the European Communities Council Directive 2010/63/EU for animal experiments. The experimental procedures were conducted following national rules on animal experiments and were approved by the Bulgarian Food Safety Agency (No. 229/No. 145/09.04.2019).

Informed Consent Statement: Not applicable.

Data Availability Statement: Not applicable.

Acknowledgments: V. Gledacheva is grateful for the financial support of this research by the Bulgarian Ministry of Education and Science under the National Program "Young Scientists and Postdoctoral Students—2".

Conflicts of Interest: The authors declare no conflict of interest.

References

1. Vardanyan, R.; Hruby, V. *Synthesis of Essential Drugs*, 1st ed.; Elsevier: Amsterdam, The Netherlands, 2006; pp. 323–335. ISBN 9780080462127. [CrossRef]
2. Kalishwaralal, K.; Deepak, V.; Pandian, S.B.R.K.; Kottaisamy, M.; Kanth, S.B.M.; Kartikeyan, B.; Gurunathan, S. Biosynthesis of Silver and Gold nanoparticles Using Brevibacterium casei. *Colloids Surf. B* **2010**, *77*, 257–262. [CrossRef] [PubMed]
3. Shrivastava, S.; Bera, T.; Singh, S.K.; Singh, G.; Ramachandrarao, P.; Dash, D. Characterisation of antiplatelet properties of silver nanoparticles. *ACS Nano* **2009**, *3*, 1357–1364. [CrossRef] [PubMed]
4. Kemp, M.M.; Kumar, A.; Mousa, S.; Park, T.J.; Ajayan, P.; Kubotera, N.; Mousa, S.A.; Linhardt, R.J. Synthesis of gold and silver nanoparticles stabilized with glycosaminoglycans having distinctive biological activities. *Biomacromolecules* **2009**, *10*, 589–595. [CrossRef] [PubMed]
5. White, H.D. Oral antiplatelet therapy for atherothrombotic disease: Current evidence and new directions. *Am. Heart J.* **2011**, *161*, 450–461. [CrossRef]
6. Latib, A.; Ielasi, A.; Ferri, L.; Chieffo, A.; Godino, C.; Carlino, M.; Montorfano, M.; Colombo, A. Aspirin intolerance and the need for dual antiplatelet therapy after stent implantation: A proposed alternative regimen. *Int. J. Cardiol.* **2013**, *165*, 444–447. [CrossRef]
7. Reiter, R.A.; Jilma, B. Platelets and new antiplatelet drugs. *Therapy* **2005**, *2*, 465–502. [CrossRef]
8. Ilas, J.; Jakopin, Z.; Borstnar, T.; Stegnar, M.; Kikelj, D. 3,4-Dihydro-2H-1,4-benzoxazine derivatives combining thrombin inhibitory and glycoprotein IIb/IIIa receptor antagonistic activity as a novel class of antithrombotic compounds with dual function. *J. Med. Chem.* **2008**, *51*, 5617–5629. [CrossRef]
9. Chan, Y.C.; Valenti, D.; Mansfield, A.O.; Stansby, G. Warfarin induced skin necrosis. *Br. J. Surg.* **2000**, *87*, 266–272. [CrossRef]
10. Liu, G.; Xu, J.; Chen, N.; Zhang, S.; Ding, Z.; Du, H. Synthesis of N6-alkyl(aryl)-2-alkyl(aryl) thioadenosines as antiplatelet agents. *Eur. J. Med. Chem.* **2012**, *53*, 114–123. [CrossRef]
11. Griffin, J.P.; D'Arcy, P.F. *A Manual of Adverse Drug Interactions*, 5th ed.; Elsevier Science: Amsterdam, The Netherlands, 1997; pp. 111, 113–175, 177–231. ISBN 9780080525839.
12. Pineo, G.F.; Hull, R.D. Chapter 43—Conventional Treatment of Deep Venous Thrombosis. In *The Vein Book*; Bergan, J.J., Ed.; Academic Press: Cambridge, MA, USA, 2007; pp. 381–394. [CrossRef]
13. Merlob, P.; Habermann, J. Chapter 7—Anticoagulants, thrombocyte aggregation inhibitors and fibrinolytics. In *Drugs During Pregnancy and Lactation*, 3rd ed.; Schaefer, C., Peters, P., Miller, R.K., Eds.; Academic Press: Cambridge, MA, USA, 2015; pp. 725–729. [CrossRef]
14. Pharmacology, B.J. Indanedione anticoagulants. In *Me'ler's Side Effects of Drugs*, 16th ed.; Aronson, J.K., Ed.; Elsevier: Amsterdam, The Netherlands, 2016; pp. 50–51. [CrossRef]
15. Vilums, M.; Heuberger, J.; Heitman, L.H.; Ijzerman, A.P. Indanes—Properties, Preparation, and Presence in Ligands for G Protein Coupled Receptors. *Med. Res. Rev.* **2015**, *35*, 1097–1126. [CrossRef]
16. Mukhtar, A.; Shah, S.; Kanwal; Hameed, S.; Khan, K.M.; Khan, S.U.; Zaib, S.; Iqbal, J.; Parveen, S. Indane-1,3-diones: As Potential and Selective α-glucosidase Inhibitors, their Synthesis, in vitro and in silico Studies. *Med. Chem.* **2021**, *17*, 887–902. [CrossRef]
17. Das, S. Annulations involving 2-arylidene-1,3-indanediones: Stereoselective synthesis of spiro- and fused scaffolds. *New J. Chem.* **2020**, *44*, 17148–17176. [CrossRef]
18. Yellappa, S. An anti-Michael route for the synthesis of indole-spiro (indene-pyrrolidine) by 1,3-cycloaddition of azomethineylide with indole-derivatised ofelins. *J. Heterocycl. Chem.* **2020**, *57*, 1083–1089. [CrossRef]
19. Abolhasani, H.; Zarghi, A.; Movahhed, T.K.; Abolhasani, A.; Daraei, B.; Dastmalchi, S. Design, synthesis and biological evaluation of novel indanone containing spiroisoxazoline derivativea with selective COX-2 inhibition as anticancer agents. *Bioorg. Med. Chem.* **2021**, *32*, 115960. [CrossRef]
20. Ahmed, N. Synthetic advances in the indane natural products scaffolds as drug candidates: A review. *Stud. Nat. Prod. Chem.* **2016**, *51*, 383–434. [CrossRef]
21. Prasher, P.; Sharma, M. Medicinal Chemistry of Indane and Its Analogues: A Mini Review. *ChemistrySelect* **2021**, *6*, 2658–2677. [CrossRef]
22. Lee, J.Y.; Spicer, A.P. Hyaluronan: A multifunctionynthesizelton, stealth molecule. *Curr. Opin. Cell Biol.* **2000**, *12*, 581–586. [CrossRef]
23. Raja, S.; Ramesh, V.; Thivaharan, V. Antibacterial and anticoagulant activity of silver nanoparticynthesizedised from a novel source–pods of Peltophorum pterocarpum. *J. Ind. Eng. Chem.* **2015**, *29*, 257–264. [CrossRef]
24. Roy, K.; Srivastwa, A.; Ghosh, C. Anticoagulant, thrombolytic and antibacterial activities of Euphorbia acruensis latex-mediated bioengineered silver nanoparticles. *Green Process. Synth.* **2019**, *8*, 590–599. [CrossRef]
25. Dakshayani, S.S.; Marulasiddeshwara, M.B.; Kumar, M.N.S.; Ramesh, G.; Kumar, P.R.; Devaraja, S.; Hosamani, R. Antimicrobial, anticoagulant and antiplatelet activities of green synthesized silver nanoparticles using Selaginella (Sanjeevini) plant extract. *Int. J. Biol. Macromol.* **2019**, *131*, 787–797. [CrossRef]

26. Talank, N.; Morad, H.; Barabadi, B.; Mojab, F.; Amidi, S.; Kobarfard, F.; Mahjoub, M.A.; Jounaki, K.; Mohammadi, N.; Salehi, G.; et al. Bioengineering of green-synthesized silver nanoparticles: In vitro physicochemical, antibacterial, biofilm inhibitory, anticoagulant, and antioxidant performance. *Talanta* **2022**, *243*, 123374. [CrossRef] [PubMed]

27. Asghar, M.A.; Yousuf, R.I.; Shoaib, M.H.; Asghar, M.A. Antibacterial, anticoagulant and cytotoxic evaluation of biocompatible nanocomposite of chitosan loaded green synthesized bioinspired silver nanoparticles. *Int. J. Biol. Macromol.* **2020**, *160*, 934–943. [CrossRef] [PubMed]

28. Namasivayam, S.K.R.; Srinivasan, S.; Samrat, K.; Priyalakshmi, B.; Kumar, R.D.; Bharani, A.; Kumar, R.G.; Kavisri, M.; Moovendhan, M. Sustainable approach to manage the vulnerable rodents using eco-friendly green rodenticides formulation through nanotechnology principles—A review. *Process Saf. Environ. Prot.* **2023**, *171*, 591–606. [CrossRef]

29. Douglass, M.; Garren, M.; Devine, R.; Mondal, A. Hitesh Handa, Bio-inspired hemocompatible surface modifications for biomedical applications. *Prog. Mater. Sci.* **2022**, *130*, 100997. [CrossRef]

30. Barabadi, H.; Noqani, H.; Ashouri, F.; Prasad, A.; Jounaki, K.; Mobaraki, K.; Mohanta, Y.K.; Mostafavi, E. Nanobiotechnological approaches in anticoagulant therapy: The role of bioengineered silver and gold nanomaterials. *Talanta* **2023**, *256*, 124279. [CrossRef]

31. Chardoli, A.; Karimi, N.; Fattahi, A. Biosynthesis, characterization, antimicrobial and cytotoxic effects of silver nanoparticles using Nigellaarvensis seed extract. *Iran. J. Pharm. Res.* **2017**, *16*, 1167–1175.

32. Amin, Z.; Khashyarmanesh, Z.; Bazzaz, B.; Noghabi, Z. Does biosynthetic silver nanoparticles are more stable with lower toxicity than their synthetic counterparts? *Iran. J. Pharm. Res.* **2019**, *18*, 210–221.

33. Salari, S.; Bahabadi, S.; Samzadeh-Kermani, A.; Yosefzaei, F. In-vitro evaluation of antioxidant and antibacterial potential of greensynthesized silver nanoparticles using prosopis farcta fruit extract. *Iran. J. Pharm. Res.* **2019**, *18*, 430–455.

34. Yaqoob, A.A.; Ahmad, H.; Parveen, T.; Ahmad, A.; Oves, M.; Ismail, I.M.I.; Qari, H.A.; Umar, K.; Ibrahim, M.N.M. Recent advances in metal decorated nanomaterials and their various biological applications: A review. *Front. Chem.* **2020**, *8*, 341. [CrossRef]

35. Barabadi, H.; Kobarfard, F.; Vahidi, H. Biosynthesis and Characterization of Biogenic Tellurium Nanoparticles by Using *Penicillium chrysogenum* PTCC 5031: A Novel Approach in Gold Biotechnology. *Iran. J. Pharm. Res.* **2018**, *17*, 87–97.

36. Barabadi, H.; Honary, S.; Ebrahimi, P.; Alizadeh, A.; Naghibi, F.; Saravanan, M. Optimization of myco-synthesized silver nanoparticles by response surface methodology employing Box-Behnken design. *Inorg. Nano Met. Chem.* **2019**, *49*, 33–43. [CrossRef]

37. De Matteis, V.; Rizzello, L.; Cascione, M.; Liatsi-Douvitsa, E.; Apriceno, A.; Rinaldi, R. Green Plasmonic Nanoparticles and Bio-Inspired Stimuli-Responsive Vesicles in Cancer Therapy Application. *Nanomaterials* **2020**, *10*, 1083. [CrossRef]

38. Sharma, V.K.; Yngard, R.A.; Lin, Y. Silver nanoparticles: Green synthesis and their antimicrobial activities. *Adv. Colloid. Interface Sci.* **2009**, *145*, 83–96. [CrossRef] [PubMed]

39. Panacek, A.; Kvítek, L.; Prucek, R.; Kolar, M.; Vecerova, R.; Pizúrova, N.; Sharma, V.; Nevecna, T.; Zboril, R. Silver colloid nanoparticles: Synthesis, characterization, and their antibacterial activity. *J. Phys. Chem. B* **2006**, *110*, 16248–16253. [CrossRef] [PubMed]

40. Sabbagh, F.; Kiarostami, K.; Khatir, N.; Rezania, S.; Muhamad, I.; Hosseini, F. Effect of zinc content on structural, functional, morphological, and thermal properties of kappa-carrageenan/NaCMC nanocomposites. *Polym. Test.* **2021**, *93*, 106922. [CrossRef]

41. Todorova, M.; Milusheva, M.; Kaynarova, L.; Georgieva, D.; Delchev, V.; Simeonova, S.; Pilicheva, B.; Nikolova, S. Drug-Loaded Silver Nanoparticles—A Tool for Delivery of a Mebeverine Precursor in Inflammatory Bowel Diseases Treatment. *Biomedicines* **2023**, *11*, 1593. [CrossRef]

42. Kamble, S.; Bhosale, K.; Mohite, M.; Navale, S. Methods of Preparation of Nanoparticles. *Int. Adv. Res. Sci. Commun. Technol.* **2022**, *2*, 2581–9429. [CrossRef]

43. Filippo, E.; Manno, D.; Serra, A. Self assembly and branching of sucrose stabilized silver nanoparticles by microwave assisted synthesis: From nanoparticles to branched nanowires structures. *Colloids Surf. A Physicochem. Eng. Asp.* **2009**, *348*, 205–211. [CrossRef]

44. Filippo, E.; Serra, A.; Buccolieri, A.; Manno, D. Green synthesis of silver nanoparticles with sucrose and maltose: Morphological and structural characterization. *J. Non-Cryst. Solids* **2010**, *356*, 344–350. [CrossRef]

45. Shervani, Z.; Yamamoto, Y. Carbohydrate-directed synthesis of silver and gold nanoparticles: Effect of the structure of carbohydrates and reducing agents on the size and morphology of the composites. *Carbohydr. Res.* **2011**, *346*, 651–658. [CrossRef]

46. Ghiyasiyan-Arani, M.; Salavati-Niasari, M.; Masjedi-Arani, M.; Mazloom, F. An Easy Sonochemical Route for Synthesis, Characterization and Photocatalytic Performance of Nanosized FeVO 4 in the Presence of Aminoacids as green Capping Agents. *J. Mater. Sci. Mater. Elect.* **2018**, *29*, 474–485. [CrossRef]

47. Caschera, D.; Toro, R.G.; Federici, F.; Montanari, R.; de Caro, T.; Al-Shemy, M.T.; Adel, A.M. Green Approach for the Fabrication of Silver-Oxidized Cellulose Nano-composite with Antibacterial Properties. *Cellulose* **2020**, *27*, 8059–8073. [CrossRef]

48. Garza-Cervantes, J.A.; Mendiola-Garza, G.; de Melo, E.M.; Dugmore, T.I.; Matharu, A.S.; Morones-Ramirez, J.R. Antimicrobial Activity of a Silver-Microfibrillated Cellulose Biocomposite against Susceptible and Resistant Bacteria. *Sci. Rep.* **2020**, *10*, 7281. [CrossRef]

49. Rather, R.A.; Sarwara, R.K.; Das, N.; Pal, B. Impact of Reducing and Capping Agents on Carbohydrates for the Growth of Ag and Cu Nanostructures and Their Antibacterial Activities. *Particuology* **2019**, *43*, 219–226. [CrossRef]

50. Shanmuganathan, R.; Edison, T.N.J.I.; LewisOscar, F.; Kumar, P.; Shanmugam, S.; Pugazhendhi, A. Chitosan Nanopolymers: An Overview of Drug Delivery against Cancer. *Int. J. Biol. Macromol.* **2019**, *130*, 727–736. [CrossRef]

51. Maiti, P.K.; Ghosh, A.; Parveen, R.; Saha, A.; Choudhury, M.G. Preparation of Carboxy-Methyl Cellulose-Capped Nanosilver Particles and Their Antimicrobial Evaluation by an Automated Device. *Appl. Nanosci.* **2019**, *9*, 105–111. [CrossRef]

52. Duan, H.; Wang, D.; Li, Y. Green chemistry for nanoparticle synthesis. *Chem. Soc. Rev.* **2015**, *44*, 5778–5792. [CrossRef]

53. Zeng, Q.; Shao, D.; Ji, W.; Li, J.; Chen, L.; Song, J.T. The nanotoxicity investigation of optical nanoparticles to cultured cells in vitro. *Toxicol. Rep.* **2014**, *2*, 137–144. [CrossRef]

54. Dieckmann, W. Zur kenntnis der homo thalsaur. *Chem. Ber.* **1914**, *47*, 1439. [CrossRef]

55. Pace, H.; Rogers, N.; Jarolimek, C.; Coleman, V.; Higgins, C.; Ranville, J. Determining Transport Efficiency for the Purpose of Counting and Sizing Nanoparticles via Single Particle Inductively Coupled Plasma Mass Spectrometry. *Anal. Chem.* **2011**, *83*, 9361–9369. [CrossRef]

56. Frisch, M.; Trucks, G.; Schlegel, H.B.; Scuseria, G.E.; Robb, M.A.; Cheeseman, J.R.; Scalmani, G.; Barone, V.; Mennucci, B.; Petersson, G. *Gaussian 09, Revision d. 01*; Gaussian, Inc.: Wallingford, CT, USA, 2009; p. 201.

57. Hadi, Z.; Milad, N.; Asal, Y.-S.; Hamedreza, J.; Saber, M.C.; Seyedeh, S.H. A DFT study on the therapeutic potential of carbon nanostructures as sensors and drug delivery carriers for curcumin molecule: NBO and QTAIM analyses. *Colloids Surf. A Physicochem. Eng. Asp.* **2022**, *651*, 129698. [CrossRef]

58. Tumbarski, Y.; Deseva, I.; Mihaylova, D.; Stoyanova, M.; Krastev, L.; Nikolova, R.; Yanakieva, V.; Ivanov, I. Isolation, Characterization and Amino Acid Composition of a Bacteriocin Produced by Bacillus methylotrophicus Strain BM47. *Food Technol. Biotechnol.* **2018**, *56*, 546–552. [CrossRef]

59. Clauss, A. Rapid physiological coagulation method in determination of fibrinogen. *Acta Haematol.* **1957**, *17*, 237–246. [CrossRef] [PubMed]

60. Kennedy, D.; Orts-Gil, G.; Lai, C.-H.; Müller, L.; Haase, A.; Luch, A.; Seeberger, P. Carbohydrate functionalization of silver nanoparticles modulates cytotoxicity and cellular uptake. *J. Nanobiotechnol.* **2014**, *12*, 59. [CrossRef]

61. Pryshchepa, O.; Pomastowski, P.; Buszewski, B. Silver nanoparticles: Synthesis, investigation techniques, and properties, Advances in Colloid and Interface Science. *Adv. Colloid Interface Sci.* **2020**, *284*, 102246. [CrossRef]

62. Asif, M.; Yasmin, R.; Asif, R.; Ambreen, A.; Mustafa, M.; Umbreen, S. Green Synthesis of Silver Nanoparticles (AgNPs), Structural Characterization, and their Antibacterial Potential. *Dose-Response* **2022**, *20*, 15593258221088709. [CrossRef]

63. De Leersnyder, I.; De Gelder, L.; Van Driessche, I.; Vermeir, P. Revealing the Importance of Aging, Environment, Size and Stabilization Mechanisms on the Stability of Metal Nanoparticles: A Case Study for Silver Nanoparticles in a Minimally Defined and Complex Undefined Bacterial Growth Medium. *Nanomaterials* **2019**, *9*, 1684. [CrossRef]

64. Du, P.; Xu, Y.; Shi, Y.; Xu, Q.; Li, S.; Gao, M. Preparation and shape change of silver nanoparticles (AgNPs) loaded on the dialdehyde cellulose by in-situ synthesis method. *Cellulose* **2022**, *29*, 6831–6843. [CrossRef]

65. Xavier, J.; Vincent, S.; Meder, F.; Vollmer, F. Advances in Optoplasmonic Sensors—Combining Optical Nano/Microcavities and Photonic Crystals with Plasmonic Nanostructures and Nanoparticles. *Nanophotonics* **2018**, *7*, 51807062. [CrossRef]

66. Martinsson, E.; Otte, M.A.; Shahjamali, M.M.; Sepulveda, B.; Aili, D. Substrate Effect on the Refractive Index Sensitivity of Silver Nanoparticles. *J. Phys. Chem. C* **2014**, *118*, 24680–24687. [CrossRef]

67. Arcas, A.S.; Jaramillo, L.; Costa, N.S.; Allil, R.C.S.B.; Werneck, M.M. Localized Surface Plasmon Resonance-Based Biosensor on Gold Nanoparticles for Taenia Solium Detection. *Appl. Opt.* **2021**, *60*, 8137–8144. [CrossRef] [PubMed]

68. Raiche-Marcoux, G.; Loiseau, A.; Maranda, C.; Poliquin, A.; Boisselier, E. Parametric Drug Release Optimization of Anti-Inflammatory Drugs by Gold Nanoparticles for Topically Applied Ocular Therapy. *Int. J. Mol. Sci.* **2022**, *23*, 16191. [CrossRef] [PubMed]

69. Zaheer, Z. Biogenic synthesis, optical, catalytic, and in vitro antimicrobial potential of Ag-nanoparticles prepared using Palm date fruit extract. *J. Photochem. Photobiol. B Biol.* **2018**, *178*, 584–592. [CrossRef] [PubMed]

70. Jiang, P.; Zhou, J.J.; Li, R.; Gao, Y.; Sun, T.L.; Zhao, X.W.; Xiang, Y.J.; Xie, S.S. PVP-capped twinned gold plates from nanometer to micrometer. *J. Nanoparticle Res.* **2006**, *8*, 927. [CrossRef]

71. Ezraa, L.; O'Della, Z.J.; Huia, J.; Riley, K.R. Emerging investigator series: Quantifying silver nanoparticle aggregation kinetics in realtime using particle impact voltammetry coupled with UV-vis spectroscopy. *Environ. Sci. Nano.* **2020**, *7*, 2509–2521. [CrossRef]

72. AbuDalo, M.; Al-Mheidat, I.; Al-Shurafat, A.; Grinham, C.; Oyanedel-Craver, V. Synthesis of silver nanoparticles using a modified Tollens' method in conjunction with phytochemicals and assessment of their antimicrobial activity. *Peer J.* **2019**, *7*, e641. [CrossRef]

73. Ahmed, K.B.A.; Mohammed, A.S.; Anbazhagan, V. Interaction of sugar stabilized silver nanoparticles with the T-antigen specific lectin, jacalin from Artocarpus Integrifolia. *Spectrochim. Acta Part A Mol. Biomol. Spectrosc.* **2015**, *145*, 110–116. [CrossRef]

74. Wiercigroch, E.; Szafraniec, E.; Czamara, K.; Pacia, M.Z.; Majzner, K.; Kochan, K.; Kaczor, A.; Baranska, M.; Malek, K. Raman and infrared spectroscopy of carbohydrates: A review. *Spectrochim. Acta A Mol. Biomol. Spectrosc.* **2017**, *185*, 317–335. [CrossRef]

75. Wells, H.; Atalla, R. An investigation of the vibrational spectra of glucose, galactose, and mannose. *J. Mol. Struct.* **1990**, *224*, 385–424. [CrossRef]

76. Sivakesava, S.; Irudayaraj, J. Prediction of inverted cane sugar adulteration of honey by Fourier transform infrared spectroscopy. *J. Food Sci.* **2001**, *66*, 972–978. [CrossRef]

77. David, M.; Hategan, A.; Berghian-Grosan, C.; Magdas, D. The Development of Honey Recognition Models Based on the Association between ATR-IR Spectroscopy and Advanced Statistical Tools. *Int. J. Mol. Sci.* **2022**, *23*, 9977. [CrossRef]

78. Iravani, S.; Korbekandi, H.; Mirmohammadi, S.; Zolfaghari, B. Synthesis of silver nanoparticles: Chemical, physical and biological methods. *Res. Pharm. Sci.* **2014**, *9*, 385–406.

79. Islam, N. Green synthesis and biological activities of gold nanoparticles functionalized with *Salix alba*. *Arab. J. Chem.* **2019**, *12*, 2914–2925. [CrossRef]

80. *Nanomaterials in Plants, Algae, and Microorganisms: Concepts and Controversies*, 1st ed.; Tripathi, D.; Ahmad, P.; Sharma, S.; Chauhan, D.; Dubey, N.K. (Eds.) Academic Press: Cambridge, MA, USA, 2018.

81. Younas, M.; Ahmad, M.A.; Jannat, F.T.; Ashfaq, T.; Ahmad, A. Chapter 18—Role of silver nanoparticles in multifunctional drug delivery. In *Micro and Nano Technologies, Nanomedicine Manufacturing and Applications*; Verpoort, F., Ahmad, I., Ahmad, A., Khan, A., Chee, C.Y., Eds.; Elsevier: Orlando, FL, USA, 2021; pp. 297–319. [CrossRef]

82. Ivask, A. Toxicity mechanisms in Escherichia coli vary for silver nanoparticles and differ from ionic silver. *ACS Nano* **2014**, *8*, 374–386. [CrossRef]

83. Stoehr, L.; Gonzalez, E.; Stampfl, A.; Casals, E.; Duschl, A.; Puntes, V.; Oostingh, G. Shape matters: Effects of silver nanospheres and wires on human alveolar epithelial cells. *Part. Fibre Toxicol.* **2011**, *8*, 36. [CrossRef]

84. Huk, A.; Izak-Nau, E.; Reidy, B.; Boyles, M.; Duschl, A.; Lynch, I.; Dušinska, M. Is the toxic potential of nanosilver dependent on its size? *Part. Fibre Toxicol.* **2014**, *11*, 65. [CrossRef]

85. Zhang, T.; Wang, L.; Chen, Q.; Chen, C. Cytotoxic potential of silver nanoparticles. *Yonsei Med. J.* **2014**, *55*, 283–291. [CrossRef]

86. León-Silva, S.; Fernández-Luqueño, F.; López-Valdez, F. Silver Nanoparticles (AgNP) in the Environment: A Review of Potential Risks on Human and Environmental Health. *Water Air Soil Pollut.* **2016**, *227*, 306. [CrossRef]

87. Sigalov, M.V. Keto–Enol Tautomerism of Phenindione and Its Derivatives: An NMR and Density Functional Theory (DFT) Reinvestigation. *J. Phys. Chem. A* **2015**, *119*, 1404–1414. [CrossRef]

88. Mulvaney, P.; Liz-Marzan, L.; Giersig, M.; Ung, T.J. Silica encapsulation of quantum dots and metal clusters. *Mater. Chem.* **2000**, *10*, 1259–1270. [CrossRef]

89. Philip, D. Biosynthesis of Au, Ag and Au-Ag nanoparticles using edible mushroom extract. *Spectrochim. Acta A Mol. Biomol. Spectrosc.* **2009**, *73*, 374–381. [CrossRef] [PubMed]

90. Feizi-Dehnayebi, M.; Dehghanian, E.; Mansouri-Torshizi, H. A novel palladium (II) antitumor agent: Synthesis, characterization, DFT perspective, CT-DNA and BSA interaction studies via in-vitro and in-silico approaches. *Spectrochim. Acta A Mol. Biomol. Spectrosc.* **2021**, *249*, 119215. [CrossRef] [PubMed]

91. Feizi-Dehnayebi, M.; Dehghanian, E.; Mansouri-Torshizi, H. Synthesis and characterization of Pd (II) antitumor complex, DFT calculation and DNA/BSA binding insight through the combined experimental and theoretical aspects. *J. Mol. Struct.* **2021**, *1240*, 130535. [CrossRef]

92. Mantasha, I.; Shahid, M.; Kumar, M.; Ansari, A.; Akhtar, M.N.; AlDamen, M.A.; Song, Y.; Ahmad, M.; Khan, I.M. Exploring solvent dependent catecholase activity in transition metal complexes: An experimental and theoretical approach. *New J. Chem.* **2020**, *44*, 1371–1388. [CrossRef]

93. Feizi-Dehnayebi, M.; Dehghanian, E.; Mansouri-Torshizi, H. Probing the biomolecular (DNA/BSA) interaction by new Pd (II) complex via in-depth experimental and computational perspectives: Synthesis, characterization, cytotoxicity, and DFT approach. *J. Iran. Chem. Soc.* **2022**, *19*, 3155–3175. [CrossRef]

94. Feizi-Dehnayebi, M.; Dehghanian, E.; Mansouri-Torshizi, H. Biological activity of bis-(morpholineacetato) palladium (II) complex: Preparation, structural elucidation, cytotoxicity, DNA-/serum albumin-interaction, density functional theory, in-silico prediction and molecular modeling. *Spectrochim. Acta A Mol. Biomol. Spectrosc.* **2022**, *281*, 121543. [CrossRef]

95. Feizi-Dehnayebi, M.; Dehghanian, E.; Mansouri-Torshizi, H. DNA/BSA binding affinity studies of new Pd (II) complex with SS and NN donor mixed ligands via experimental insight and molecular simulation: Preliminary antitumor activity, lipophilicity and DFT perspective. *J. Mol. Liq.* **2021**, *344*, 117853. [CrossRef]

96. Séguy, L.; Groo, A.-C.; Malzert-Fréon, A. How nano-engineered delivery systems can help marketed and repurposed drugs in Alzheimer's disease treatment? *Drug Discov. Today* **2022**, *27*, 1575–1589. [CrossRef]

97. Prakash, P.; Gnanaprakasam, P.; Emmanuel, R.; Arokiyaraj, M.; Saravanan, M. Green synthesis of silver nanoparticles from leaf extract of *Mimusops elengi*, Linn. for enhanced antibacterial activity against multi drug resistant clinical isolates. *Colloids Surf. B Biointerfaces* **2013**, *108*, 255–259. [CrossRef]

98. Solomon, D.; Gupta, N.; Mulla, N.S.; Shukla, S.; Guerrero, Y.A.; Gupta, V. Role of In Vitro Release Methods in Liposomal Formulation Development: Challenges and Regulatory Perspective. *AAPSJ* **2017**, *19*, 1669–1681. [CrossRef]

99. Yu, M.; Yuan, W.; Li, D.; Schwendeman, A.; Schwendeman, S.P. Predicting Drug Release Kinetics from Nanocarriers inside Dialysis Bags. *J. Control. Release* **2019**, *315*, 23–30. [CrossRef]

100. Kumar, B.; Jalodia, K.; Kumar, P.; Gautam, H.K. Recent Advances in Nanoparticle-Mediated Drug Delivery. *J. Drug Deliv. Sci. Technol.* **2017**, *41*, 260–268. [CrossRef]

101. Sabbagh, F.; Khatir, N.M.; Kiarostami, K. Synthesis and Characterization of *k*-Carrageenan/PVA Nanocomposite Hydrogels in Combination with MgZnO Nanoparticles to Evaluate the Catechin Release. *Polymers* **2023**, *15*, 272. [CrossRef]

102. Jain, A.S.; Pawar, P.S.; Sarkar, A.; Junnuthula, V.; Dyawanapelly, S. Bionanofactories for Green Synthesis of Silver Nanoparti-cles: Toward Antimicrobial Applications. *Int. J. Mol. Sci.* **2021**, *22*, 11993. [CrossRef]

103. Guzman, M.; Arcos, M.; Dille, J.; Godet, S.; Rousse, C. Effect of the Concentration of NaBH4 and N2H4 as Reductant Agent on the Synthesis of Copper Oxide Nanoparticles and its Potential Antimicrobial Applications. *Nano Biomed. Eng.* **2018**, *10*, 392–405. [CrossRef]
104. Narayanan, S.; Gupta, P.; Nazim, U.; Ali, M.; Karadkhelkar, N.; Ahmad, M.; Chen, Z.-S. Anti-cancer effect of Indanone-based thiazolyl hydrazone derivative on colon cancer cell lines. *Int. J. Biochem. Cell Biol.* **2019**, *110*, 21–28. [CrossRef]
105. Chanda, D.; Bhushan, S.; Guru, S.K.; Shanker, K.; Wani, Z.A.; Rah, B.A.; Luqman, S.; Mondhe, D.M.; Pal, A.; Negi, A.S. Anticancer activity, toxicity and pharmacokinetic profile of an indanone derivative. *Eur. J. Pharm. Sci.* **2012**, *47*, 988–995. [CrossRef]
106. Balasubramaniam, M.; Lakkaniga, N.R.; Dera, A.A.; Fayi, M.A.; Abohashrh, M.; Ahmad, I.; Chandramoorthy, H.C.; Nalini, G.; Rajagopalan, P. FCX-146, a potent allosteric inhibitor of Akt kinase in cancer cells: Lead optimization of the second-generation arylidene indanone scaffold. *Biotechnol. Appl. Biochem.* **2021**, *68*, 82–91. [CrossRef]
107. Cohen, A.T.; Agnelli, G.; Anderson, F.A.; Arcelus, J.I.; Bergqvist, D.; Brecht, J.G. Venousthromboembolism (VTE) in Europe. The number of VTE events and associated morbidityand mortality. *Thromb. Haemost.* **2007**, *98*, 756–764. [PubMed]
108. Heitt, J.A. The epidemiology of venous thromboembolism in the community. *Arterioscler. Thromb. Vasc. Biol.* **2008**, *28*, 370–372. [CrossRef]
109. Biss, T.T.; Brandão, L.R.; Kahr, W.H.; Chan, A.K.; Williams, S. Clinical features andoutcome of pulmonary embolism in children. *Brit. J. Haematol.* **2008**, *142*, 808–818. [CrossRef] [PubMed]
110. Ede, O.A.; Bindá, F.A.; Silva, A.M.; Costa, T.D.; Fernandes, M.C.; Fernandes, M.C. Risk factors andprophylaxis for venous thromboembolism in hospitals in the city of Manaus. *J. Bras. Pneumol.* **2009**, *35*, 114–121. [CrossRef]
111. Santos Fernandes, C.J.S.; AlvesJúnior, J.L.; Gavilanes, F.; Prada, L.F.; Morinaga, L.K.; Rogerio, S. New anticoagulants for the treatment of venous thromboembolism. *J. Bras. Pneumol.* **2016**, *42*, 146–154. [CrossRef] [PubMed]
112. Azeez, M.A.; Durodola, F.A.; Lateef, A.; Yekeen, T.A.; Adubi, A.O.; Oladipo, I.C.; Adebayo, E.A.; Badmus, J.A.; Abawulem, A.O. Green synthesized novel silver nanoparticles and their application as anticoagulant and thrombolytic agents: A perspective. *IOP Conf. Ser. Mater. Sci. Eng.* **2020**, *805*, 012043. [CrossRef]
113. Quinn, A.; Bellamy, M. Chapter 53—Hemostasis and coagulation. In *Foundations of Anesthesia*, 2nd ed.; Hemmings, H.C., Hopkins, P.M., Eds.; Mosby: Maryland Heights, MO, USA, 2006; pp. 635–645. [CrossRef]
114. Xiang, J.; Wang, H.; Ma, C.; Zhou, M.; Wu, Y.; Wang, L.; Guo, S.; Chen, T.; Shaw, C. Ex Vivo Smooth Muscle Pharmacological Effects of a Novel Bradykinin-Related Peptide, and Its Analogue, from Chinese Large Odorous Frog, Odorrana livida Skin Secretions. *Toxins* **2016**, *8*, 283. [CrossRef]
115. Wright, D.; Sharma, P.; Ryu, M.H.; Rissé, P.A.; Ngo, M.; Maarsingh, H.; Koziol-White, C.; Jha, A.; Halayko, A.J.; West, A.R. Models to study airway smooth muscle contraction in vivo, ex vivo and in vitro: Implications in understanding asthma. *Pulm. Pharmacol. Ther.* **2013**, *26*, 24–36. [CrossRef]
116. Liu, S.; Lin, Z. Vascular Smooth Muscle Cells Mechanosensitive Regulators and Vascular Remodeling. *J. Vasc. Res.* **2022**, *59*, 90–113. [CrossRef]
117. Hashitani, H.; Brading, A.F.; Suzuki, H. Correlation between spontaneous electrical, calcium and mechanical activity in detrusor smooth muscle of the guinea-pig bladder. *Br. J. Pharmacol.* **2004**, *141*, 183–193. [CrossRef]
118. Durnin, L.; Kwok, B.; Kukadia, P.; McAvera, R.; Corrigan, R.D.; Ward, S.M.; Zhang, Y.; Chen, Q.; Koh, S.D.; Sanders, K.M.; et al. An ex vivo bladder model with detrusor smooth muscle removed to analyze biologically active mediators released from the suburothelium. *J. Physiol.* **2019**, *597*, 1467–1485. [CrossRef]
119. Tarhan, F.; Duman, N.Ç.; Özkulab, S.; Karaalp, A.; Cangüven, Ö. In vitro contractile responses of human detrusor smooth muscle to oxytocin: Does it really have effect? *Aging Male* **2020**, *23*, 1141–1145. [CrossRef]
120. Upchurch, W.J.; Iaizzo, P.A. In vitro contractile studies within isolated tissue baths: Translational research from Visible Heart®Laboratories. *Exp. Biol. Med.* **2022**, *247*, 584–597. [CrossRef]
121. Brozovich, F.V.; Nicholson, C.J.; Degen, C.V.; Gao, Y.Z.; Aggarwal, M.; Morgan, K.G. Mechanisms of Vascular Smooth Muscle Contraction. *Pharmacol. Rev.* **2016**, *68*, 476–532. [CrossRef]
122. Jespersen, B.; Tykocki, N.R.; Watts, S.W.; Cobbett, P.J. Measurement of smooth muscle function in the isolated tissue bathapplications to pharmacology research. *J. Vis. Exp.* **2015**, *95*, 52324. [CrossRef]
123. Ivanov, I.; Nikolova, S.; Aladjov, D.; Stefanova, I.; Zagorchev, P. Synthesis and Contractile Activity of Substituted 1,2,3,4-Tetrahydroisoquinolines. *Molecules* **2011**, *16*, 7019–7042. [CrossRef]
124. Stefanova, I.; Argirova, M.; Krustev, A. Influence of model melanoidins on calciumdependent transport mechanisms in smooth muscle tissue. *Mol. Nutr. Food Res.* **2007**, *51*, 468–472. [CrossRef]
125. Argirova, M.; Stefanova, I.; Krustev, A. New biological properties of coffee melanoidins. *Food Funct.* **2013**, *4*, 1204–1208. [CrossRef]
126. Sagorchev, P.; Lukanov, J.; Beer, A.-M. Effects of 1.8-cineole (eucalyptol) on the spontaneous contractile activity of smooth muscles fibre. *J. Med. Plants Res.* **2015**, *9*, 486–493. [CrossRef]
127. Beer, A.M.; Lukanov, J.; Sagorchev, P. Effect of Thymol on the spontaneous contractile activity of the smooth muscles. *Phytomedicine* **2007**, *14*, 65–69. [CrossRef]

128. González, C.; Salazar-García, S.; Palestino, G.; Martínez-Cuevas, P.P.; Ramírez-Lee, M.A.; Jurado-Manzano, B.B.; Rosas-Hernández, H.; Gaytán-Pacheco, N.; Martel, G.; Espinosa-Tanguma, R.; et al. Effect of 45 nm silver nanoparticles (AgNPs) upon the smooth muscle of rat trachea: Role of nitric oxide. *Toxicol. Lett.* **2011**, *207*, 306–313. [CrossRef]
129. Sheridan, H.; Lemon, S.; Frankish, N.; McArdle, P.; Higgins, T.; James, J.P.; Bhandari, P. Synthesis and antispasmodic activity of nature identical substituted indanes and analogues. *Eur. J. Med. Chem.* **1990**, *25*, 603–608. [CrossRef]

 *biomedicines*

Communication

Delivery of Lipid Nanoparticles with ROS Probes for Improved Visualization of Hepatocellular Carcinoma

Vera S. Shashkovskaya [1,2,*], Polina I. Vetosheva [3], Arina G. Shokhina [2,4,5], Ilya O. Aparin [3,5], Tatiana A. Prikazchikova [6], Arsen S. Mikaelyan [7], Yuri V. Kotelevtsev [1], Vsevolod V. Belousov [2,4,5], Timofei S. Zatsepin [6] and Tatiana O. Abakumova [2,*]

[1] Vladimir Zelman Center for Neurobiology and Brain Rehabilitation, Skolkovo Institute of Science and Technology, 121205 Moscow, Russia
[2] Institute of Translational Medicine, Pirogov Russian National Research Medical University, 117997 Moscow, Russia
[3] Center for Molecular and Cellular Biology, Skolkovo Institute of Science and Technology, 121205 Moscow, Russia
[4] Federal Center of Brain Research and Neurotechnologies, Federal Medical Biological Agency, 119435 Moscow, Russia
[5] Shemyakin-Ovchinnikov Institute of Bioorganic Chemistry, 117997 Moscow, Russia
[6] Department of Chemistry, M. V. Lomonosov Moscow State University, 119991 Moscow, Russia
[7] Koltsov Institute of Developmental Biology of Russian Academy of Sciences, 152742 Moscow, Russia
* Correspondence: verashashkovskaya@gmail.com (V.S.S.); sandalovato@gmail.com (T.O.A.)

Abstract: Reactive oxygen species (ROS) are highly reactive products of the cell metabolism derived from oxygen molecules, and their abundant level is observed in many diseases, particularly tumors, such as hepatocellular carcinoma (HCC). In vivo imaging of ROS is a necessary tool in preclinical research to evaluate the efficacy of drugs with antioxidant activity and for diagnosis and monitoring of diseases. However, most known sensors cannot be used for in vivo experiments due to low stability in the blood and rapid elimination from the body. In this work, we focused on the development of an effective delivery system of fluorescent probes for intravital ROS visualization using the HCC model. We have synthesized various lipid nanoparticles (LNPs) loaded with ROS-inducible hydrocyanine pro-fluorescent dye or plasmid DNA (pDNA) with genetically encoded protein sensors of hydrogen peroxide (HyPer7). LNP with an average diameter of 110 ± 12 nm, characterized by increased stability and pDNA loading efficiency ($64 \pm 7\%$), demonstrated preferable accumulation in the liver compared to 170 nm LNPs. We evaluated cytotoxicity and demonstrated the efficacy of hydrocyanine-5 and HyPer7 formulated in LNP for ROS visualization in mouse hepatocytes (AML12 cells) and in the mouse xenograft model of HCC. Our results demonstrate that obtained LNP could be a valuable tool in preclinical research for visualization ROS in liver diseases.

Keywords: reactive oxygen species; hepatocellular carcinoma; HyPer7; hydrocyanine; lipid nanoparticles

Citation: Shashkovskaya, V.S.; Vetosheva, P.I.; Shokhina, A.G.; Aparin, I.O.; Prikazchikova, T.A.; Mikaelyan, A.S.; Kotelevtsev, Y.V.; Belousov, V.V.; Zatsepin, T.S.; Abakumova, T.O. Delivery of Lipid Nanoparticles with ROS Probes for Improved Visualization of Hepatocellular Carcinoma. *Biomedicines* **2023**, *11*, 1783. https://doi.org/10.3390/biomedicines11071783

Academic Editors: Yongtai Zhang and Zhu Jin

Received: 26 May 2023
Revised: 13 June 2023
Accepted: 19 June 2023
Published: 21 June 2023

1. Introduction

Reactive oxygen species (ROS) are chemically reactive molecules with various essential functions in living organisms and include hydroxyl radical ($\cdot$OH), superoxide anion ($O_2 \cdot^-$), singlet oxygen ($^1O_2$), and hydrogen peroxide (H_2O_2) [1,2]. Under normal physiological conditions, the production of ROS is essential for the cells to defend against pathogens and to promote growth and death. However, the increased production of ROS leads to the damage of many biomolecules, apoptosis, and cell arrest [3–5]. Most cellular H_2O_2 is produced spontaneously or catalytically by superoxide dismutase from ($O_2 \cdot^-$) [6]. H_2O_2 is relatively stable in the aqueous solution. It is a potent redox signaling molecule and also can facilitate transmembrane transport [7].

Various methods have been developed to visualize ROS, based on spectrophotometry, fluorescence, chemiluminescence, and electron spin resonance [8,9]. Fluorescence imaging methods are widely used and allow continuous monitoring of the ROS-associated molecules in situ in living cells on a real-time scale using fluorescence-based probes [9]. The most common probes in biomedical research are non-selective sensors based on a leuco-form of fluorescent dyes, which can light up when oxidized by ROS to parental fluorescein, rhodamine, or hydroethidine. However, there are numerous limitations for in vivo applications of these probes critically reviewed in publications [10–13]. Alternative fluorochromes with emission maxima in the near-infrared region offer low phototoxicity to the cells, low autofluorescence, and good tissue penetration, making them attractive for imaging in tissues [11]. Among these, cyanine dyes are particularly interesting due to their physical and chemical properties and their ability to react with intracellular and extracellular ROS. Genetically encoded sensors are another class of ROS-sensitive probes, e.g., HyPer7. It is a selective ratiometric sensor that allows monitoring H_2O_2 concentration, and it can be easily visualized by fluorescence microscopy. Although these probes were used to detect ROS in vivo in several studies, it is still challenging to deliver them to specific organs or tissues [14–16]. To prevent early oxidation and increase blood circulation time, cyanine dyes and genetically encoded sensors, such as HyPer7, can be loaded or conjugated to nanoparticles for better visualization of ROS in vivo. Kim et al. showed that hydrocyanine conjugated to chitosan-functionalized pluronic-based nanocarriers can detect ROS in tumor sites by fluorescence and show fluorescence up to two days after injection [14]. Another study showed that the nanoencapsulation of sensors could improve detection or imaging of H_2O_2 in biological systems [17]. Hammond et al. developed a novel nanoprobe based on dye encapsulation with improved sensor function for intracellular measurement of H_2O_2 [18]. Lipid nanoparticles (LNPs), possessing excellent biocompatibility and low toxicity, are a clinically approved small interfering RNA (siRNA) delivery system for liver diseases, so we decided to use this technology to deliver fluorescent probes for detection of ROS in hepatocellular carcinoma (HCC).

HCC is one of the most malignant and common diseases and is the third leading cause of cancer death [19]. The common mechanism of hepatocarcinogenesis is chronic inflammation associated with severe oxidative stress [20]. It is evident that ROS play a pathogenetic role in the progression of HCC, as they can stimulate the growth of cancer cells [21]. To assess the HCC, noninvasive imaging tools such as magnetic resonance imaging (MRI), ultrasound (US), and computed tomography (CT) are currently used. US is a well-suited tool for diagnosing HCC due to its cost-effectiveness and accuracy in detecting the focal liver lesions [22]. The introduction of microbubbles contrast agents leads to the improved diagnostic capability of HCC by conventional US [23]. MRI and CT are considered as powerful tools for detection of HCC. Recently, several functioning MRI techniques, such as hepatobiliary contrast agents, have been developed to improve the evaluation of liver lesions [23]. However, early diagnosis of HCC still remains challenging, and more studies are required to develop the imaging tools for detection of HCC. Visualization of ROS in HCC models could help to diagnose the disease at an early stage and investigate new drug delivery systems.

Here, we report on the development of an effective delivery system of HyPer7 and hydrocyanine-5 (hydro-Cy5) formulated in LNPs for ROS visualization in HCC. LNPs have demonstrated successful accumulation in the liver and were able to deliver ROS sensors. We expect that it will become an effective system for the detection of ROS in liver diseases.

2. Materials and Methods

2.1. Cell Culture

AML12 and Huh-7 cell lines were obtained from ATCC. AML12 cells and the Huh-7 cell line were maintained in a 37 °C humidified incubator with 5% CO_2 in DMEM/F12 (PanEco, Moscow, Russia) supplemented with 10% fetal bovine serum (FBS) (Capricorn, Ebsdorfergrund, Germany).

2.2. Synthesis of Hydro-Cy5-siRNA-LNPs

Hydro-Cy5 dye was synthesized based on starting sulfo-Cy5 dicarboxylic acid (Lumiprobe, Moscow, Russia) based on a previously established procedure [24] and formulated in LNPs as described previously [25].

First, hydro-Cy5 was dissolved in ethanol (10 mg/mL) and mixed with lipids. LNP were obtained by mixing a solution of small interfering RNA to firefly luciferase (siLuc) (0.4 mg/mL) in a microfluidic device in a citrate buffer (10 mM, pH 3.0) and an ethanol solution of a mixture of hydrocyanine (3:1 by volume) and lipids and lipidoids (C12-200, 1,2-dystearoyl-CH-glycero-3-phosphocholine, 1,2-dimyristoyl-CH-glycero-3-phosphoethanolamine-N-[methoxy(polyethylene glycol)-2000] and cholesterol) as described previously [25]. The molar ratio of hydro-Cy5:siLuc was 1:1 (hydro-Cy5-LNP_1) or 2:1 (hydro-Cy5-LNP_2).

For biodistribution studies, we used cyanine 5.5 dye (instead of leuco-form of hydrocyanine 5) to obtain fluorescently labeled LNP with different diameters (110 nm and 170 nm); the molar ratio for cyanine 5.5:Luc was 1:10.

2.3. Synthesis of LNPs with Plasmid Encoding HyPer7

For HyPer7 construct generation, EGFP was removed from pAAV.TBG.PI.eGFP.WPRE.bGH (addgene #105535, Watertown, MA, USA) using NotI and HindIII restriction enzyme sites. Then, amplified from pCS2+HyPer7-NES (addgene #136467, forward primer: 5′-ATATATGCGGCCGCGCCACCATGCACCTGGC-3′, reverse primer: 5′-ATATATAAGCTTT-TACAGGGTCAGCCGCTCCA-3′), a cytosolic version of HyPer7 with an added nuclear exclusion sequence (NES) was cloned into the prepared backbone.

Two methods were used to obtain lipid nanoparticles with plasmid DNA (pDNA): manual formulation and microfluidics. In manual formulation, LNPs were obtained by mixing solutions of pDNA (0.1 mg/mL) in citrate buffer (10 mM, pH 3) and lipids in ethanol (3:1 by volume) by rapid addition of ethanol in the buffer and further suspension. The lipids (C12-200, dioleoylphosphatidylethanolamine—DOPE or dioleoylphosphatidylcholine—DOPC, cholesterol, PEG-lipid) were mixed in ethanol in a molar ratio of 35:16:46.5:2.5. The weight ratio of DNA/C12-200 is 1:10. The mixture was incubated for 10 min at room temperature to complete particle self-assembly. The nanoparticle suspension was diluted in a PBS, dialyzed overnight in 2000 molecular weight dialysis cassettes against 500 times the volume of PBS at pH 7.4, and filtered through a polyethersulfone syringe filter (0.22 μm) (Table S1).

Another formulation method of LNPs was focused on mixing solutions of pDNA and lipids (3:1 by volume) using the NanoAssemblr® Benchtop microfluidic device (Precision NanoSystems, Vancouver, BC, Canada). The molar ratio of lipids in the mixture, the mass ratio of DNA/C12-200, and the concentration of pDNA were used the same as in the manual formulation. The solution mixing range was 10 mL/min. The use of microfluidic formulation allows the mixing rate to be adjusted by the flow rate of each of the solutions. The aqueous and ethanol phases containing dissolved components of nanoparticles are injected into each inlet of the NanoAssemblr cartridge using syringes. Computer control allows the input mixing parameters to be controlled to optimize particle characteristics such as particle size and encapsulation efficiency and eliminate variability.

2.4. Characterization of Obtained LNP

Particle sizes were measured and calculated by intensity using Zetasizer Nano ZSP (Malvern Panalytical, Malvern, UK) according to the manufacturer's protocol [26]. The loading efficiency of pDNA was analyzed by using the Quant-iT™ RiboGreen® reagent (Thermo Fisher Scientific R11491, Waltham, MA, USA) as described earlier [27]. First, a calibration curve was obtained using standard pDNA in the concentration range 0.03–0.2 μg/mL. After that, DNA concentration was measured in obtained nanoparticles before and after disruption with the Triton X-100 buffer. The difference in these values shows the portion of loaded pDNA.

2.5. Cytotoxicity of Obtained Lipid Nanoparticles (MTT Assay)

The cytotoxicity of selected LNPs was measured on a culture of mouse hepatocytes (AML12). Cells were seeded in a 96-well plate and grown to 70% confluent monolayer (48 h, 5% CO_2, 37 °C). The LNPs were diluted in DMEM/F12 (PanEco, Moscow, Russia) with 10% FBS (Capricorn, Ebsdorfergrund, Germany), added to the cells, and incubated for 24 h (5% CO_2, 37 °C). The next day, the cells were washed with PBS and MTS-reagent was added (Promega, Madison, WI, USA) according to the manufacturer's protocol. Absorbance at 490 nm was then analyzed. Unexposed cells were used as a control.

2.6. Analysis of ROS Activity In Vitro

Analysis of ROS activity in vitro was performed using a fluorescent microplate reader (Varioskan Lux, ThermoFisher, Waltham, MA, USA). We incubated hydro-Cy5-LNPs with 10 μM H_2O_2 for 10 min at room temperature and quantified fluorescence at 640/680 emission/excitation filters.

For imaging experiments, AML12 cells were seeded in a 35 mm glass confocal dish. Twenty-four hours after incubation with LNP, cells were washed with 1 mL PBS (PanEco, Moscow, Russia). To determine the response of LNP- hydro-Cy5 to exogenous H_2O_2, a solution of H_2O_2 was added to a final concentration of 10 μM. Cell imaging was performed using a Nikon Eclipse Ti2 microscope. Fluorescence of LNP- hydro-Cy5 probes was analyzed at 640/680 excitation/emission filters.

2.7. Delivery of LNPs to the Liver and Analysis of ROS Activity in Vivo in HCC Model

All experiments were approved and performed in accordance with institutional guidelines of the Koltsov Institute of Developmental Biology of the Russian Academy of Sciences (Approval #60) and Pirogov Russian National Research Medical University (Moscow, Russia, Approval 06/2021).

Biodistribution. For biodistribution experiments we intravenously injected cyanine 5.5-LNP (size 110 nm and 170 nm) and analyzed whole body fluorescence using IVIS Spectrum CT (Perkin Elmer, Waltham, MA, USA) at 5 min, 1, 4, 6, and 24 h after injection at 675/720 excitation/emission filters. Spectral unmixing was performed using Living Imaging 4.5.3 software to separate the tissue autofluorescence signal. For the biodistribution experiment of LNPs loaded with pDNA, we used plasmid-encoding firefly luciferase (Fluc) with an average diameter 110 nm and injected it intravenously (500 μg DNA/kg); after that, we analyzed its luminescence using IVIS Spectrum CT (Perkin Elmer, Waltham, MA, USA) at 3, 6, 9, and 14 days after injection.

HCC model. To induce ROS in the liver, we used an HCC xenograft model. We injected Huh-7-Luc cells (2×10^6 cells in 50% Matrigel®) to the nu/nu mice (male, 6–8 weeks old) in the median lobe of liver. Seven days after tumor implantation, we injected the obtained LNPs in PBS, and analyzed its fluorescence using IVIS Spectrum CT. For ex vivo visualization of ROS, animals were anesthetized (Zoletil® (tiletamine + zolazepam) 20 mg/kg, xylazine 0.2 mg/kg) and livers were imaged using 640/680 excitation/emission filters. Photons of tumor and healthy tissue were quantified using Living Image software (Xenogen Corp., Alameda, CA, USA).

2.8. Statistical Analysis

GraphPad Prism 8.2.1 (GraphPad Prism Software, Inc., San Diego, CA, USA) was used for statistical analysis. Tukey's multiple comparison test with 95 % confidence level or two-way ANOVAs were used for statistical analysis. A *p*-value less than 0.05 was considered as significant.

3. Results

3.1. Synthesis and Characterisation of Lipid Nanoparticles and ROS Analysis In Vitro

Previously, we synthesized two different hydro-Cy5 derivatives and analyzed their sensitivity for ROS detection [7]. In this study, we investigated the possibility of using the free dye after intravenous bolus injection into the blood of healthy animals and animals with HCC, as was previously described in other studies [28,29]. Unfortunately, we observed rapid hydro-Cy5 dye excretion from the body and a strong fluorescence signal in the bladder within 5 min after intravenous injection (Figure S1). Therefore, we decided to encapsulate hydro-Cy5 in LNP to improve targeted delivery to the liver. The zeta potential of hydro-Cy5-LNP is -7.58 mV.

One of the newly developed molecular probes for ROS visualization is the pDNA HyPer7, which encodes the HyPer protein sensitive to H_2O_2 and capable of responding to the ROS changes in the cell within seconds [30]. In order to efficiently deliver HyPer7 to the liver in vivo, we also decided to encapsulate it in LNPs, as publications show that LNPs can serve as an efficient delivery system for different molecules [25]. While LNP are clinically approved for siRNA and mRNA delivery, efficient encapsulation and delivery of pDNA requires optimization of composition and molar ratio of lipids. For the synthesis of LNPs with HyPer7, we decided to compare the LNP formulation by using two different helper lipids, DOPE and DOPC. The other lipids used and their proportions were not changed. The manual formulation leads to the formation of particles of uncontrolled size in the diameter range of 120 to 250 nm with a high degree of sample heterogeneity. The main disadvantage of this method is the lack of the possibility to control the parameters of the LNPs using the mixing rate in manual formation. In the case of microfluidic cartridges, the particles have a smaller size, significantly higher efficiency of pDNA loading, and their characteristics are much more reproducible. The polydispersity of LNPs is comparable to that of particles obtained by manual formulation, but the polydispersity index (PDI) values are not critical for using such nanoparticles in vivo (Table 1).

Table 1. The characteristics of LNPs obtained by manual and microfluidic formulation.

Formulation	Cationic Lipid	Particle Size, nm	PDI	DNA Concentration in the Particles, ng/μL	DNA Loading Efficiency, %
Manual	DOPE	177.7 ± 48.1	0.18	6.3 ± 1.1	$10.8 \pm 1.9\%$
Manual	DOPC	214.3 ± 15	0.2	8	13.6%
Microfluidic	DOPE	120	0.2	34	62%
Microfluidic	DOPC	108 ± 12	0.22 ± 0.02	36 ± 5	$64 \pm 7\%$

Cytotoxicity and stability of the obtained LNPs were then evaluated. The particles were shown to retain their size at 4 °C in the sterile solution (Figure 1B). We evaluated the cytotoxicity of LNP-pDNAs (IC_{50} = 0.429 ng DNA/ml) that is comparable with other studies (Figure S2). We used the similar LNP concentration as in our previous study that demonstrates that the administration of LNP did not result in pathological changes in liver tissue [31]. It is also comparable to the IC_{50} data of previously studied LNPs with siRNAs [32].

We obtained two types of particles: LNP_1, LNP_2 ($1\times$ and $2\times$ dye excess to siLuc during synthesis, respectively) with an average diameter of 80 nm. It was shown that the hydro-Cy5 encapsulated in particles didn't change the fluorescence intensity when interacting with the H_2O_2. In contrast, LNP_2 showed a slight fluorescence increase when reacted with highly reactive hydroxyl radical (Figure 1C). These data suggest a high stability of the LNPs and a high degree of dye protection against oxidation after intravenous injection.

Next, we decided to determine optimal amount of dye incoroporated in LNP for the detection using IVIS Spectrum CT. For this purpose, we conjugated LNP with different amount of cyanine 5.5 and showed that the amount of incorporated dye impacts LNP detection in organs (Figure S3) and further efficacy evaluation. So, we proceeded experiments with LNP_2 and tested their efficacy in vitro.

Next, we analyzed the efficacy of hydro-Cy5-LNP for ROS visualization on the AML12 cell line in vitro. For this purpose, AML12 were incubated with hydro-Cy5-LNP for 24 h, and then 10 μmol H_2O_2 was added to the cells. Confocal images were taken of the AML12 before and after H_2O_2 addition. Cell nuclei were stained with Hoechst 33342 (in blue), and the LNPs were imaged in red to monitor oxidation of hydro-Cy5-LNP in the cells. We demonstrated in vitro that hydro-Cy5-LNPs fluorescence increased in the presence of 10 μmol H_2O_2, which is relevant to ROS changes in vivo during liver inflammation (Figure 2).

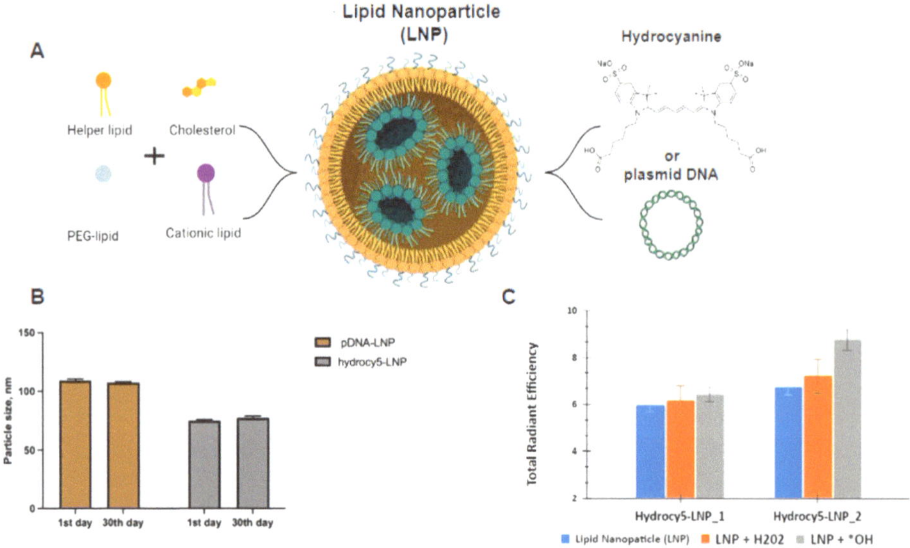

Figure 1. Synthesis and characterization of LNP. (**A**) Schematic representation of the LNPs composition; (**B**) the stability of obtained LNPs; (**C**) fluorescence intensity analysis of LNP with hydrocyanine (LNP_1, LNP_2) before and after interaction with ROS (H_2O_2 and hydroxyl radical (*OH)).

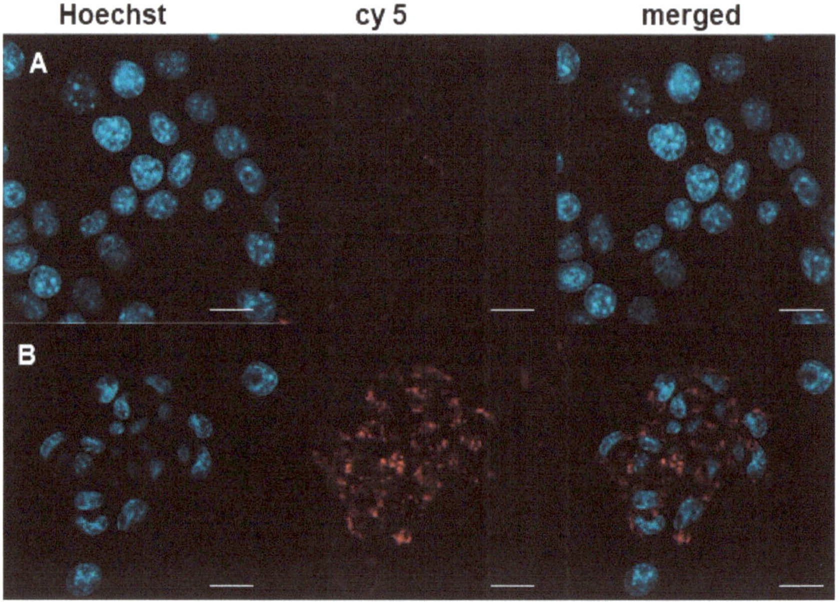

Figure 2. Confocal microscopy images of the AML12 cells incubated with hydro-Cy5-LNP before (**A**) and after (**B**) H_2O_2. Nuclei are stained with Hoechst, and hydro-Cy5 is indicated in red color. The scale bar is 50 μm.

3.2. Intravital Delivery of ROS Sensor-Lipid Nanoparticles and Visualization of Hepatocellular Carcinoma

To preliminarily evaluate the size effect on particle delivery efficiency, we intravenously injected cyanine 5.5-labeled LNPs to the FvB/N mice. Whole-body fluorescence images were analyzed using the IVIS Imaging CT system in 15 min, 1, 4, and 24 h of injection of the 110 and 170 nm particles (Figure 3C). It was shown that LNPs with an average diameter of 110 nm demonstrated preferable accumulation in the liver after 24 h (Figure 3A), while LNP with size of 170 nm mainly accumulated in the spleen (Figure 3B). So, we decided to use 110 nm LNP for further experiments.

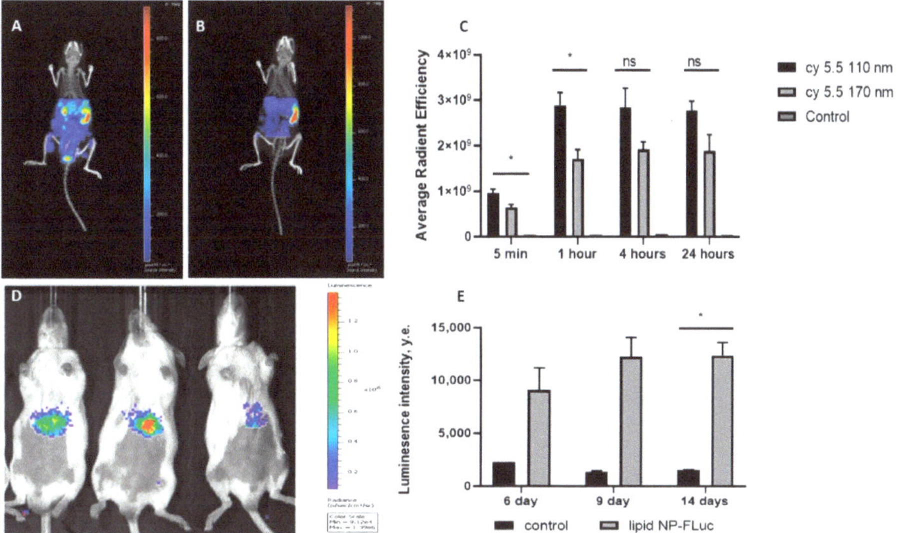

Figure 3. Biodistribution of 110 nm (**A**) and 170 nm (**B**) cy5.5-labeled LNPs in 24 h after injection and the quantification of fluorescence intensity in the liver in 5 min, 1, 4, and 24 h after injection (**C**); the luminescence intensity (**D**) in mice in 9 days after intravenous injection of LNPs loaded with pDNA Fluc and quantification of luminescence intensity at 6th, 9th, and 14th days after injection of LNPs (**E**). Control indicates a healthy mouse. Data are presented as mean $\pm$ SD, $n = 4$ (ns—not significant, *—$p < 0.05$).

Next, we evaluated the efficiency of pDNA-LNP delivery to mouse liver cells. LNPs containing pDNA to Fluc with a diameter 110 nm were injected intravenously and the luminescence intensity was analyzed 3, 6, 9, and 14 days after injection. It was shown that the particles effectively transfected the liver after 6 days up to 14 days without reinjection (Figure 3D,E). These results confirm the possibility of effective pDNA delivery using LNPs.

To evaluate the efficacy of ROS imaging using the developed LNPs, we decided to use the HCC mouse model. We demonstrated the efficacy of ROS imaging in a mouse model of HCC after the administration of hydrocyanine nanoparticles. It was shown that these particles effectively accumulate in liver tissue and specifically fluoresce in the area of the pathology (Figure 4).

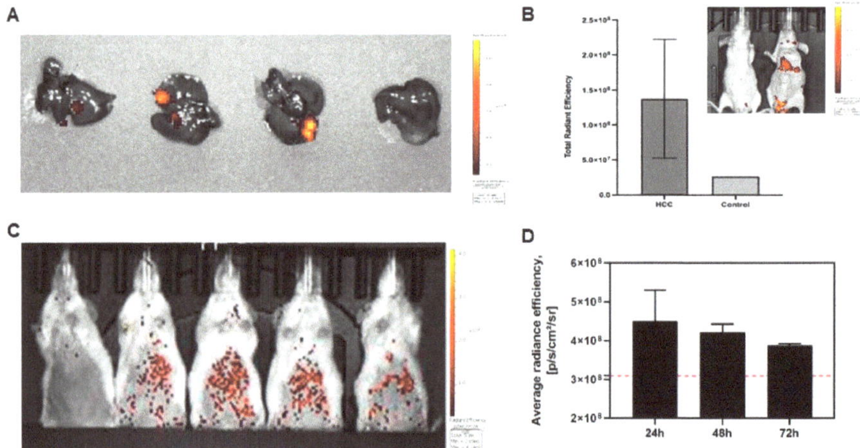

Figure 4. Images of liver tissues of the mice with HCC after administration of LNPs loaded with hydrocyanine: ex vivo imaging using IVIS Spectrum CT 3 h after injection (**A**) and quantification of fluorescence intensities (**B**); (**C**). Images of mice after injection of LNPs loaded with HyPer7; (**D**). Quantification of fluorescence intensities after injection of LNPs loaded with HyPer7; red line indicates control. Data are presented as mean $\pm$ SD, n = 3.

We observed an insignificant increase in fluorescence intensity (1.5 times in comparison with control) in 48 h after LNP-HyPer7 injection (Figure 4C). Similar results were observed 10 days after injection of HyPer7-LNP (Figure S4). However, the injection of obtained LNPs into the HCC-bearing mouse did not significantly change the fluorescent signal. These results demonstrate persistent limitations of HyPer7 for in vivo imaging that requires either different pDNA delivery methods (viral or hydrodynamic) or other detection methods.

4. Discussion

ROS play an important role in many biological pathways, but long-term inflammation leads to the overproduction of ROS, causing oxidative stress and the development of many pathologies, including liver diseases [11,33]. Cancer cells have elevated levels of ROS that promote tumor growth and are responsible for the development of numerous liver diseases, including alcoholic liver diseases, nonalcoholic fatty liver diseases, hepatic fibrosis, and hepatitis C virus [5]. ROS activate hepatic stellate cells, increasing the inflammation and initiating liver fibrosis [34]. One of the most common malignant liver diseases is HCC, which is associated with many cellular processes and affects the metabolism of liver cells, resulting in increased ROS production. For diagnosis and monitoring of different liver diseases and namely HCC, visualization of ROS remains an important tool for preclinical study. Here, we demonstrate the development of two types of LNPs loaded with ROS sensors—pDNA HyPer7 and the hydro-Cy5 dye, which can be used for the detection and monitoring of ROS in HCC.

LNP technology has played the central role in the development of the first three approved drugs, namely Patisiran as siRNA medicine and Tozinameran and Elasomeranas as mRNA vaccines [35]. Patisiran (ONPATTRO®) is used for the treatment of polyneuropathies associated with the hereditary disease transthyretin-mediated amyloidosis [36]. The two recent FDA approval of COVID-19 based modified mRNA vaccines are formulated in LNPs [37]. The LNPs formulation is a promising strategy for DNA/RNA delivery (pDNA, mRNA, siRNA) to the different tissues, and it can be effective for the accumulation of ROS-sensitive dyes in the liver and improve the visualization of ROS [38,39]. In our work, we compared the manual formulation of LNPs and formulation by microfluidic device. Manual formulation can be easily conducted while microfluidic technologies allow to produce homogenous-sized LNPs and have accelerated the development of LNP-based nanomedicine [40]. Lopes et al. outlined that microfluidic technologies promote the more stable and uniform LNPs [41]. Unsurprisingly, microfluidic device technology allows production of LNP with high loading efficiency and low PDI, which we observed by pDNA and hydro-Cy5 formulation. We demonstrated that obtained LNPs are stable and retain their size within 30 days, which means that the LNPs do not aggregate over time, which is essential for using them both in vitro and in vivo. Composition of lipids can impact LNP characteristics; however, we did not observe any difference in DNA loading efficiency using two different helper lipids, DOPE and DOPC. This could be explained by the similar chemical structure of both lipids. However, selection between DOPE and DOPC by formulation in LNPs can affect the transfection efficiency both in vitro and in vivo. Zhang et al. found that DOPE-containing LNP formulation enhances the mRNA delivery to the liver [42]. While Kulkarni et al. showed that replacement of DSPC with DOPC increased transfection efficacy with pDNA-LNP in vitro more than 40x times [43]. However, we observed only slight changes in DOPE- and DOPC-LNP-pDNA transfection efficacy in vivo in 24–72 h after intravenous injection. Danaei et al. outlined that the nanoparticles with the PDI value 0.2 and below are most commonly accepted in practice for polymer-based nanoparticle materials [44]. In our work, we used LNPs with PDI values no more than 0.22 obtained by a microfluidic device, which is consistent with the previous studies [45]. Multiple formulation parameters, lipid components and composition, lipid ration, and processing parameters are crucial for LNPs with PDI values less than 0.2 [46].

Genetically encoded probes that are sensitive to the H_2O are promising approaches for monitoring the production of intracellular H_2O_2. The main problem of HyPer and other probes containing the circular permutation is their high sensitivity to the pH. Recently, a new member of the HyPer family, HyPer7, was introduced that is non sensitive to the changes in pH and more sensitive to the H_2O_2 [47]. We used HyPer7 in this study for visualization of ROS in vivo that was formulated in LNP for the efficient delivery to the liver. Although, we demonstrated that we can successfully deliver pDNA Fluc to the liver cells, we did not observe the significant fluorescence after HyPer7-LNP injection in comparison with control. Possibly, it might be due to the fact that the one of main disadvantages of green fluorescence probes to be used in vivo models is the low tissue penetration [48]. The low fluorescence intensity of the liver after intravenous injection of HyPer7-loaded LNPs could be also explained by the insufficient delivery method that could be optimized by the other composition of the lipids or amount of pDNA or detection methods. Godbout et al. outlined that it is crucial to modify the composition of the lipids used for formulation in LNPs that will lead to the specific delivery of LNPs to the organs [49]. Additionally, adenovirus-mediated delivery, polymeric nanoparticles, or hydrodynamic transfection could increase the efficacy of HyPer7 delivery. Zamboni et al. demonstrated that polymeric biodegradable poly(beta-amino-ester) nanoparticles enabled high DNA delivery to HCC cells [50]. Additionally, it should be noted that HyPer is generally used for ratiometric imaging of the living cells, so detection method plays an essential role in the efficacy of H_2O_2 measurement; in particular, intravital microscopy should be considered [51].

One of the probes that can be used for visualization of ROS are hydrocyanine dyes that are nonfluorescent in leuco form but can be oxidized by hydroxyl radicals and become fluorescent cyanine molecules [52]. Different probes from the hydrocyanine family have different emission wavelengths and can be used for different purposes for in vitro and in vivo studies. Hydrocyanines were used for measuring the ROS production in vivo in different animal models, such as inflammation, but their application is limited due to the low tissue penetration of optical probes [52]. In our work, we used hydro-Cy5 dye that was formulated in LNP to protect it from oxidation in the blood stream and increase accumulation in vivo. To prove the high stability of obtained LNPs, we decided to compare the efficacy of LNPs loaded with two different excess levels of hydro-Cy5. We did not observe significant changes in the fluorescence intensities by adding H_2O_2, which proves the high stability of LNPs. We also tested efficacy of hydro-Cy5-LNP in

AML12 cells and demonstrated that LNPs increased fluorescence already 5 min after H_2O_2 addition compared to non-treated cells, which indicates that hydro-Cy5 encapsulated in LNPs is sensitive to the intracellular H_2O_2 and can be used for further in vivo experiments. Therefore, we decided to study the biodistribution of hydro-Cy5-LNPs of two sizes—110 and 170 nm. Although intravenous studies show that the liver is the major organ depot for nanoparticles [42], we observed the accumulation of hydro-Cy5-LNP 170 nm in the spleen but not in the liver compared to the hydro-Cy5-LNP 110 nm, which does not allow us to use it for detection of ROS in HCC. We successfully demonstrated ROS-sensitive detection of tumor tissue in an orthotopic HCC model after intravenous injection of hydro-Cy5-LNP. It should also be noted that this delivery platform could also be used for simultaneous siRNA delivery and ROS visualization.

In conclusion, we synthesized LNPs with ROS-sensitive probes and demonstrated the successful delivery of pDNA-LNP and hydro-Cy5-LNP to the liver. We showed that accumulation of hydro-Cy5-LNP results in ROS-specific visualization of tumor tissue in the mouse model of HCC compared to the control. We believe that our results present a valuable tool to visualize ROS in liver diseases by LNPs loaded with hydro-Cy5 that can be used in preclinical research.

Supplementary Materials: The following supporting information can be downloaded at: https://www.mdpi.com/article/10.3390/biomedicines11071783/s1, Figure S1: The fluorescent images of mice in 5 min (A) and 15 min (B) after administration of unconjugated hydro-Cy5; Figure S2: Cytotoxicity of 110 nm LNPs; Figure S3: Images of organs (organs (heart, lungs, liver, kidneys, spleen) after cy5.5-labeled LNP administration with $1\times$ or $2\times$ dye excess in 4 h and 24 h after injection. Figure S4: Whole body fluorescent images in 10 days after LNP-HyPer7 administration; Table S1: The parameters of LNPs synthesis.

Author Contributions: T.O.A. and V.S.S. wrote the main manuscript and prepared most of the figures; T.O.A. performed design of study, in vivo experiment, funding acquisition; V.S.S. performed in vitro experiments; V.V.B. and A.G.S. provided the plasmids for formulation in LNP and further administration; I.O.A. synthesized hydrocyanine dyes; P.I.V. synthesized pDNA-LNP and hydro-Cy5-LNP; T.A.P. formulated cy 5.5-LNP for biodistribution studies; T.S.Z.—conceptualization, design of studies, editing of manuscript; A.S.M. and Y.V.K. designed the experiments and edited the manuscript. All authors have read and agreed to the published version of the manuscript.

Funding: This work was supported by the grant of President MK 1128.2020.4 (ROS measurements in vitro/in vivo) and RSF 22-75-101 (LNP synthesis).

Institutional Review Board Statement: Biodistribution studies were approved by the Institute of Developmental Biology (Moscow, Russia, Approval #60, 16 June 2022), Experiments with HCC model was approved by Pirogov National Research Medical University (Approval #6/2021, 1 April 2021).

Data Availability Statement: The datasets used and analyzed during the study are available from the corresponding author upon request.

Acknowledgments: We would like to thank Arsen Mikaelyan and Yuri Kotelevtsev for helpful comments on this project. We also thank Vsevolod Belousov for providing HyPer7 and Veronika Usatova for the help in the imaging of the cells. The body fluorescence was performed using the equipment of the "Bioimaging and Spectroscopy" Core Facility of the Skolkovo Institute of Science and Technology IVIS Spectrum CT. Biodistribution experiments were conducted using equipment of the Core Centrum of Institute of Developmental Biology RAS.

Conflicts of Interest: The authors declare no conflict of interest.

References

1. Zou, Z.; Chang, H.; Li, H.; Wang, S. Induction of Reactive Oxygen Species: An Emerging Approach for Cancer Therapy. *Apoptosis* **2017**, *22*, 1321–1335. [CrossRef]
2. Trachootham, D.; Alexandre, J.; Huang, P. Targeting Cancer Cells by ROS-Mediated Mechanisms: A Radical Therapeutic Approach? *Nat. Rev. Drug Discov.* **2009**, *8*, 579–591. [CrossRef] [PubMed]
3. Boonstra, J.; Post, J.A. Molecular Events Associated with Reactive Oxygen Species and Cell Cycle Progression in Mammalian Cells. *Gene* **2004**, *337*, 1–13. [CrossRef] [PubMed]

4. Xu, Y.; Ji, Y.; Li, X.; Ding, J.; Chen, L.; Huang, Y.; Wei, W. Uri1 Suppresses Irradiation-Induced Reactive Oxygen Species (Ros) by Activating Autophagy in Hepatocellular Carcinoma Cells. *Int. J. Biol. Sci.* **2021**, *17*, 3091–3103. [CrossRef] [PubMed]
5. Cichoz-Lach, H.; Michalak, A. Oxidative Stress as a Crucial Factor in Liver Diseases. *World J. Gastroenterol.* **2014**, *20*, 8082–8091. [CrossRef]
6. Rezende, F.; Brandes, R.P.; Schröder, K. Detection of Hydrogen Peroxide with Fluorescent Dyes. *Antioxid. Redox. Signal.* **2018**, *29*, 585–602. [CrossRef] [PubMed]
7. Abakumova, T.; Prikazchikova, T.; Aparin, I.; Vaneev, A.; Gorelkin, P.; Erofeev, A.; Zatsepin, T. ROS-Sensitive Dyes in Lipid Nanoparticles for in Vivo Imaging. In Proceedings of the 2020 IEEE 10th International Conference on "Nanomaterials: Applications and Properties", NAP 2020, Sumy, Ukraine, 9–13 November 2020; Institute of Electrical and Electronics Engineers Inc.: Piscataway, NJ, USA, 2020.
8. Wang, H.; Wang, X.; Li, P.; Dong, M.; Yao, S.Q.; Tang, B. Fluorescent Probes for Visualizing ROS-Associated Proteins in Disease. *Chem. Sci.* **2021**, *12*, 11620–11646. [CrossRef] [PubMed]
9. Zhang, Y.; Dai, M.; Yuan, Z. Methods for the Detection of Reactive Oxygen Species. *Anal. Methods* **2018**, *10*, 4625–4638. [CrossRef]
10. Bai, X.; King-Hei Ng, K.; Jacob Hu, J.; Ye, S.; Yang, D. Small-Molecule-Based Fluorescent Sensors for Selective Detection of Reactive Oxygen Species in Biological Systems. *Annu. Rev. Biochem.* **2019**, *88*, 605–633. [CrossRef] [PubMed]
11. Oushiki, D.; Kojima, H.; Terai, T.; Arita, M.; Hanaoka, K.; Urano, Y.; Nagano, T. Development and Application of a Near-Infrared Fluorescence Probe for Oxidative Stress Based on Differential Reactivity of Linked Cyanine Dyes. *J. Am. Chem. Soc.* **2010**, *132*, 2795–2801. [CrossRef]
12. Wardman, P. Fluorescent and Luminescent Probes for Measurement of Oxidative and Nitrosative Species in Cells and Tissues: Progress, Pitfalls, and Prospects. *Free Radic. Biol. Med.* **2007**, *43*, 995–1022. [CrossRef]
13. Winterbourn, C.C. The Challenges of Using Fluorescent Probes to Detect and Quantify Specific Reactive Oxygen Species in Living Cells. *Biochim. Biophys. Acta Gen. Subj.* **2014**, *1840*, 730–738. [CrossRef]
14. Kim, J.Y.; Choi, W.I.; Kim, Y.H.; Tae, G. Highly Selective In-Vivo Imaging of Tumor as an Inflammation Site by ROS Detection Using Hydrocyanine-Conjugated, Functional Nano-Carriers. *J. Control. Release* **2011**, *156*, 398–405. [CrossRef] [PubMed]
15. Erard, M.; Dupré-Crochet, S.; Nüße, X.O. Biosensors for Spatiotemporal Detection of Reactive Oxygen Species in Cells and Tissues. *Am. J. Physiol. Regul. Integr. Comp. Physiol.* **2018**, *314*, 667–683. [CrossRef] [PubMed]
16. Guo, H.; Aleyasin, H.; Dickinson, B.C.; Haskew-Layton, R.E.; Ratan, R.R. Recent Advances in Hydrogen Peroxide Imaging for Biological Applications. *Cell Biosci.* **2014**, *4*, 64. [CrossRef]
17. Kim, G.; Koo, Y.E.L.; Xu, H.; Philbert, M.A.; Kopelman, R. Nanoencapsulation Method for High Selectivity Sensing of Hydrogen Peroxide inside Live Cells. *Anal. Chem.* **2010**, *82*, 2165–2169. [CrossRef]
18. Hammond, V.J.; Aylott, J.W.; Greenway, G.M.; Watts, P.; Webster, A.; Wiles, C. An Optical Sensor for Reactive Oxygen Species: Encapsulation of Functionalised Silica Nanoparticles into Silicate Nanoprobes to Reduce Fluorophore Leaching. *Analyst* **2008**, *133*, 71–75. [CrossRef] [PubMed]
19. Muriel, P. Role of Free Radicals in Liver Diseases. *Hepatol. Int.* **2009**, *3*, 526–536. [CrossRef] [PubMed]
20. Seitz, H.K.; Stickel, F. Risk Factors and Mechanisms of Hepatocarcinogenesis with Special Emphasis on Alcohol and Oxidative Stress. *Biol. Chem.* **2006**, *387*, 349–360. [CrossRef]
21. Marx, J. Cancer research. Inflammation and Cancer: The Link Grows Stronger. *Science* **2004**, *306*, 966–968. [CrossRef] [PubMed]
22. Chartampilas, E.; Rafailidis, V.; Georgopoulou, V.; Kalarakis, G.; Hatzidakis, A.; Prassopoulos, P. Current Imaging Diagnosis of Hepatocellular Carcinoma. *Cancers* **2022**, *14*, 3997. [CrossRef]
23. Jiang, H.Y.; Chen, J.; Xia, C.C.; Cao, L.K.; Duan, T.; Song, B. Noninvasive Imaging of Hepatocellular Carcinoma: From Diagnosis to Prognosis. *World J. Gastroenterol.* **2018**, *24*, 2348–2362. [CrossRef] [PubMed]
24. Kundu, K.; Knight, S.F.; Willett, N.; Lee, S.; Taylor, W.R.; Murthy, N. Hydrocyanines: A Class of Fluorescent Sensors That Can Image Reactive Oxygen Species in Cell Culture, Tissue, and in Vivo. *Angew. Chem.—Int. Ed.* **2009**, *48*, 299–303. [CrossRef] [PubMed]
25. Whitehead, K.A.; Dorkin, J.R.; Vegas, A.J.; Chang, P.H.; Veiseh, O.; Matthews, J.; Fenton, O.S.; Zhang, Y.; Olejnik, K.T.; Yesilyurt, V.; et al. Degradable Lipid Nanoparticles with Predictable in Vivo SiRNA Delivery Activity. *Nat. Commun.* **2014**, *5*, 4277. [CrossRef] [PubMed]
26. *Zetasizer Nano User Manual (English)*; MAN0485; Malvern Instruments Ltd.: Malvern, UK, 2013.
27. *Quant-IT ^{TM} RiboGreen ^{TM} RNA Reagent and Kit*; MAN0002073; Thermo Fisher Scientific Inc.: Eugene, OR, USA, 2022.
28. Asghar, M.N.; Emani, R.; Alam, C.; Helenius, T.O.; Grönroos, T.J.; Sareila, O.; Din, M.U.; Holmdahl, R.; Hänninen, A.; Toivola, D.M. In Vivo Imaging of Reactive Oxygen and Nitrogen Species in Murine Colitis. *Inflamm. Bowel Dis.* **2014**, *20*, 1435–1447. [CrossRef] [PubMed]
29. Fan, N.; Silverman, S.M.; Liu, Y.; Wang, X.; Kim, B.J.; Tang, L.; Clark, A.F.; Liu, X.; Pang, I.H. Rapid Repeatable in Vivo Detection of Retinal Reactive Oxygen Species. *Exp. Eye Res.* **2017**, *161*, 71–81. [CrossRef]
30. Belousov, V.V.; Fradkov, A.F.; Lukyanov, K.A.; Staroverov, D.B.; Shakhbazov, K.S.; Terskikh, A.V.; Lukyanov, S. Genetically Encoded Fluorescent Indicator for Intracellular Hydrogen Peroxide. *Nat. Methods* **2006**, *3*, 281–286. [CrossRef] [PubMed]
31. Sergeeva, O.; Abakumova, T.; Kurochkin, I.; Ialchina, R.; Kosyreva, A.; Prikazchikova, T.; Varlamova, V.; Shcherbinina, E.; Zatsepin, T. Level of Murine Ddx3 Rna Helicase Determines Phenotype Changes of Hepatocytes In Vitro and In Vivo. *Int. J. Mol. Sci.* **2021**, *22*, 6958. [CrossRef]

32. Leboeuf, D.; Abakumova, T.; Prikazchikova, T.; Rhym, L.; Anderson, D.G.; Zatsepin, T.S.; Piatkov, K.I. Downregulation of the Arg/N-Degron Pathway Sensitizes Cancer Cells to Chemotherapy In Vivo. *Mol. Ther.* **2020**, *28*, 1092–1104. [CrossRef] [PubMed]
33. Agita, A.; Thaha, M. Inflammation, Immunity, and Hypertension. *Acta Med. Indones.* **2017**, *49*, 158–165.
34. Sánchez-Valle, V.; Chávez-Tapia, N.C.; Uribe, M.; Méndez-Sánchez, N. Role of Oxidative Stress and Molecular Changes in Liver Fibrosis: A Review. *Curr. Med. Chem.* **2012**, *19*, 4850–4860. [CrossRef] [PubMed]
35. Suzuki, Y.; Ishihara, H. Difference in the Lipid Nanoparticle Technology Employed in Three Approved SiRNA (Patisiran) and MRNA (COVID-19 Vaccine) Drugs. *Drug Metab Pharm.* **2021**, *41*, 100424. [CrossRef] [PubMed]
36. Akinc, A.; Maier, M.A.; Manoharan, M.; Fitzgerald, K.; Jayaraman, M.; Barros, S.; Ansell, S.; Du, X.; Hope, M.J.; Madden, T.D.; et al. The Onpattro Story and the Clinical Translation of Nanomedicines Containing Nucleic Acid-Based Drugs. *Nat. Nanotechnol.* **2019**, *14*, 1084–1087. [CrossRef]
37. Blakney, A.K.; McKay, P.F.; Hu, K.; Samnuan, K.; Jain, N.; Brown, A.; Thomas, A.; Rogers, P.; Polra, K.; Sallah, H.; et al. Polymeric and Lipid Nanoparticles for Delivery of Self-Amplifying RNA Vaccines. *J. Control. Release* **2021**, *338*, 201–210. [CrossRef] [PubMed]
38. Oberli, M.A.; Reichmuth, A.M.; Dorkin, J.R.; Mitchell, M.J.; Fenton, O.S.; Jaklenec, A.; Anderson, D.G.; Langer, R.; Blankschtein, D. Lipid Nanoparticle Assisted MRNA Delivery for Potent Cancer Immunotherapy. *Nano. Lett.* **2017**, *17*, 1326–1335. [CrossRef] [PubMed]
39. Kaczmarek, J.C.; Patel, A.K.; Kauffman, K.J.; Fenton, O.S.; Webber, M.J.; Heartlein, M.W.; DeRosa, F.; Anderson, D.G. Polymer–Lipid Nanoparticles for Systemic Delivery of MRNA to the Lungs. *Angew. Chem.—Int. Ed.* **2016**, *55*, 13808–13812. [CrossRef]
40. Maeki, M.; Uno, S.; Niwa, A.; Okada, Y.; Tokeshi, M. Microfluidic Technologies and Devices for Lipid Nanoparticle-Based RNA Delivery. *J. Control. Release* **2022**, *344*, 80–96. [CrossRef]
41. Lopes, C.; Cristóvão, J.; Silvério, V.; Lino, P.R.; Fonte, P. Microfluidic Production of MRNA-Loaded Lipid Nanoparticles for Vaccine Applications. *Expert Opin. Drug Deliv.* **2022**, *19*, 1381–1395. [CrossRef]
42. Zhang, R.; El-Mayta, R.; Murdoch, T.J.; Warzecha, C.C.; Billingsley, M.M.; Shepherd, S.J.; Gong, N.; Wang, L.; Wilson, J.M.; Lee, D.; et al. Helper Lipid Structure Influences Protein Adsorption and Delivery of Lipid Nanoparticles to Spleen and Liver. *Biomater. Sci.* **2021**, *9*, 1449–1463. [CrossRef] [PubMed]
43. Kulkarni, J.A.; Myhre, J.L.; Chen, S.; Tam, Y.Y.C.; Danescu, A.; Richman, J.M.; Cullis, P.R. Design of Lipid Nanoparticles for in Vitro and in Vivo Delivery of Plasmid DNA. *Nanomedicine* **2017**, *13*, 1377–1387. [CrossRef]
44. Danaei, M.; Dehghankhold, M.; Ataei, S.; Hasanzadeh Davarani, F.; Javanmard, R.; Dokhani, A.; Khorasani, S.; Mozafari, M.R. Impact of Particle Size and Polydispersity Index on the Clinical Applications of Lipidic Nanocarrier Systems. *Pharmaceutics* **2018**, *10*, 57. [CrossRef] [PubMed]
45. Jaradat, E.; Weaver, E.; Meziane, A.; Lamprou, D.A. Microfluidic Paclitaxel-Loaded Lipid Nanoparticle Formulations for Chemotherapy. *Int. J. Pharm.* **2022**, *628*, 122320. [CrossRef]
46. Prakash, G.; Shokr, A.; Willemen, N.; Bashir, S.M.; Shin, S.R.; Hassan, S. Microfluidic Fabrication of Lipid Nanoparticles for the Delivery of Nucleic Acids. *Adv. Drug Deliv. Rev.* **2022**, *184*, 114197. [CrossRef] [PubMed]
47. Pak, V.V.; Ezerina, D.; Lyublinskaya, O.G.; Pedre, B.; Tyurin-Kuzmin, P.A.; Mishina, N.M.; Thauvin, M.; Young, D.; Wahni, K.; Martínez Gache, S.A.; et al. Ultrasensitive Genetically Encoded Indicator for Hydrogen Peroxide Identifies Roles for the Oxidant in Cell Migration and Mitochondrial Function. *Cell Metab.* **2020**, *31*, 642–653. [CrossRef] [PubMed]
48. Ermakova, Y.G.; Bilan, D.S.; Matlashov, M.E.; Mishina, N.M.; Markvicheva, K.N.; Subach, O.M.; Subach, F.V.; Bogeski, I.; Hoth, M.; Enikolopov, G.; et al. Red Fluorescent Genetically Encoded Indicator for Intracellular Hydrogen Peroxide. *Nat. Commun.* **2014**, *5*, 5222. [CrossRef]
49. Godbout, K.; Tremblay, J.P. Delivery of RNAs to Specific Organs by Lipid Nanoparticles for Gene Therapy. *Pharmaceutics* **2022**, *14*, 2129. [CrossRef]
50. Zamboni, C.G.; Kozielski, K.L.; Vaughan, H.J.; Nakata, M.M.; Kim, J.; Higgins, L.J.; Pomper, M.G.; Green, J.J. Polymeric Nanoparticles as Cancer-Specific DNA Delivery Vectors to Human Hepatocellular Carcinoma. *J. Control. Release* **2017**, *263*, 18–28. [CrossRef]
51. Huang, B.K.; Ali, S.; Stein, K.T.; Sikes, H.D. Interpreting Heterogeneity in Response of Cells Expressing a Fluorescent Hydrogen Peroxide Biosensor. *Biophys. J.* **2015**, *109*, 2148–2158. [CrossRef]
52. Prunty, M.C.; Aung, M.H.; Hanif, A.M.; Allen, R.S.; Chrenek, M.A.; Boatright, J.H.; Thule, P.M.; Kundu, K.; Murthy, N.; Pardue, M.T. In Vivo Imaging of Retinal Oxidative Stress Using a Reactive Oxygen Species-Activated Fluorescent Probe. *Investig. Ophthalmol. Vis. Sci.* **2015**, *56*, 5862–5870. [CrossRef]

Article

Drug-Loaded Silver Nanoparticles—A Tool for Delivery of a Mebeverine Precursor in Inflammatory Bowel Diseases Treatment

Mina Todorova [1], Miglena Milusheva [1,2], Lidia Kaynarova [3], Deyana Georgieva [3], Vassil Delchev [4], Stanislava Simeonova [5,6], Bissera Pilicheva [5,6] and Stoyanka Nikolova [1,*]

1 Department of Organic Chemistry, Faculty of Chemistry, University of Plovdiv, 4000 Plovdiv, Bulgaria; minatodorova@uni-plovdiv.bg (M.T.); miglena.milusheva@uni-plovdiv.bg or miglena.milusheva@mu-plovdiv.bg (M.M.)

2 Department of Bioorganic Chemistry, Faculty of Pharmacy, Medical University of Plovdiv, 4002 Plovdiv, Bulgaria

3 Department of Analytical Chemistry and Computer Chemistry, Faculty of Chemistry, University of Plovdiv, 4000 Plovdiv, Bulgaria; l.kainarova@uni-plovdiv.bg (L.K.); georgieva@uni-plovdiv.bg (D.G.)

4 Department of Physical Chemistry, Faculty of Chemistry, University of Plovdiv, 4000 Plovdiv, Bulgaria

5 Department of Pharmaceutical Sciences, Faculty of Pharmacy, Medical University of Plovdiv, 4002 Plovdiv, Bulgaria

6 Research Institute, Medical University of Plovdiv, 4002 Plovdiv, Bulgaria

* Correspondence: tanya@uni-plovdiv.bg

Citation: Todorova, M.; Milusheva, M.; Kaynarova, L.; Georgieva, D.; Delchev, V.; Simeonova, S.; Pilicheva, B.; Nikolova, S. Drug-Loaded Silver Nanoparticles—A Tool for Delivery of a Mebeverine Precursor in Inflammatory Bowel Diseases Treatment. *Biomedicines* **2023**, *11*, 1593. https://doi.org/10.3390/biomedicines11061593

Academic Editor: Yongtai Zhang

Received: 15 April 2023
Revised: 26 May 2023
Accepted: 29 May 2023
Published: 30 May 2023

Abstract: Chronic, multifactorial illnesses of the gastrointestinal tract include inflammatory bowel diseases. One of the greatest methods for regulated medicine administration in a particular region of inflammation is the nanoparticle system. Silver nanoparticles (Ag NPs) have been utilized as drug delivery systems in the pharmaceutical industry. The goal of the current study is to synthesize drug-loaded Ag NPs using a previously described 3-methyl-1-phenylbutan-2-amine, as a mebeverine precursor (MP). Methods: A green, galactose-assisted method for the rapid synthesis and stabilization of Ag NPs as a drug-delivery system is presented. Galactose was used as a reducing and capping agent forming a thin layer encasing the nanoparticles. Results: The structure, size distribution, zeta potential, surface charge, and the role of the capping agent of drug-loaded Ag NPs were discussed. The drug release of the MP-loaded Ag NPs was also investigated. The Ag NPs indicated a very good drug release between 80 and 85%. Based on the preliminary results, Ag NPs might be a promising medication delivery system for MP and a useful treatment option for inflammatory bowel disease. Therefore, future research into the potential medical applications of the produced Ag NPs is necessary.

Keywords: silver nanoparticles; mebeverine precursor; green method; galactose; inflammatory bowel diseases

1. Introduction

Inflammatory bowel diseases (IBDs) are functional chronic multifactorial diseases of the gastrointestinal tract with a complex pathogenesis and multifaceted therapy approaches, aimed at alleviating clinical symptoms and improving the life quality of patients. IBDs are becoming more widespread globally, although the actual cause is still unknown. Their treatment includes dietary changes and uptake of drugs from various pharmacological groups such as antidiarrheals, prokinetic agents, bulk-forming laxatives, tricyclic antidepressants, anticholinergics, serotonin receptor antagonists, and newer pharmacological classes targeting chloride ion channels and guanylate cyclase C [1]. Chronic inflammation most often affects the colon or small and large intestines [2]. Despite the fact that some of the currently available treatments for ulcerative colitis are helpful, there are still certain restrictions on their use due to their nonspecific distribution, gastrointestinal drug metabolism, and considerable side effects [3]. A well-designed drug delivery system is beneficial to enhance therapeutic efficacy, such as the use of drug-release system, prodrugs, system affected by pH, micro- and nanoparticulate systems, etc. [4,5]. A number of medications, including corticosteroids, antibiotics, and immunosuppressants, have recently been shown to support mucosal healing. Innovative platforms based on biomaterials with greater efficacy and fewer adverse effects are required to deliver pharmaceutical medications to the wounded location. They can be used as effective drug delivery systems to facilitate colon-specific administration and stable drug release [6]. Drug delivery methods to the micro- or nanometer scale might prolong colonic residency duration but also have more advantages for treating IBD. A size-dependent accumulation of micro- and nanoparticles, especially in the inflamed intestinal regions, can be observed [7,8]. That makes them one of the best options for drug delivery regulation.

The synthesis of nanomaterials is an essential aspect of nanotechnology. Nanoparticles (NPs) of noble metals have been known for the last decades [9] due to their expanding number of pharmaceutical and biomedical applications, such as drug delivery, photothermal therapy, radiotherapy, imaging, etc. [10,11]. Amongst the different noble metal NPs, silver and gold NPs are mostly synthesized so far due to their remarkable biological activities and unique physicochemical properties [12–27]. Typically, they are synthesized using hazardous chemicals which affect the environment and human health. This necessitated the development of environment-friendly methods and gave rise to the green synthesis methodology [28,29].

Critical objectives in modern synthetic organic chemistry include the improvement of reaction efficiency, the avoidance of toxic reagents, the reduction of waste, and the responsible utilization of our resources. Starting from this assumption, the search for a green alternative to produce NPs or the discovery of green molecules is a challenge in order to obtain safe materials [30]. The biological techniques include the use of different biological resources as reducing and stabilizing agents to change the noble metal cations to their nanoforms [31–34].

Recently, the application of carbohydrates has become a popular area in NPs synthesis. Many interesting methods are being applied currently to the green preparation of nanosized silver nanoparticles (Ag NPs) such as the biosynthesis of nanoparticles by starch [35–37], by plant leaf broth [38], by edible mushroom extract [39], etc. Green syntheses of Ag NPs with polyoxometallates, polysaccharides, Tollens, Fehling's, irradiation, etc., were reported as well [40,41]. Polymer-based NPs were synthesized due to the spectrum of biological interest [42]. Sugars as reducing agents for the green synthesis of metal NPs were applied in the last ten years [43–47]. Many mono-, di-, and polysaccharides were used as reducing and capping agents for the preparation of noble metals NPs due to their sustainability, abundance, low cost, harmlessness, renewability, biodegradability, and compatibility with biological systems [48–52]. To prevent aggregation, the capping agents were applied. The capping agents are most often organic molecules that bind the metallic core, developing a layer on the surface of NPs. Many capping agents have been used in conventional syntheses, such as cetyltrimethylammonium bromide, polyvinylpyrrolidone, oleic acid, sodium dodecyl sulfate, tetraethyl ammonium bromide, etc. [53–55]. These reagents have an effective role in NPs growth and control of their size but could be hazardous [56], causing many doubts about their use in biological applications [57]. The reducing agents commonly used in a chemical route are hydrazine, formaldehyde, and sodium tetrahydridoborate [58], which are toxic to the environment and living organisms. Other typical capping and reducing agents are sodium citrate and tannic acid, which are used to obtain highly monodispersed NPs [30,59].

The major challenge of using Ag NPs as drug-delivery systems, particularly in the treatment of inflammatory bowel disease, is carrying out site-specific drug delivery to the colon. Contrarily, long-term medicine has numerous adverse effects that are detrimental to the patients' quality of life, causing diarrhea, ulcers, or vomiting [60–63]. Targeting the medications directly to the colonic region of the gastrointestinal tract is the main problem facing researchers [64]. Another obstacle to overcome is the retardation of skeletal and growth development brought on by an inadequate diet. Cytokines and other inflammatory mediators lead to changes in the hormonal axis that have an immediate impact on growth. A proper diet and the correct anti-inflammatory treatment can help with these issues.

The quest for NPs synthesis and their application as drug-delivery systems prompted us to obtain Ag NPs as drug-delivery systems for recently described antispasmodics in new therapeutic approaches for irritable bowel syndrome treatment [65]. Recently, we explored the synthesis of a mebeverine precursor (MP, 3-methyl-1-phenylbutan-2-amine, Figure 1b) based on its pharmacological properties. Our decision was also driven by the structural similarity to mebeverine (Figure 1a), a second-generation papaverine analogue with a history of pharmacological applications as an IBD treatment.

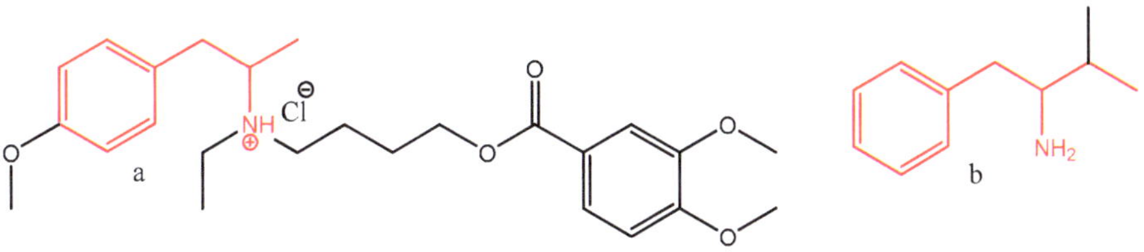

Figure 1. Structure of mebeverine hydrochloride (**a**) and the mebeverine precursor (MP) (**b**) (MP common structure in red).

In order to explore their biological activity, we synthesized many MPs with a C-3 substituent [65] (Scheme 1).

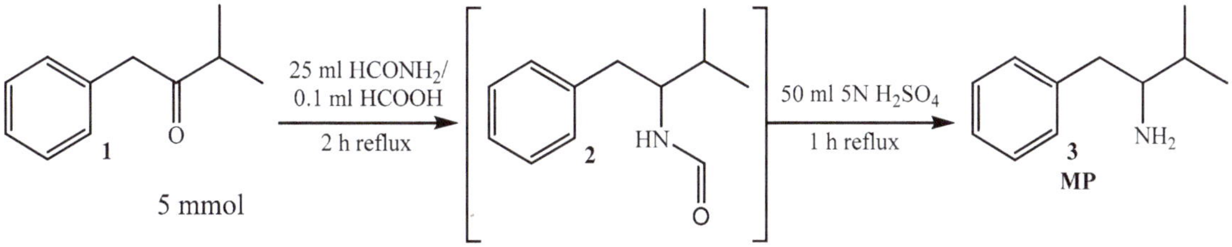

Scheme 1. Reaction pathway for the synthesis of MP.

In the present study, we focused our investigation on the synthesis of drug-loaded Ag NPs with the most active MP (**3**, Figure 1b). The MP structure was chosen as the leader compound with favorable antimicrobial, immunohistochemical, and spasmolytic activity, and a reasonable effective choice for orally active long-term therapy of chronic IBD [65].

2. Materials and Methods

All solvents and reagents were purchased from Merck (Merck Bulgaria EAD, Darmstadt, Germany). Melting point was determined on a Boetius hot stage apparatus and was uncorrected. IR spectra were determined on a VERTEX 70 FT-IR spectrometer (Bruker Optics, Ettlingen, Germany). MP was characterized by ^{1}H NMR, ^{13}CNMR, IR, and HRESIMS. The purity was determined by TLC using several solvent systems of different polarity. TLC was carried out on precoated 0.2 mm Fluka silica gel 60 plates (Merck Bulgaria EAD), using chloroform: diethyl ether: n-hexane = 6:3:1 as a chromatographic system. Neutral Al_2O_3 was used for column chromatographic separation. The product, after evaporation of the solvent, was purified by recrystallization from diethyl ether.

2.1. Synthetic Protocol

2.1.1. Synthesis of MP [65]

To a solution of 5 mmol of the starting ketone 3-methyl-1-phenylbutan-2-one **1** in 25 mL formamide, a catalytic amount of methanoic acid was added. The mixture was refluxed for 2 h at 180 °C, then poured into water and extracted with CH_2Cl_2 (3 × 20 mL). The combined extracts were washed with Na_2CO_3 solution, water, and dried using anhydrous Na_2SO_4, filtered on the short column filled with neutral Al_2O_3, and then concentrated. The obtained formamide was directly hydrolyzed with 50 mL 5N H_2SO_4 and 1 h reflux at 100 °C to 3-methyl-1-phenylbutan-2-amine (**3**, MP). The mixture then was poured into water and extracted with CH_2Cl_2 (2 × 20 mL). The water layer was alkalized with NH_4OH and extracted with CH_2Cl_2 (3 × 20 mL). The combined extracts were dried using anhydrous Na_2SO_4, filtered on the short column filled with basic Al_2O_3, and then concentrated. Spectral data confirmed the structure of the MP.

3-methyl-1-phenylbutan-2-amine (MP) ^{1}H-NMR: 0.98 (dd, J = 9.8, 6.8, 6H, 2×CH_3), 1.35 (broad s, 2H, NH_2), 1.67 (dq, J = 10.8, 6.8, 1H, $CH(CH_3)_2$), 2.40–2.43 (m, 1H, $CHNH_2$), 2.81–2.86 (m,2H, CH_2), 7.19–7.22 (m, 3H, Ar), 7.28–7.31 (m, 2H, Ar); ^{13}C-NMR: 140.3 (Ar), 129.2 (Ar), 128.5 (Ar), 126.1 (Ar), 58.2 ($CHNH_2$), 41.3 (CH_2), 33.1 ($CH(CH_3)_2$), 19.4 ($CH(CH_3)_2$), 17.5 ($CH(CH_3)_2$). IR(KBr) ν_{max}, cm^{-1}: 3419 ν(N–H), 3071, 3061, 3030 ν(C-H, Ph), 2960 ν^{as}(C–H, CH_3), 2938 ν^{as}(C–H, CH_2), 2898 ν^{s}(C–H, CH_3), 1627 δ(NH_2), 1590, 1571 ν(C=C, Ph), 1494 δ(CH_2), 1464 δ^{as}(CH_3), 1375 δ(CH_3, i-Pr); HRMS (ESI) m/z 164.14323.

2.1.2. Synthesis of Galactose-Assisted Ag NPs

A total of 1.25 g (0.007 mol) galactose was dissolved in 25 mL water and refluxed for 5 min; then 0.63 mL of 0.01 M $AgNO_3$ solution was added, and the mixture was kept at boiling. In about 5 min, the color of the solution turned pale yellow, indicating the formation of Ag NPs.

2.1.3. Synthesis of Galactose-Assisted MP-Loaded Ag NPs

A total of 1.25 g (0.007 mol) galactose was dissolved in 25 mL water and refluxed for 2 min; then 0.63 mL of 0.01 M $AgNO_3$ solution was added. To find out how the ratio of Ag NPs to the drug molecule would affect the drug release, different amounts of MP were used in the preparation of the solutions. Throughout the trials, Ag NP concentration was constant, whereas drug concentration changed depending on the ratio. The chosen ratios of drug molecules to Ag NPs were 1:1, 1:3, and 1:6, respectively. The solution's color changed to a light yellow after roughly 5 min of reflux, signifying the generation of Ag NPs.

2.2. Characterization of the Ag NPs. Analytical Techniques

After NPs preparation, the solution was used for UV-Vis, transmission electron microscopy (TEM), single-particle ICP-MS (sp-ICP-MS), dynamic light scattering (DLS), and zeta potential. For Fourier transform infrared spectra (FTIR) and X-ray diffraction (XRD) analysis, the suspension was centrifuged at 5000 rpm for 15 min and filtered (0.22 μm, Chromafil®, Macherey-Nagel, Düren, Germany), and the precipitate was used.

2.2.1. UV-Vis Spectra

UV–Vis spectra were recorded in the range between 320 and 800 nm using a Cary-60 UV-Vis spectrophotometer (Agilent Technologies, Santa Clara, CA, USA). Solution spectra were obtained by measuring the absorption of the prepared nanoparticle dispersions in a quartz cuvette with a 1 cm optical path.

2.2.2. FTIR Spectra

IR spectra were determined on a VERTEX 70 FT-IR spectrometer (Bruker Optics, Ettlingen, Germany). The spectra were collected in the range from 600 cm^{-1} to 4000 cm^{-1} with a resolution of 4 nm and 20 scans. The instrument was equipped with a diamond attenuated total reflection (ATR) accessory. The IR spectra were analyzed with the OPUS-Spectroscopy Software, Bruker (Version 7.0, Bruker, Ettlingen, Germany).

2.2.3. spICP-MS

A 7700 Agilent ICP-MS spectrometer (Agilent Technologies, Tokyo, Japan) equipped with MicroMist™ nebulizer and Peltier cooled double-pass spray chamber was used for the characterization of silver nanoparticles at 107 amu. The ICP-MS operating parameters were as follows: RF power—1.55 kW; sample flow rate—0.336 mL min^{-1}; carrier Ar gas flow rate—1.2 L min^{-1}; acquisition time 60 s; dwell time 5 ms; and transport efficiency 0.032. The transport efficiency was determined by the particle-size method [66]. Ultrapure water (UPW) was used throughout the experiments (PURELAB Chorus 2+ (ELGA Veolia) water purification system). For the sonication of silver colloids, an ultrasonic bath (Kerry US) was used. Reference materials (RMs) of citrate-stabilized silver dispersions Ag NPs (Merck Bulgaria EAD) with mean size 40 $\pm$ 4 nm and total mass concentration of silver 0.02 mg mL^{-1} were used in this study for transport efficiency determination and calibration.

A standard solution of Ag 9.974 $\pm$ 0.041 mg L^{-1} in 2% HNO$_3$ (CPAchem Ltd., Bogomilovo, Bulgaria) was used for the preparation of ionic standards.

2.2.4. TEM

The morphology, shape, and size of the nanoparticles were investigated by TEM (Talos 1.15.3, Thermo Fisher Scientific, Waltham, MA, USA). The nanoparticles suspension was added dropwise onto a formvar/carbon-coated copper grid, and then the TEM observation of the samples was performed at an operating voltage of 200 kV.

2.2.5. DLS and Zeta Potential

DLS measurements were performed on a Brookhaven BI-200 goniometer with vertically polarized incident light at a wavelength l = 632.8 nm supplied by a He–Ne laser operating at 35 mW and equipped with a Brookhaven BI-9000 AT digital autocorrelator. The scattered light was measured for dilute aqueous dispersions in the concentration range 0.056–0.963 mg mL^{-1} at 25, 37, and 65 °C. Measurements were made at θ angles in the range of 50–130°. The system allows measurements of ζ-potential in the range from -200 mV to +200 mV. All analyses were performed in triplicate at 25 °C.

2.2.6. X-ray Diffraction (XRD)

The degree of crystallinity of the synthesized nanoparticles was studied by X-ray powder diffractometry. The diffraction patterns of Ag NPs (blank) and drug-loaded Ag NPs were recorded at a 2θ range from 10° to 80° using a SIEMENS D500 X-ray powder diffractometer (KS Analytical Systems, Aubrey, TX, USA). All the measurements were performed at a voltage of 35 kV and a current of 25 mA. The monochromatic X-rays (1.5406 Å) were generated by Cu-anticathode (K$_{\alpha1}$).

2.3. In Vitro Drug Release

In vitro release of MP from the Ag NPs was performed using the dialysis bag method. A dialysis membrane (MWCO 12 kDa, Sigma-Aldrich, St. Louis, MO, USA) was hydrated in distilled water for 24 h. An accurately weighed amount of drug-loaded NPs (equivalent to the amount of MP in ratios of 1:1, 1:3, and 1:6) was dispersed in 10 mL of phosphate buffered saline (PBS) and then transferred to the dialysis bag, which was closed using a plastic clamp. Each bag was placed into a beaker containing 40 mL PBS (pH 7.4, dialysis medium). Aliquots (2 mL) from the dialysis medium were taken for measurements and subsequently replaced at predetermined time intervals with fresh medium. The drug-release study was performed for 24 h. The mean results of triplicate measurements and standard deviations were reported. The solution in the Falcon tube was shaken before each evaluation of UV–visible absorbance. For data analysis, the MP's maximum absorption band values (λ = 192 nm) were employed. Additionally, drug-only controls (one drug control for each ratio) were made.

3. Results and Discussion

The accumulation of nanoparticles in the colon region is one of the most important features of an effective nanomedicine for colon diseases. In the present study, galactose-assisted drug-loaded Ag NPs were synthesized using a direct rapid, entirely green one-pot method. Ag NPs were chosen due to a variety of biological activities [67–74]. In addition, more important is their protective effect on gastrointestinal tract [75]. Galactose was used as a reducing and capping agent. It can quickly reduce Ag^+ to the ground state at the boiling point of water (100 °C) [45]. Galactose-capped Ag NPs were chosen for the lower toxicity [76,77], so they can be recognized as more biocompatible and therefore are more suitable for exploitation in medical applications. With favorable spasmolytic, antibacterial, immunohistochemical activity and lack of cytotoxicity, the MP structure was selected as a viable and effective option for orally active long-term therapy of chronic IBD [65].

3.1. Characteristics of Galactose-Assisted Ag NPs Compared to Galactose-Assisted Drug-Loaded Ag NPs

Initially, the synthesis of the Ag NPs in an aqueous solution was monitored by recording the absorption spectra at a wavelength range of 190–600 nm (Figure 2). The absorption maximum in galactose solution was detected at 287 nm and MP was observed a 192 nm. In the early 10 min of the Ag NPs formation process, the UV–Vis spectrum included the appearance of a single, strong, and broad surface plasmon absorbance at 412 nm. The surface plasmon resonance effect causes an absorbance band in the electromagnetic spectrum, usually in the visible light range. Particle size and shape, coating, particle spacing, and other variables can all have an impact on the peak [78–81]. The fact that we observed the peak at the region 410–450 nm corresponds with the literature data for spherical nanoparticles [45,82,83]. The absorption spectra obtained from all the prepared solutions after 10 min are shown in Figure 2.

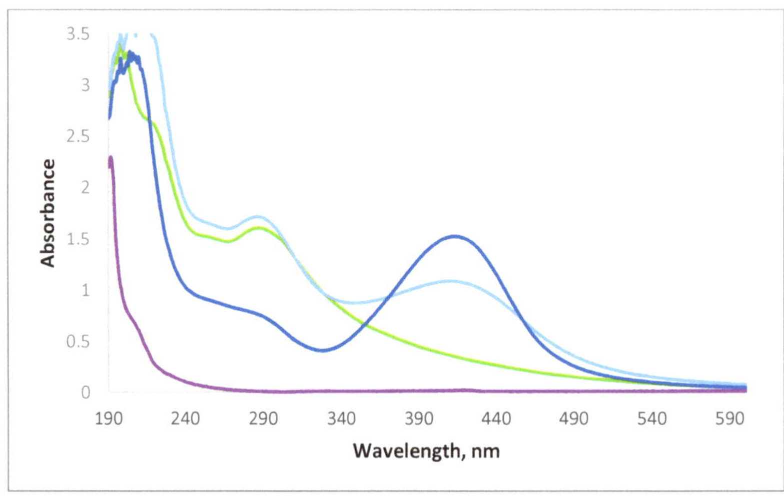

Figure 2. Absorption spectra of galactose (green), MP (purple), Ag NPs (blue), and drug-loaded Ag NPs (dark blue).

A hypochromic effect of the absorption maximum of galactose at 287 nm was observed for MP-loaded Ag NPs. In the meanwhile, a hyperchromic effect of the absorption maximum at 412 nm was also seen. The observed effects were proportional to the increased concentration of Ag NPs in the presence of the MP [84].

The UV–visible spectra showed an application of a simple one-pot galactose-assisted green method for the synthesis and stabilization of Ag NPs as drug-delivery systems. The symmetrical plasmon band of MP-loaded Ag NPs indicated the low degree of aggregation [85]. We presume that the lack of aggregation is due to the presence of an isopropyl group in the structure of the MP. Since the intensity of an absorption peak is proportional to the concentration of Ag NPs in the colloidal solution, the appearance of such well-defined peaks suggests that the production yield of our method was quite good.

The stability of drug-loaded Ag NPs was also investigated by FTIR and compared to MP, galactose, and galactose-assisted Ag NPs spectrum (Figure 3). IR analysis depicted shifts in the characteristic peaks of galactose, Ag NPs, and MP-loaded Ag NPs, which indicates the interactions between the molecules. The shifts were observed for stretching vibration ν(O–H) from 3386, 3205, and 3131 cm^{-1} in galactose compared to 3421 cm^{-1} for Ag NPs and 3382, 3203, and 3124 cm^{-1} for MP-loaded Ag NPs. Shifts could be found also for the stretching C–H vibration ν(C–H) at 2949, 2937, and 2916 cm^{-1} for galactose compared to 2939–2849 cm^{-1} region Ag NPs and 2972 cm^{-1} for drug-loaded Ag NPs. The deformation vibrations (scissoring δ, wagging ω, and torsion τ) of the CH_2 group were also shifted from 1457–1248 cm^{-1} in IR of galactose to 1500–1238 cm^{-1} for the IR of Ag NPs, and 1456–1245 cm^{-1} for MP-loaded Ag NPs. The stretching vibrations ν(C–O), ν(C–C), and in-plane bending β(COH) in the pyranose ring of galactose were observed at 1151, 1104, 1067, and 1045 cm^{-1} [86], while the Ag NPs showed peaks at 1119 and 1049 cm^{-1} and MP-loaded Ag NPs—at 1103 cm^{-1}. The IR spectrum of galactose and MP-loaded Ag NPs showing the band at 837 cm^{-1} indicates the presence of the α-anomer, while for galactose-assisted Ag NPs, two bands appeared at 836 cm^{-1} and 883 cm^{-1}, corresponding to both the anomers—α and β [87,88]. The shifts of the bands in the IR spectra of galactose-assisted Ag NPs and MP-loaded Ag NPs relative to the bands in the IR spectrum of galactose are most indicative of the adsorption and modification of Ag NPs on the surface of the carbohydrate [89].

The depicted shifts in the IR spectra of galactose and MP compared to Ag NPs and MP-loaded Ag NPs confirmed that galactose molecules participate with their C–O and glycosidic OH groups in the formation and stabilization of the Ag NPs [90].

Carbohydrate-assisted (by reduction of glucose, galactose, maltose, and lactose) synthesis of silver NPs in the presence of $[Ag(NH_3)_2]^+$ obtained Ag NPs with controllable sizes [91]. It is well known that, irrespective of the approach, biomolecules are involved in the reduction of silver nanoparticles, and also have an interaction with the upper face of silver, which is their initial connection of originating particles [92,93]. The coating of silver nanoparticles decreases the agglomeration rate and size [94]. The dynamic surface site of the silver nanoparticles persisted through the particle size, shape, and accumulation rate. The reaction conditions such as temperature, pH, extract volume, reactant concentration, and time describe the size and gradation of the growing particles and, thus, influence the shape and size of the silver accumulates [95]. Silver nanoparticles less than 10 nm may elapse by the nuclear opening and interact with genetic material. These crystals are appropriate for diagnostics and gene therapy, but they are genotoxic. The shape of the nanoparticles is shown to affect cytotoxicity, e.g., plate-shaped silver nanoparticles have revealed higher toxicity as compared to wires or spherical nanoparticles [96–99].

To establish the size and shape of the Ag NPs, spICP-MS, DLS, and zeta potential were used.

The nanocolloid suspensions were examined by spICP-MS to determine the size and distribution by size of the silver core in synthesized galactose-assisted Ag NPs and MP-loaded Ag NPs. The information about Ag core size is obtained by single-point calibration with RM of citrate-stabilized Ag NPs with size 40 ± 4 nm and one ionic standard of Ag^+ (1 ng L^{-1}) for sensitivity determination.

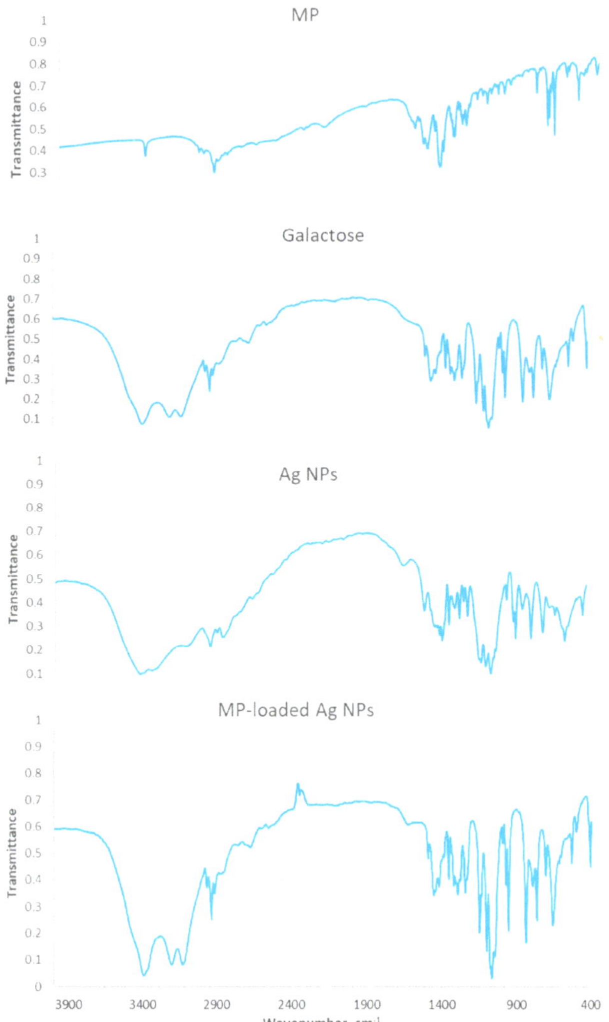

Figure 3. IR spectra of MP, galactose, galactose-assisted Ag NPs, and MP-loaded Ag NPs in the region 4000–400 cm^{-1}.

The introduction of a sufficient number of particles and a 10% likelihood of particle coincidence at the chosen dwell duration were ensured by gravimetrically diluting nanocolloid suspensions with UPW to DF = 5 × 105. Before each dilution step and during the instrumental measurements, the NPs were subjected to an ultrasound treatment for 10 min in order to achieve a uniform dispersion in the solutions.

The histogram size distribution for galactose-assisted Ag NPs, presented in Figure 4, shows that the particle size distribution is asymmetric with a high fraction of small particles (19–22 nm), which is close to the method's detection limit—LODsize 18 nm. The estimated mean diameter for this sample is 42 ± 1 nm, while the most frequent diameter is 20 nm. The results received for MP-loaded Ag NPs show that most detected particles have a size near or below the determined method detection limit (LODsize 18 nm).

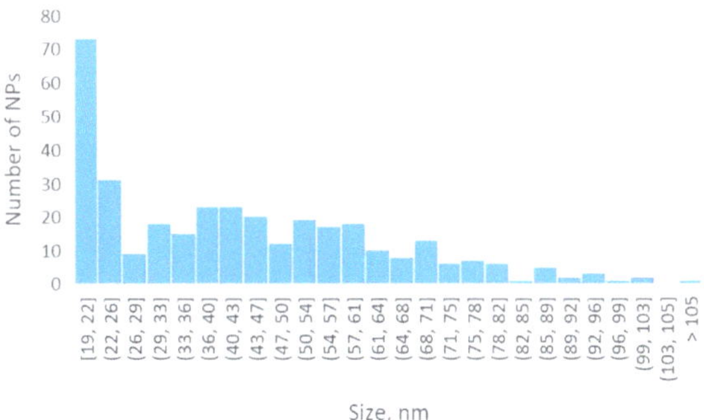

Figure 4. Particle size distribution histogram of galactose-assisted Ag NPs (~400 NPs).

The results from the size distribution histograms (Figure 4) were confirmed by the TEM images (Figure 5a,b). The obtained results clearly indicate the shape and size of the nanoparticles and illustrate the individual nanoparticles as well. The TEM image confirms the synthesis of smaller spherical particles with different size between 10 and 26 nm for galactose-assisted Ag NPs and between 10 and 17 nm size for the drug-loaded Ag NPs. This explains the fact that we cannot quantify the MP-loaded Ag NPs using the spICP-MS.

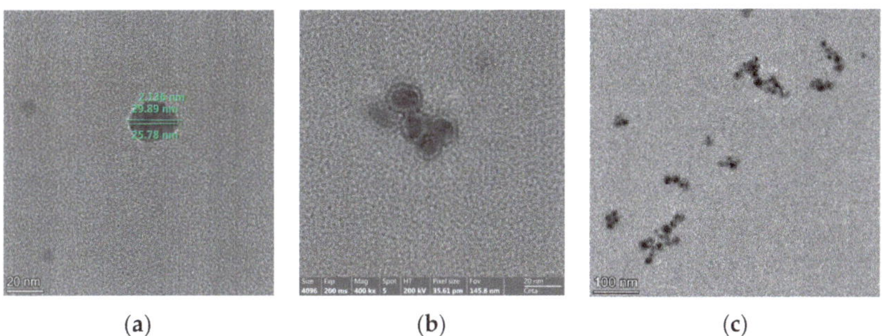

(a) (b) (c)

Figure 5. TEM images of galactose-assisted Ag NPs (**a**,**b**); MP-loaded Ag NPs (**c**).

The TEM images show that the galactose participated as a capping and protecting agent. A thin, galactose-based coat with an average size of 2.136 nm can be observed in TEM images (Figure 5b,c). Many groups such as hydroxyl, carboxyl, phenol, and carbonyl are linked with oxygen and nitrogen with covalent bond linkage for the complex formation of silver, and so they are probably absorbed on its surface [94]. We assume that galactose in galactose-assisted Ag NPs (Figure 5a,b), as well as the isopropyl group of MP in MP-loaded Ag NPs (Figure 5c,d), prevent the aggregation of the particles.

The dynamic light scattering histograms show that the median average size of the obtained particles is 27.9 nm for galactose-assisted Ag NPs and 18 nm for MP-loaded Ag NPs (Figure 6).

The zeta potential of galactose-assisted Ag NPs was -21.99 ± 1.53 mV and for drug-loaded Ag NPs it was −10.72 ± 0.74 mV. The negative charge of the zeta potential could refer to carboxylic groups of galactonic acid produced. Carboxylic acids (in this case, the oxidation products of the sugars) are used to provide a negative surface charge density in order to counteract the van der Waals forces responsible for particle coalescence. Self-assembled carboxylic acids ensure dense coating on the metal surfaces and therefore stabilize them [100].

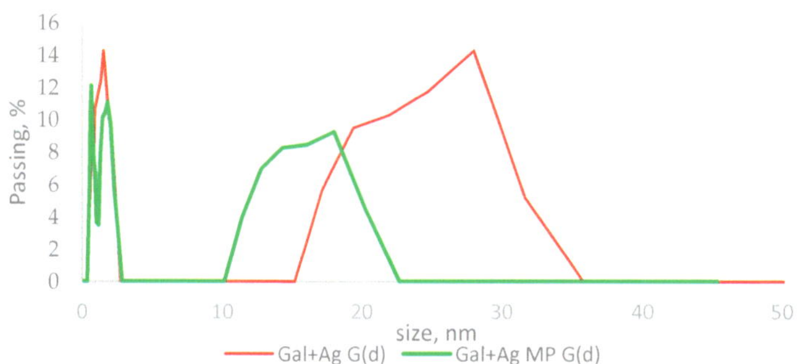

Figure 6. Dynamic light scattering histograms of galactose-assisted Ag NPs and MP-loaded Ag NPs.

The XRD analysis supports galactose's role in the synthesis of Ag NPs (Figure 7). Pure galactose 's XRD pattern (Figure 7a) and MP's structure (Figure 7d) revealed multiple distinctive peaks in the 2θ area because of their crystalline structure.

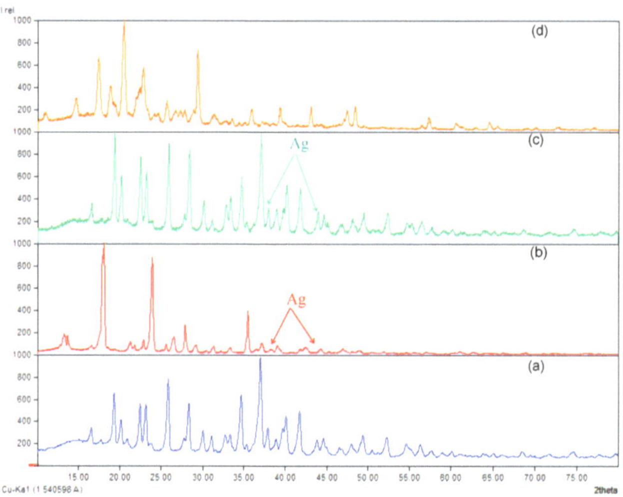

Figure 7. X-ray diffraction patterns of galactose (**a**), Ag NPs (**b**), MP-loaded Ag NPs (**c**), and MP (**d**).

The Ag NPs pattern modified the normal galactose XRD pattern (Figure 7b). At 38.16° and 44.36° of 2θ, the detected signals of Ag NPs are quite weak. The 1 1 1 plane, which can be indexed to the spherical structure of silver [45], corresponds to the strong Bragg reflection at 38.16°.

The silver reflection of MP-loaded Ag NPs (Figure 7c) can be observed at 37.92° and 44.02° of 2θ.

3.2. Parametric Drug-Release Optimization

The most popular method of medicine administration is oral. Oral medication is self-administrable, painless, easy to take, and less expensive than intravenous methods. A drug's oral bioavailability will be hampered if its water solubility is too low. Nanocarriers can be used to resolve such solubility issues [70]. The influence of Ag NPs-to-drug ratio was evaluated, utilizing an in vitro approach with dialysis bags in order to obtain insights into the drug-release trends of MP-loaded Ag NPs [101,102]. Due to a lack of universal, accepted practices, this is a typical technique to learn about innovative drug delivery systems [103].

To determine the in vivo–in vitro correlation of nanoparticle formulations, as well as to direct the development and quality control, drug-release profiles from dialysis-based assays are used [104]. Additionally, they can be utilized to distinguish between different release patterns, such as fast vs. gradual releases [105].

Profiles of MP-molecules only (controls) were compared to those seen in the presence of Ag NPs in order to determine the drug-release properties of the MP-loaded Ag NPs. Different Ag NPs-to-drug ratios (1:1, 1:3, and 1:6) were utilized to examine how the ratio might impact the drug's release. One drug control per ratio was made using an ethanol solution of MP.

Under normal physiological conditions, the pH value from the stomach to the intestine basically showed an increasing trend, specifically, acidic stomach (pH 1.5 to 3.5), duodenum (pH 6), terminal ileum (pH 7.4), terminal cecum (pH 6), and colon (pH 6.7) [106,107]. To simulate the physiological conditions, the drug-release study was carried out in a neutral medium (PBS, pH 7.4). A cellulose dialysis bag with a molecular weight cut-off of 12,400 Da was used to hold the solution of drug-containing Ag NPs. The pores in this cut-off were large enough to let the medication out of the dialysate while still keeping the Ag NPs inside the membrane. Since the calibration curve for MP was linear, it was possible to conduct the quantification tests.

The length of the intestine, the composition of the gut microbiota, and individual variances in stomach and intestinal fluid and peristalsis may all have a role in the transport time of the gastrointestinal system. There are considerable variations in gastrointestinal emptying periods, not just between healthy participants but also between IBD patients and healthy people, which may raise the uncertainty of when the medicine will reach the colon. Although the small intestine's average transport time is thought to be 4 h, individual differences often range from 2 to 6 h [107]. Therefore, determining the drug release at the first 6 h was crucial for our investigation.

During the 24 h period, the drug release was measured. We found that the absorbance intensity rose with time during the initial five hours of the drug-release experiment, indicating an increasing concentration of the MP in the dialysate (Figure 8).

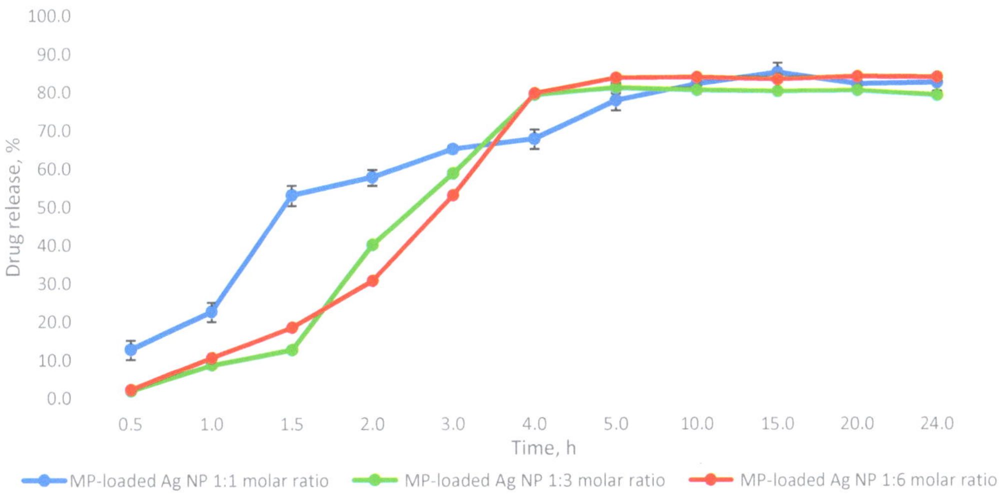

Figure 8. Drug-release concentrations for MP-loaded Ag NPs in molar ratio 1:1, 1:3, and 1:6.

After the fifth hour, the pattern of drug release changed slightly as the rate of increase in absorbance intensity keep unchanged until the 24th hour. The UV–visible spectra of the released pharmaceuticals in the dialysate, therefore, overlapped in the 24th hour, indicating that the drug release was complete under these circumstances and that the concentration of the released drug in the dialysate was stable. All experiments showed that the dialysate drug concentrations of MP only were a bit higher. That could refer to the protective coat of the galactose and indicates the effectively loaded MP, which slows down the drug release. Nevertheless, MP-loaded Ag NPs indicate a very good drug release between 80 and 85% for all drug ratios at the 6th hour of the experiment. According to the preliminary findings, Ag NPs might be a useful medication delivery mechanism for MP and a potent tool for the treatment of inflammatory bowel disease. For this reason, additional research on long-acting delivery formulations and potential medical uses for MP-loaded Ag NPs will also be required to add to the benefits of this study.

4. Conclusions

The rapid synthesis and stabilization of silver nanoparticles using galactose was accomplished utilizing a straightforward, entirely green, one-pot approach. Without the use of microwave irradiation or any other intermediate procedures, the nanoparticles were synthesized in a short amount of time at relatively low temperatures at the water boiling point. The size range for the produced Ag NPs, as determined by various approaches, was 10 to 26 nm for galactose-assisted Ag NPs and 10 to 17 nm for Ag NPs loaded with MP. Galactose was employed as a reducing and capping agent. A thin galactose layer encasing the nanoparticles was visible in the TEM pictures. Additionally established were the Ag NPs' drug-release capabilities. A valuable perspective for our future investigation is the ex vivo and in vivo determination of the long-acting delivery systems and possible medicinal applications of drug-loaded Ag NPs, namely, in the field of gastrointestinal inflammatory diseases.

Author Contributions: Conceptualization, M.T. and S.N.; methodology, M.T., S.N., L.K. and D.G.; investigation, M.T., M.M. and S.N.; synthesis of Ag NPs, M.T.; UV, FTIR spectra, spICP-MS, L.K. and D.G.; TEM, S.S. and B.P.; XRD, V.D.; writing—original draft preparation, S.N., M.M. and M.T.; writing—review and editing, S.N. and M.M.; visualization, S.N., M.M., M.T. and S.S.; supervision, S.N.; project administration, S.N. and M.T. All authors have read and agreed to the published version of the manuscript.

Funding: National Program "Young Scientists and Postdoctoral Students—2" by the Bulgarian Ministry of Education and Science.

Acknowledgments: M. Milusheva and L. Kaynarova are grateful for the financial support of this research by the Bulgarian Ministry of Education and Science under the National Program "Young Scientists and Postdoctoral Students—2".

References

1. Talaei, F.; Atyabi, F.; Azhdarzadeh, M.; Dinarvand, R.; Saadatzadeh, A. Overcoming therapeutic obstacles in inflammatory bowel diseases: A comprehensive review on novel drug delivery strategies. *Eur. J. Pharm. Sci.* **2013**, *49*, 712–722. [CrossRef]
2. Friend, D.R. New oral delivery systems for treatment of inflammatory bowel disease. *Adv. Drug Deliv. Rev.* **2005**, *57*, 247–265. [CrossRef]
3. Vong, L.; Mo, J.; Abrahamsson, B.; Nagasaki, Y. Specific accumulation of orally administered redox nanotherapeutics in the inflamed colon reducing inflammation with dose–response efficacy. *J. Control. Release* **2015**, *210*, 19–25. [CrossRef] [PubMed]
4. Tiwari, A.; Verma, A.; Kumar Panda, P.; Saraf, S.; Jain, A.; Jain, S. Stimuli-responsive polysaccharides for colon-targeted drug delivery. In *Biomaterials, Stimuli Responsive Polymeric Nanocarriers for Drug Delivery Applications*; Makhlouf, A.S.H., Abu-Thabit, N.Y., Eds.; Woodhead Publishing: Sawston, UK, 2019; pp. 547–566. [CrossRef]
5. Wang, C.-P.; Ji Byun, M.; Kim, S.-N.; Park, W.; Park, H.; Kim, T.-H.; Lee, J.; Park, C. Biomaterials as therapeutic drug carriers for inflammatory bowel disease treatment. *J. Control. Release* **2022**, *345*, 1–19. [CrossRef]
6. Collnot, E.-M.; Ali, H.; Lehr, C.-M. Nano- and microparticulate drug carriers for targeting of the inflamed intestinal mucosa. *J. Control. Release* **2012**, *161*, 235–246. [CrossRef] [PubMed]
7. Nidhi; Rashid, M.; Kaur, V.; Hallan, S.S.; Sharma, S.; Mishra, N. Microparticles as controlled drug delivery carrier for the treatment of ulcerative colitis: A brief review. *Saudi Pharm. J.* **2016**, *24*, 458–472. [CrossRef] [PubMed]

8. Katas, H.; Moden, N.Z.; Lim, C.S.; Celesistinus, T.; Chan, J.Y.; Ganasan, P.; Abdalla, S.S.I. Biosynthesis and potential applications of silver and gold nanoparticles and their chitosan-based nanocomposites in nanomedicine. *J. Nanotechnol.* **2018**, *2018*, 4290705. [CrossRef]

9. Jeevanandam, J.; Barhoum, A.; Chan, Y.; Dufresne, A.; Danquah, M. Review on nanoparticles and nanostructured materials: History, sources, toxicity and regulations. *Beilstein J. Nanotechnol.* **2018**, *9*, 1050. [CrossRef] [PubMed]

10. Ordikhani, F.; Erdem Arslan, M.; Marcelo, R.; Sahin, I.; Grigsby, P.; Schwarz, J.K.; Azab, A.K. Drug Delivery Approaches for the Treatment of Cervical Cancer. *Pharmaceutics* **2016**, *8*, 23. [CrossRef] [PubMed]

11. Kodiha, M.; Wang, Y.; Hutter, E.; Maysinger, D.; Stochaj, U. Off to the Organelles—Killing Cancer Cells with Targeted Gold Nanoparticles. *Theranostics* **2015**, *5*, 357–370. [CrossRef]

12. Petros, R.; DeSimone, J. Strategies in the design of nanoparticles for therapeutic applications. *Nat. Rev. Drug Discov.* **2010**, *9*, 615–627. [CrossRef]

13. Rai, A.; Prabhune, A.; Perry, C. Antibiotic mediated synthesis of gold nanoparticles with potent antimicrobial activity and their application in antimicrobial coatings. *J. Mater. Chem.* **2010**, *20*, 6789. [CrossRef]

14. Peng, G.; Tisch, U.; Adams, O.; Hakim, M.; Shehada, N.; Broza, Y.; Billan, S.; Abdah-Bortnyak, R.; Kuten, A.; Haick, H. Diagnosing lung cancer in exhaled breath using gold nanoparticles. *Nat. Nanotech.* **2009**, *4*, 669–673. [CrossRef] [PubMed]

15. Saravanan, M.; Asmalash, T.; Gebrekidan, A.; Gebreegziabiher, D.; Araya, T.; Hilekiros, H.; Barabadi, H.; Ramanathan, K. Nano-Medicine as a Newly Emerging Approach to Combat Human Immunodeficiency Virus (HIV). *Pharm. Nanotechnol.* **2018**, *6*, 17. [CrossRef] [PubMed]

16. Barabadi, H.; Alizadeh, Z.; Rahimi, M.T.; Barac, A.; Maraolo, A.E.; Robertson, L.J.; Masjedi, A.; Shahrivar, F.; Ahmadpour, E. Nanobiotechnology as an Emerging Approach to Combat Malaria: A Systematic Review. *Nanomedicine* **2019**, *18*, 221–233. [CrossRef]

17. Virmani, I.; Sasi, C.; Priyadarshini, E.; Kumar, R.; Kumar Sharma, S.; Pratap Singh, G.; Babu Pachwarya, R.; Paulraj, R.; Barabadi, H.; Saravanan, M.; et al. Comparative Anticancer Potential of Biologically and Chemically Synthesized Gold Nanoparticles. *J. Clust. Sci.* **2020**, *31*, 867–876. [CrossRef]

18. Mostafavi, E.; Zarepour, A.; Barabadi, H.; Zarrabi, A.; Truong, L.; Medina-Cruz, D. Antineoplastic activity of biogenic silver and gold nanoparticles to combat leukemia: Beginning a new era in cancer theragnostic. *Biotechnol. Rep.* **2022**, *34*, e00714. [CrossRef] [PubMed]

19. Saravanan, M.; Vahidi, H.; Cruz, D.M.; Vernet-Crua, A.; Mostafavi, E.; Stelmach, R.; Webster, T.J.; Mahjoub, M.A.; Rashedi, M.; Barabadi, H. Emerging antineoplastic biogenic gold nanomaterials for breast cancer therapeutics: A systematic review. *Int. J. Nanomed.* **2020**, *15*, 3577–3595. [CrossRef] [PubMed]

20. Barabadi, H.; Webster, T.J.; Vahidi, H.; Sabori, H.; Damavandi Kamali, K.; Jazayeri Shoushtari, F.; Mahjoub, M.A.; Rashedi, M.; Mostafavi, E.; Medina Cruz, D.; et al. Green nanotechnology-based gold nanomaterials for hepatic cancer therapeutics: A systematic review. *Iran. J. Pharm. Res.* **2020**, *3*, 19. [CrossRef]

21. Jain, A.; Pawar, P.; Sarkar, A.; Junnuthula, V.; Dyawanapelly, S. Bionanofactories for Green Synthesis of Silver Nanoparticles: Toward Antimicrobial Applications. *Int. J. Mol. Sci.* **2021**, *22*, 11993. [CrossRef] [PubMed]

22. Wei, L.; Lu, J.; Xu, H.; Patel, A.; Chen, Z.-S.; Chen, G. Silver nanoparticles: Synthesis, properties, and therapeutic applications. *Drug Discov. Today* **2015**, *20*, 595–601. [CrossRef] [PubMed]

23. Khorrami, S.; Abdollahi, Z.; Eshaghi, G.; Khosravi, A.; Bidram, E.; Zarrabi, A. An improved method for fabrication of Ag-GO nanocomposite with controlled anticancer and anti-bacterial behavior; a comparative study. *Sci. Rep.* **2019**, *9*, 9167. [CrossRef]

24. Yadi, M.; Mostafavi, E.; Saleh, B.; Davaran, S.; Aliyeva, I.; Khalilov, R.; Nikzamir, M.; Nikzamir, N.; Akbarzadeh, A.; Panahi, Y.; et al. Current developments in green synthesis of metallic nanoparticles using plant extracts: A review. *Artif. Cells Nanomed. Biotechnol.* **2018**, *46* (Suppl. S3), S336–S343. [CrossRef]

25. Kalantari, K.; Mostafavi, E.; Afifi, A.; Izadiyan, Z.; Jahangirian, H.; Rafiee Moghaddam, R.; Webster, T. Wound dressings functionalized with silver nanoparticles: Promises and pitfalls. *Nanoscale* **2020**, *4*, 12. [CrossRef]

26. Lotfollahzadeh, R.; Yari, M.; Sedaghat, S.; Delbari, A. Biosynthesis and characterization of silver nanoparticles for the removal of amoxicillin from aqueous solutions using Oenothera biennis water extract. *J. Nanostructure Chem.* **2021**, *4*, 693–706. [CrossRef]

27. Abbas, Q.; Saleem, M.; Phull, A.; Rafiq, M.; Hassan, M.; Lee, K.-H.; Seo, S.-Y. Green synthesis of silver nanoparticles using Bidens frondosa extract and their tyrosinase activity. *Iran. J. Pharm. Res.* **2017**, *16*, 760.

28. Karimi, N.; Chardoli, A.; Fattahi, A. Biosynthesis, characterization, antimicrobial and cytotoxic effects of silver nanoparticles using Nigellaarvensis seed extract. *Iran. J. Pharm. Res.* **2017**, *16*, 1167.

29. Amin, Z.; Khashyarmanesh, Z.; Bazzaz, B.; Noghabi, Z. Does biosynthetic silver nanoparticles are more stable with lower toxicity than their synthetic counterparts? *Iran. J. Pharm. Res.* **2019**, *18*, 210–221.

30. Salari, S.; Bahabadi, S.; Samzadeh-Kermani, A.; Yosefzaei, F. In-vitro evaluation of antioxidant and antibacterial potential of greensynthesized silver nanoparticles using prosopis farcta fruit extract. *Iran. J. Pharm. Res.* **2019**, *18*, 430–455. [PubMed]

31. Yaqoob, A.A.; Ahmad, H.; Parveen, T.; Ahmad, A.; Oves, M.; Ismail, I.M.I.; Qari, H.A.; Umar, K.; Ibrahim, M.N.M. Recent advances in metal decorated nanomaterials and their various biological applications: A review. *Front. Chem.* **2020**, *8*, 341. [CrossRef]

32. Barabadi, H.; Kobarfard, F.; Vahidi, H. Biosynthesis and Characterization of Biogenic Tellurium Nanoparticles by Using *Penicillium chrysogenum* PTCC 5031: A Novel Approach in Gold Biotechnology. *Iran. J. Pharm. Res.* **2018**, *17*, 87–97.

33. Barabadi, H.; Honary, S.; Ebrahimi, P.; Alizadeh, A.; Naghibi, F.; Saravanan, M. Optimization of myco-synthesized silver nanoparticles by response surface methodology employing Box-Behnken design. *Inorg. Nano Met. Chem.* **2019**, *49*, 33–43. [CrossRef]

34. De Matteis, V.; Rizzello, L.; Cascione, M.; Liatsi-Douvitsa, E.; Apriceno, A.; Rinaldi, R. Green Plasmonic Nanoparticles and Bio-Inspired Stimuli-Responsive Vesicles in Cancer Therapy Application. *Nanomaterials* **2020**, *10*, 1083. [CrossRef] [PubMed]

35. Chairam, S.; Poolperm, C.; Somsook, E. Starch vermicelli template-assisted synthesis of size/shape-controlled nanoparticles. *Carbohydr. Polym.* **2009**, *75*, 694–704. [CrossRef]

36. Kassaee, M.Z.; Akhavan, A.; Sheikh, N.; Beteshobabrud, R. γ-Ray synthesis of starch-stabilized silver nanoparticles with antibacterial activities. *Radiat. Phys. Chem.* **2008**, *77*, 1074–1078. [CrossRef]

37. Vigneshwaran, N.; Nachane, R.P.; Balasubramanya, R.H.; Varadarajan, P.V. A novel one-pot green synthesis of stable silver nanoparticles using soluble starch. *Carbohydr. Res.* **2006**, *341*, 2012–2018. [CrossRef] [PubMed]

38. Shankar, S.; Rai, A.; Ahmad, A.; Sastry, M. Rapid synthesis of Au, Ag, and bimetallic Au core-Ag shell nanoparticles using Neem (Azadirachta indica) leaf broth. *J. Colloid Interface Sci.* **2004**, *275*, 496–502. [CrossRef]

39. Philip, D. Biosynthesis of Au, Ag and Au-Ag nanoparticles using edible mushroom extract. *Spectrochim. Acta A Mol. Biomol. Spectrosc.* **2009**, *73*, 374–381. [CrossRef]

40. Sharma, V.K.; Yngard, R.A.; Lin, Y. Silver nanoparticles: Green synthesis and their antimicrobial activities. *Adv. Colloid Interface Sci.* **2009**, *145*, 83–96. [CrossRef]

41. Panacek, A.; Kvítek, L.; Prucek, R.; Kolar, M.; Vecerova, R.; Pizúrova, N.; Sharma, V.; Nevecna, T.; Zboril, R. Silver colloid nanoparticles: Synthesis, characterization, and their antibacterial activity. *J Phys. Chem. B* **2006**, *110*, 16248–16253. [CrossRef]

42. Sabbagh, F.; Kiarostami, K.; Khatir, N.; Rezania, S.; Muhamad, I.; Hosseini, F. Effect of zinc content on structural, functional, morphological, and thermal properties of kappa-carrageenan/NaCMC nanocomposites. *Polym. Test.* **2021**, *93*, 106922. [CrossRef]

43. Kamble, S.; Bhosale, K.; Mohite, M.; Navale, S. Methods of Preparation of Nanoparticles. *Int. Adv. Res. Sci. Commun. Technol.* **2022**, *2*, 2581–9429. [CrossRef]

44. Filippo, E.; Manno, D.; Serra, A. Self assembly and branching of sucrose stabilized silver nanoparticles by microwave assisted synthesis: From nanoparticles to branched nanowires structures. *Colloids Surf. A Physicochem. Eng. Asp.* **2009**, *348*, 205–211. [CrossRef]

45. Filippo, E.; Serra, A.; Buccolieri, A.; Manno, D. Green synthesis of silver nanoparticles with sucrose and maltose: Morphological and structural characterization. *J. Non-Cryst. Solids* **2010**, *356*, 344–350. [CrossRef]

46. Shervani, Z.; Yamamoto, Y. Carbohydrate-directed synthesis of silver and gold nanoparticles: Effect of the structure of carbohydrates and reducing agents on the size and morphology of the composites. *Carbohydr. Res.* **2011**, *346*, 651–658. [CrossRef]

47. Ghiyasiyan-Arani, M.; Salavati-Niasari, M.; Masjedi-Arani, M.; Mazloom, F. An Easy Sonochemical Route for Synthesis, Characterization and Photocatalytic Performance of Nanosized FeVO 4 in the Presence of Aminoacids as green Capping Agents. *J. Mater. Sci. Mater. Elect.* **2018**, *29*, 474–485. [CrossRef]

48. Caschera, D.; Toro, R.G.; Federici, F.; Montanari, R.; de Caro, T.; Al-Shemy, M.T.; Adel, A.M. Green Approach for the Fabrication of Silver-Oxidized Cellulose Nano-composite with Antibacterial Properties. *Cellulose* **2020**, *27*, 8059–8073. [CrossRef]

49. Garza-Cervantes, J.A.; Mendiola-Garza, G.; de Melo, E.M.; Dugmore, T.I.; Matharu, A.S.; Morones-Ramirez, J.R. Antimicrobial Activity of a Silver-Microfibrillated Cellulose Biocomposite against Susceptible and Resistant Bacteria. *Sci. Rep.* **2020**, *10*, 7281. [CrossRef]

50. Rather, R.A.; Sarwara, R.K.; Das, N.; Pal, B. Impact of Reducing and Capping Agents on Carbohydrates for the Growth of Ag and Cu Nanostructures and Their Antibacterial Activities. *Particuology* **2019**, *43*, 219–226. [CrossRef]

51. Shanmuganathan, R.; Edison, T.N.J.I.; LewisOscar, F.; Kumar, P.; Shanmugam, S.; Pugazhendhi, A. Chitosan Nanopolymers: An Overview of Drug Delivery against Cancer. *Int. J. Biol. Macromol.* **2019**, *130*, 727–736. [CrossRef]

52. Maiti, P.K.; Ghosh, A.; Parveen, R.; Saha, A.; Choudhury, M.G. Preparation of Carboxy-Methyl Cellulose-Capped Nanosilver Particles and Their Antimicrobial Evaluation by an Automated Device. *Appl. Nanosci.* **2019**, *9*, 105–111. [CrossRef]

53. Javed, R.; Zia, M.; Naz, S.; Aisida, S.O.; Ain, N.U.; Ao, Q. Role of Capping Agents in the Application of Nanoparticles in Biomedicine and Environmental Remediation: Recent Trends and Future Prospects. *J. Nanobiotechnology* **2020**, *18*, 172. [CrossRef]

54. Kumar, A.; Das, N.; Satija, N.K.; Mandrah, K.; Roy, S.K.; Rayavarapu, R.G. A Novel Approach towards Synthesis and Characterization of Non-cytotoxic Gold Nanoparticles Using Taurine as Capping Agent. *Nanomaterials* **2020**, *10*, 45. [CrossRef]

55. Priya, N.; Kaur, K.; Sidhu, A.K. Green Synthesis: An Eco-Friendly Route for the Synthesis of Iron Oxide Nanoparticles. *Front. Nanotechnol.* **2021**, *3*, 47. [CrossRef]

56. Duan, H.; Wang, D.; Li, Y. Green chemistry for nanoparticle synthesis. *Chem. Soc. Rev.* **2015**, *44*, 5778–5792. [CrossRef]

57. Zeng, Q.; Shao, D.; Ji, W.; Li, J.; Chen, L.; Song, J.T. The nanotoxicity investigation of optical nanoparticles to cultured cells in vitro. *Toxicol. Rep.* **2014**, *2*, 137–144. [CrossRef]

58. Maribel Guzman, M.A.; Dille, J.; Godet, S.; Rousse, C. Effect of the Concentration of NaBH4 and N2H4 as Reductant Agent on the Synthesis of Copper Oxide Nanoparticles and its Potential Antimicrobial Applications. *Nano Biomed. Eng.* **2018**, *10*, 392–405. [CrossRef]

59. Ranoszek-Soliwoda, K.; Tomaszewska, E.; Socha, E.; Krzyczmonik, P.; Ignaczak, A.; Orlowski, P.; Krzyzowska, M.; Celichowski, G.; Grobelny, J. The role of tannic acid and sodium citrate in the synthesis of silver nanoparticles. *J. Nanopart Res.* **2017**, *19*, 273. [CrossRef]

60. Buchman, A.L. Side effects of corticosteroid therapy. *J. Clin. Gastroenterol.* **2001**, *33*, 289–294. [CrossRef]
61. Ransford, R.; Langman, M. Sulphasalazine and mesalazine: Serious adverse reactions re-evaluated on the basis of suspected adverse reaction reports to the Committee on Safety of Medicines. *Gut* **2002**, *51*, 536–539. [CrossRef]
62. Rutgeerts, P.; Sandborn, W.J.; Feagan, B.G.; Reinisch, W.; Olson, A.; Johanns, J.; Travers, S.; Rachmilewitz, D.; Hanauer, S.B.; Lichtenstein, G.R.; et al. Infliximab for induction and maintenance therapy for ulcerative colitis. *N. Engl. J. Med.* **2005**, *353*, 2462–2476. [CrossRef]
63. Papadakis, K.A.; Shaye, O.A.; Vasiliauskas, E.A.; Ippoliti, A.; Dubinsky, M.C.; Birt, J.; Paavola, J.; Lee, S.K.; Price, J.; Targan, S.R.; et al. Safety and efficacy of adalimumab (D2E7) in Crohn's disease patients with an attenuated response to infliximab. *Am. J. Gastroenterol.* **2005**, *100*, 75–79. [CrossRef]
64. Sowmya, C.; Reddy, C.S.; Priya, N.V.; Sandhya, R.; Keerthi, K. Colon specific drug delivery systems: A review on pharmaceutical approaches with current trends. *Int. Res. J. Pharm.* **2012**, *3*, 45–55.
65. Milusheva, M.; Gledacheva, V.; Stefanova, I.; Pencheva, M.; Mihaylova, R.; Tumbarski, Y.; Nedialkov, P.; Cherneva, E.; Todorova, M.; Nikolova, S. In Silico, In Vitro, and Ex Vivo Biological Activity of Some Novel Mebeverine Precursors. *Biomedicines* **2023**, *11*, 605. [CrossRef] [PubMed]
66. Pace, H.; Rogers, N.; Jarolimek, C.; Coleman, V.; Higgins, C.; Ranville, J. Determining Transport Efficiency for the Purpose of Counting and Sizing Nanoparticles via Single Particle Inductively Coupled Plasma Mass Spectrometry. *Anal. Chem.* **2011**, *83*, 9361–9369. [CrossRef]
67. Sukirtha, R.; Priyanka, K.; Antony, J.; Kamalakkannan, S.; Thangam, R.; Gunasekaran, P.; Krishnan, M.; Achiraman, S. Cytotoxic effect of green synthesized silver nanoparticles using melia azedarach against in vitro hela cell lines and lymphoma mice model. *Process Biochem.* **2012**, *47*, 273–279. [CrossRef]
68. Yesilot, S.; Aydin, C. Silver nanoparticles: A new hope in cancer therapy? *Eastern J. Med.* **2019**, *24*, 111–116. [CrossRef]
69. Khorrami, S.; Zarepour, A.; Zarrabi, A. Green synthesis of silver nanoparticles at low temperature in a fast pace with unique DPPH radical scavenging and selective cytotoxicity against MCF-7 and BT-20 tumor cell lines. *Biotechnol. Rep.* **2019**, *24*, e00393. [CrossRef]
70. Séguy, L.; Groo, A.-C.; Malzert-Fréon, A. How nano-engineered delivery systems can help marketed and repurposed drugs in Alzheimer's disease treatment? *Drug Discov. Today* **2022**, *27*, 1575–1589. [CrossRef]
71. Muntimadugu, E.; Dhommati, R.; Jain, A.; Challa, V.G.S.; Shaheen, M.; Khan, W. Intranasal delivery of nanoparticle encapsulated tarenflurbil: A potential brain targeting strategy for Alzheimer's disease. *Eur. J. Pharm. Sci.* **2016**, *92*, 224–234. [CrossRef]
72. Jojo, G.; Kuppusamy, G. Scope of new formulation approaches in the repurposing of pioglitazone for the management of Alzheimer's disease. *J. Clin. Pharm. Ther.* **2019**, *44*, 337–348. [CrossRef] [PubMed]
73. Jojo, G.; Kuppusamy, G.; De, A.; Karri, V.V.S.N.R. Formulation and optimization of intranasal nanolipid carriers of pioglitazone for the repurposing in Alzheimer's disease using Box-Behnken design. *Drug. Dev. Ind. Pharm.* **2019**, *45*, 1061–1072. [CrossRef] [PubMed]
74. Angelopoulou, E.; Piperi, C. DPP-4 inhibitors: A promising therapeutic approach against Alzheimer's disease. *Ann. Transl. Med.* **2018**, *6*, 255. [CrossRef]
75. Ibrahim, A.; Abbas, I.; Hussein, M.; Muter, S.; Abdulalah, M.; Morteta, A.; Suhayla, S.; Peshawa, A.; Nabaz, A.; Mahmood, A. Effect of nano silver on gastroprotective activity against ethanol-induced stomach ulcer in rats. *Biomed. Pharmacother.* **2022**, *154*, 113550. [CrossRef]
76. Kennedy, D.; Orts-Gil, G.; Lai, C.-H.; Müller, L.; Haase, A.; Luch, A.; Seeberger, P. Carbohydrate functionalization of silver nanoparticles modulates cytotoxicity and cellular uptake. *J. Nanobiotechnol.* **2014**, *12*, 59. [CrossRef]
77. Pryshchepa, O.; Pomastowski, P.; Buszewski, B. Silver nanoparticles: Synthesis, investigation techniques, and properties. *Adv. Colloid Interface Sci.* **2020**, *284*, 102246. [CrossRef]
78. Xavier, J.; Vincent, S.; Meder, F.; Vollmer, F. Advances in Optoplasmonic Sensors—Combining Optical Nano/Microcavities and Photonic Crystals with Plasmonic Nanostructures and Nanoparticles. *Nanophotonics* **2018**, *7*, 1–38. [CrossRef]
79. Martinsson, E.; Otte, M.A.; Shahjamali, M.M.; Sepulveda, B.; Aili, D. Substrate Effect on the Refractive Index Sensitivity of Silver Nanoparticles. *J. Phys. Chem. C* **2014**, *118*, 24680–24687. [CrossRef]
80. Arcas, A.S.; Jaramillo, L.; Costa, N.S.; Allil, R.C.S.B.; Werneck, M.M. Localized Surface Plasmon Resonance-Based Biosensor on Gold Nanoparticles for Taenia Solium Detection. *Appl. Opt.* **2021**, *60*, 8137–8144. [CrossRef]
81. Raiche-Marcoux, G.; Loiseau, A.; Maranda, C.; Poliquin, A.; Boisselier, E. Parametric Drug Release Optimization of Anti-Inflammatory Drugs by Gold Nanoparticles for Topically Applied Ocular Therapy. *Int. J. Mol. Sci.* **2022**, *23*, 16191. [CrossRef]
82. Zaheer, Z. Biogenic synthesis, optical, catalytic, and in vitro antimicrobial potential of Ag-nanoparticles prepared using Palm date fruit extract. *J. Photochem. Photobiol. B Biol.* **2018**, *178*, 584–592. [CrossRef] [PubMed]
83. Jiang, P.; Zhou, J.J.; Li, R.; Gao, Y.; Sun, T.L.; Zhao, X.W.; Xiang, Y.J.; Xie, S.S. PVP-capped twinned gold plates from nanometer to micrometer. *J. Nanoparticle Res.* **2006**, *8*, 927. [CrossRef]
84. Singh, S.; Bharti, A.; Meena, V.K. Green synthesis of multi-shaped silver nanoparticles: Optical, morphological and antibacterial properties. *J. Mater. Sci Mater. Electron.* **2015**, *26*, 3638–3648. [CrossRef]
85. Huang, H.; Yang, X. Synthesis of polysaccharide-stabilized gold and silver nanoparticles: A green method. *Carbohydr. Res.* **2004**, *339*, 2627–2631. [CrossRef] [PubMed]

86. Wells, H.; Atalla, R. An investigation of the vibrational spectra of glucose, galactose and mannose. *J. Mol. Struct.* **1990**, *224*, 385–424. [CrossRef]
87. Sivakesava, S.; Irudayaraj, J. Prediction of inverted cane sugar adulteration of honey by Fourier transform infrared spectroscopy. *J. Food Sci.* **2001**, *66*, 972–978. [CrossRef]
88. David, M.; Hategan, A.; Berghian-Grosan, C.; Magdas, D. The Development of Honey Recognition Models Based on the Association between ATR-IR Spectroscopy and Advanced Statistical Tools. *Int. J. Mol. Sci.* **2022**, *23*, 9977. [CrossRef]
89. AbuDalo, M.; Al-Mheidat, I.; Al-Shurafat, A.; Grinham, C.; Oyanedel-Craver, V. Synthesis of silver nanoparticles using a modified Tollens' method in conjunction with phytochemicals and assessment of their antimicrobial activity. *Peer J.* **2019**, *7*, e6413. [CrossRef]
90. Ahmed, K.B.A.; Mohammed, A.S.; Anbazhagan, V. Interaction of sugar stabilized silver nanoparticles with the T-antigen specific lectin, jacalin from *Artocarpus Integrifolia*. *Spectrochim. Acta Part A Mol. Biomol. Spectrosc.* **2015**, *145*, 110–116. [CrossRef]
91. Iravani, S.; Korbekandi, H.; Mirmohammadi, S.; Zolfaghari, B. Synthesis of silver nanoparticles: Chemical, physical and biological methods. *Res. Pharm. Sci.* **2014**, *9*, 385–406.
92. Islam, N. Green synthesis and biological activities of gold nanoparticles functionalized with *Salix Alba Arab. J. Chem.* **2019**, *12*, 2914–2925. [CrossRef]
93. Tripathi, D. *Nanomaterials in Plants, Algae, and Microorganisms: Concepts and Controversies*; Academic Press: Cambridge, MA, USA, 2018; Volume 1.
94. Younas, M.; Ahmad, M.A.; Jannat, F.T.; Ashfaq, T.; Ahmad, A. 18—Role of silver nanoparticles in multifunctional drug delivery. In *Micro and Nano Technologies, Nanomedicine Manufacturing and Applications*; Verpoort, F., Ahmad, I., Ahmad, A., Khan, A., Chee, C.Y., Eds.; Elsevier: Orlando, FL, USA, 2021; pp. 297–319. [CrossRef]
95. Ivask, A. Toxicity mechanisms in Escherichia coli vary for silver nanoparticles and differ from ionic silver. *ACS Nano* **2014**, *8*, 374–386. [CrossRef]
96. Stoehr, L.; Gonzalez, E.; Stampfl, A.; Casals, E.; Duschl, A.; Puntes, V.; Oostingh, G. Shape matters: Effects of silver nanospheres and wires on human alveolar epithelial cells. *Part Fibre Toxicol.* **2011**, *8*, 36. [CrossRef]
97. Huk, A.; Izak-Nau, E.; Reidy, B.; Boyles, M.; Duschl, A.; Lynch, I.; Dušinska, M. Is the toxic potential of nanosilver dependent on its size? *Part Fibre Toxicol.* **2014**, *11*, 65. [CrossRef] [PubMed]
98. Zhang, T.; Wang, L.; Chen, Q.; Chen, C. Cytotoxic potential of silver nanoparticles. *Yonsei Med. J.* **2014**, *55*, 283–291. [CrossRef]
99. León-Silva, S.; Fernández-Luqueño, F.; López-Valdez, F. Silver Nanoparticles (AgNP) in the Environment: A Review of Potential Risks on Human and Environmental Health. *Water Air Soil Pollut.* **2016**, *227*, 306. [CrossRef]
100. Mulvaney, P.; Liz-Marzan, L.; Giersig, M.; Ung, T.J. Silica encapsulation of quantum dots and metal clusters. *Mater. Chem.* **2000**, *10*, 1259. [CrossRef]
101. Prakash, P.; Gnanaprakasam, P.; Emmanuel, R.; Arokiyaraj, M.; Saravanan, M. Green synthesis of silver nanoparticles from leaf extract of *Mimusops elengi*, Linn for enhanced antibacterial activity against multi drug resistant clinical isolates. *Colloids Surf. B Biointerfaces* **2013**, *108*, 255–259. [CrossRef]
102. Solomon, D.; Gupta, N.; Mulla, N.S.; Shukla, S.; Guerrero, Y.A.; Gupta, V. Role of In Vitro Release Methods in Liposomal Formulation Development: Challenges and Regulatory Perspective. *AAPSJ* **2017**, *19*, 1669–1681. [CrossRef]
103. Yu, M.; Yuan, W.; Li, D.; Schwendeman, A.; Schwendeman, S.P. Predicting Drug Release Kinetics from Nanocarriers inside Dialysis Bags. *J. Control. Release* **2019**, *315*, 23–30. [CrossRef]
104. Kumar, B.; Jalodia, K.; Kumar, P.; Gautam, H.K. Recent Advances in Nanoparticle-Mediated Drug Delivery. *J. Drug Deliv. Sci. Technol.* **2017**, *41*, 260–268. [CrossRef]
105. Shetab Boushehri, M.A.; Lamprecht, A. Nanoparticles as Drug Carriers: Current Issues with in Vitro Testing. *Nanomedicine* **2015**, *10*, 3213–3230. [CrossRef] [PubMed]
106. Chambin, O.; Dupuis, G.; Champion, D.; Voilley, A.; Pourcelot, Y. Colonspecific drug delivery: Influence of solution reticulation properties upon pectin beads performance. *Int. J. Pharm.* **2006**, *321*, 86–93. [CrossRef] [PubMed]
107. Cui, M.; Zhang, M.; Liu, K. Colon-targeted drug delivery of polysaccharide-based nanocarriers for synergistic treatment of inflammatory bowel disease: A review. *Carbohydr. Polym.* **2021**, *272*, 118530. [CrossRef]

 *biomedicines*

Article

Integrating Chinese Herbs and Western Medicine for New Wound Dressings through Handheld Electrospinning

Jianfeng Zhou [1,†], Liangzhe Wang [2,†], Wenjian Gong [1], Bo Wang [2], Deng-Guang Yu [1,*] and Yuanjie Zhu [2,*]

[1] School of Materials & Chemistry, University of Shanghai for Science and Technology, Shanghai 200093, China; 221550217@st.usst.edu.cn (J.Z.); 223353279@st.usst.edu.cn (W.G.)
[2] Department of Dermatology, Naval Special Medical Center, Naval Medical University, Shanghai 200052, China; lzwang@hotmail.com (L.W.); m18817365409@163.com (B.W.)
*   Correspondence: ydg017@usst.edu.cn (D.-G.Y.); 13918338960@126.com (Y.Z.)
†   These two authors contributed equally to this work.

Abstract: In this nanotechnology era, nanostructures play a crucial role in the investigation of novel functional nanomaterials. Complex nanostructures and their corresponding fabrication techniques provide powerful tools for the development of high-performance functional materials. In this study, advanced micro-nanomanufacturing technologies and composite micro-nanostructures were applied to the development of a new type of pharmaceutical formulation, aiming to achieve rapid hemostasis, pain relief, and antimicrobial properties. Briefly, an approach combining a electrohydrodynamic atomization (EHDA) technique and reversed-phase solvent was employed to fabricate a novel beaded nanofiber structure (BNS), consisting of micrometer-sized particles distributed on a nanoscale fiber matrix. Firstly, Zein-loaded Yunnan Baiyao (YB) particles were prepared using the solution electrospraying process. Subsequently, these particles were suspended in a co-solvent solution containing ciprofloxacin (CIP) and hydrophilic polymer polyvinylpyrrolidone (PVP) and electrospun into hybrid structural microfibers using a handheld electrospinning device, forming the EHDA product E3. The fiber-beaded composite morphology of E3 was confirmed through scanning electron microscopy (SEM) images. Fourier-transform infrared (FTIR) spectroscopy and X-ray diffraction (XRD) analysis revealed the amorphous state of CIP in the BNS membrane due to the good compatibility between CIP and PVP. The rapid dissolution experiment revealed that E3 exhibits fast disintegration properties and promotes the dissolution of CIP. Moreover, in vitro drug release study demonstrated the complete release of CIP within 1 min. Antibacterial assays showed a significant reduction in the number of adhered bacteria on the BNS, indicating excellent antibacterial performance. Compared with the traditional YB powders consisting of Chinese herbs, the BNS showed a series of advantages for potential wound dressing. These advantages include an improved antibacterial effect, a sustained release of active ingredients from YB, and a convenient wound covering application, which were resulted from the integration of Chinese herbs and Western medicine. This study provides valuable insights for the development of novel multiscale functional micro-/nano-composite materials and pioneers the developments of new types of medicines from the combination of herbal medicines and Western medicines.

Keywords: electrospinning; electrospraying; Yunnan Baiyao; ciprofloxacin; antimicrobial; combined medicine

Citation: Zhou, J.; Wang, L.; Gong, W.; Wang, B.; Yu, D.-G.; Zhu, Y. Integrating Chinese Herbs and Western Medicine for New Wound Dressings through Handheld Electrospinning. *Biomedicines* **2023**, *11*, 2146. https://doi.org/10.3390/biomedicines11082146

Academic Editors: Zhang Yongtai and Nokhodchi Ali

Received: 28 June 2023
Revised: 22 July 2023
Accepted: 27 July 2023
Published: 30 July 2023

1. Introduction

Controlled drug release is a hot topic in the field of pharmaceutical technology. Conventional drug delivery systems, as the most widely used formulations, suffer from several drawbacks: large fluctuations in blood drug concentration, leading to the occurrence of "peak-trough" phenomena, where adverse effects may arise during peak drug concentration, while subtherapeutic drug levels during trough concentration can compromise therapeutic efficacy [1–8]. Hence, achieving fast or pulsatile drug release and sustained release can not only enhance treatment effectiveness but also reduce the frequency of administration, thus improving patient convenience [9–14].

Ciprofloxacin (CIP) is a synthetic third-generation fluoroquinolone antibacterial medication with broad-spectrum antibacterial action and excellent bactericidal characteristics. It has been reported to be many times more effective than norfloxacin and enoxacin. CIP also exhibits good antibacterial action and pharmacokinetic qualities with little adverse effects [15,16]. Traditional Chinese herbs are unique medicinal substances used in traditional Chinese medicine and serve as a significant hallmark distinguishing it from other medical practices. These herbs are abundant in resources and have a long history of use. Many of them contain natural active ingredients such as organic acids, flavonoids, terpenes, saponins, alkaloids, etc. These herbs exhibit antibacterial, antiviral, anti-inflammatory, antioxidant, anti-tumor, analgesic, immune-modulating, and tissue-regenerating activities, among others [17,18]. However, due to the standard pharmaceutical procedures utilized in the manufacturing of Chinese medications, there are obstacles in establishing controllable quality, safety, efficiency, and patient compliance in traditional Chinese medicine (TCM) [19]. Yunnan Baiyao (YB), a commercially available herbal preparation, has been used to treat bruises, injuries, and bleeding wounds. However, in practical use, the powdered form of YB is difficult to apply to the surface of wounds and cannot effectively eliminate bacteria around the wound. Therefore, it is possible to design a drug delivery system that combines CIP with YB to address these limitations.

In recent decades, the rapid advancement of nanotechnology and study of new excipients has led to significant improvements in the bioavailability of drugs, while simultaneously promoting safe, effective, and convenient drug delivery systems [20–32]. Hence, many newly proven safe and biocompatible excipients have been introduced, such as PVA, PVP, PEO, PCL, gelatin, etc. [33–39]. These excipients are designed to be combined with drug molecules to create nanomedicine products for disease treatment [40]. Traditional powders prepared as nanofiber membranes offer expanded dosage forms that are easy to administer and can combine both traditional Chinese and Western medicine to enhance therapeutic efficacy [41–44]. Incorporating traditional particles and capsules into electrospun nanofiber membranes facilitates drug delivery, improves taste, and enhances patient tolerance and treatment outcomes. The handheld electrospinning device is an emerging medical product that integrates electrospinning technology with existing medical techniques, providing functions such as immediate disinfection and wound management. Additionally, it is considered one of the smallest electrospinning apparatuses in the world, characterized by its compact size, light weight, and practicality. Driven by batteries, it allows for continuous production of nanofibers with a simple press of a button, enabling single-handed operation that seamlessly conforms to the user's hand shape. It provides the flexibility to achieve in situ and directed deposition of various polymer nanofiber materials, marking a significant leap in the development of electrospinning devices.

In this study, electrospinning and electrospray coating were combined (both belonging to EHDA technique) to fabricate a novel hybrid material consisting of electrospun nanofibers and electrospray particles by using a handheld electrospinning device [45,46]. This material was developed for wound antimicrobial and healing applications. A series of characterization (SEM, XRD and FTIR), rapid disintegration, in vitro drug release, and antimicrobial study was conducted to evaluate the prepared hybrid nanofibers.

2. Materials and Methods

2.1. Materials

Yunnan Baiyao was purchased from a local Laobaixing Drugstore (Shanghai, China). Ciprofloxacin and ethanol were purchased from China National Pharmaceutical Group Corporation. Zein was bought from Shanghai Xingye Biological Technology Co., Ltd. (Shanghai, China). Polyvinylpyrrolidone (PVP K60) was purchased from Merck Chemical Reagent Co., Ltd. (Shanghai, China). Acetic acid and anhydrous ethanol were obtained from Shanghai Fitst reagent Factory (Shanghai, China). Milli-Q water was used throughout the study. All other chemical solvents were analytical grade.

2.2. Preparation of Precursor Solutions

Two different EHDA techniques were employed for the sequential preparation of Zein-YB microparticles via electrospray (single-fluid blending electrospraying) and CIP-PVP nanofibers via electrospinning (single-fluid blending electrospinning). Subsequently, a hybrid structure of nanofibers and microparticles, referred to as YB-CIP beaded nanofibers, was fabricated. Hence, three precursor solutions were prepared, corresponding to the electrostatic spray and electrostatic spinning processes. The specific parameters for the two EHDA techniques and the composition and concentrations of the three solutions are presented in Table 1.

Table 1. Parameters for the EHDA processes.

No.	EHDA Process	Working Fluid	Experimental Conditions		Drug Contents	Morphology
			V (kV)	D (cm)		
E1	Electrospraying	Fluid 1 [a]	20	20	20.0% YB	Particles
E2	Electrospinning	Fluid 2 [b]	Cell	20	20.0% CIP	Fibers
E3	Sequential EHDA process	Fluid 3 [c]	Cell	20	15% (YB) & 5% (CIP)	Hybrids

[a] **Fluid 1**: An amount of 3.0 g YB and 12.0 g Zein were co-dissolved in 100 mL 75% ethanol solution. [b] **Fluid 2**: An amount of 8.0 g PVP and 2.0 g CIP were co-dissolved into 100 mL mixture of ethanol and acetic acid with a volume ratio of 90:10. [c] **Fluid 3**: An amount of 15.0 g microparticles E1 from electrospraying were suspended into 50 mL of Fluid 2 uniformly through continuous stirring.

2.3. Preparation of YB-CIP Beaded Fiber by Electrospinning

Firstly, Zein and YB were co-dissolved in a 100 mL solution of 75% ethanol. Microparticles, denoted as E1, were prepared using a simple electrospray process. Subsequently, 15 g E1 was uniformly dispersed into 200 mL of Fluid 2 with continuous stirring, forming a suspended working fluid denoted as Fluid 3. Then, the HHE-1 handheld electrospinning apparatus (Qingdao Junada Technology Co., Ltd., Qingdao, China) was used to fabricate the YB-CIP beaded fiber E3. The specific experimental conditions were as follows: the EHDA process is powered by 7th-grade Nanfu alkaline batteries (AAA), with a voltage range of $0 \sim 10 \pm 1$ kV (rated current generally below 90 mA) and a particle deposition distance of 20 cm. The ambient temperature and humidity were maintained at $21 \pm 5\,^\circ$C and $47 \pm 7\%$, respectively.

2.4. Characterization
2.4.1. Analysis of Morphology

The YB particles and nanofibers were used for morphology characterization by SEM (FEI Quanta G450 FEG, Inc., Hillsboro, OR, USA). Briefly, steps of fixing the prepared YB particles and nanofiber membrane to a sample holder with conductive adhesive, assuring a level surface under N_2 environment, and subsequently applying a 1.5 min gold sputtering treatment using an ion sputter coater were conducted. A thin layer of 5 nm gold was deposited on the sample surface for SEM observation. ImageJ software v1.48 (National Institutes of Health, Bethesda, MD, USA) was used to measure the average diameter of the nanofibers.

2.4.2. XRD and FTIR Analysis

X-ray diffraction (XRD) analysis of YB particles and nanofiber membrane was conducted using the D8 ADVANCE diffractometer (Bruker, Carteret, NJ, USA). The operating conditions included a working voltage of 40 kV and a tube current of 30 mA. The diffraction patterns were recorded in continuous mode within an angular range of 5° to 60° with a step size of 0.02° and a scanning speed of 5°/min.

FTIR spectroscopy of samples was recorded on a PerkinElmer FTIR Spectrometer (Spectrum 100, Billerica, MA, USA) by KBr method. The spectral curve was performed in the range 500–4000 cm^{-1} with a resolution of 2 cm^{-1}.

2.5. Fast Dissolution Performance

The rapid disintegration and drug dissolution processes of the prepared drug-loaded nanofiber membrane E3 were evaluated by two self-developed methods. Briefly, one method included dropping a drop of water over a nanofiber mat collected on a glass slide, while the other method entailed depositing a piece of electrospun thin film on wet paper. All of procedures were photographed with a digital camera (Canon PowerShot SX50HS, Tokyo, Japan).

2.6. In Vitro Drug Release

The paddle method was used to assess the CIP release profiles of EHDA products in accordance with the Chinese Pharmacopoeia (2020 Edition). An amount of 25 mg EHDA product E2 and 0.1 g E3 were added to six vessels, which contained 500 mL phosphate-buffered solution (0.1 M, pH = 7.0). The dissolution media were maintained at 37 ± 1 °C with a rotation rate of 50 rpm. At a preset time point, 5.0 mL of the solution was withdrawn and filtered through a 0.22 μm film (Millipore, MA, USA). To keep the volume of the bulk solution constant, 5.0 mL of fresh PBS was added after sampling. A UV–vis spectrophotometer (UV-2102PC, Youke Instrument Co., Ltd., Shanghai, China) was used to measure the absorbance of CIP at λ_{max} = 276 nm. The experimental data were reported as mean ± standard deviation, and experiments were repeated 6 times.

There are multiple active ingredients in YB. The UV–vis spectrophotometer detection method cannot be exploited for quantitative analyses of YB sustained-release performance from the insoluble Zein microparticles. Thus, a direct observation with the eyes was conducted to assess the YB sustained release effect.

2.7. Antibacterial Performances of Nanofibers

The in vitro antibacterial effects of E1, E2, and E3 were evaluated using a plate count method. Gram-positive *Bacillus subtilis* (*Wb800*) and Gram-negative *Escherichia coli dh5α* (*E.coli dh5α*) were used as experimental microorganisms. The detailed process was as follows: (1) 5 mL of sterilized Luria–Bertani (LB) broth was loaded into an Erlenmeyer flask. (2) An amount of 50 mg EHDA products was placed into the LB broth solution, which contained about 1.5×10^5 CFU of *E. coli dh5α* and *Wb800*, respectively. (3) The mixtures were incubated in a shaking incubator for 12 h at a constant temperature of 37 °C. (4) A total of 100 μL cell solution was seeded onto LB agar by surface spread plate technique. (5) After plates were incubated at 37 °C for 24 h, the number of CFU was counted.

Pure phosphate-buffered saline (PBS) was used as a blank control, and the antibacterial efficacy (ABE, %) of the specimen could be calculated according to the equation:

$$ABE\ (\%) = (Np - Nt)/Np \times 100\%$$

where Np and Nt represent the numbers of viable bacterial colonies of the blank control (pure PBS buffer added) and experimental group, respectively. All the experiments were performed six times, and the data were expressed as the mean values.

2.8. Statistical Analysis

All experiments were performed in triplicate throughout the study. The data were analyzed and plotted using Origin Pro 2021 (Origin Lab Co., Northampton, MA, USA).

3. Results and discussion

3.1. The Sequential EHDA Process

Electrohydrodynamic atomization (EHDA) utilizes electrical energy to directly evaporate solvents, rapidly drying and solidifying microfluids within a timescale of 10^{-2} s, resulting in the generation of materials at the micro/nanoscale [47,48]. EHDA mainly comprises two processes: electrospinning and electrospray. The principles of electrospinning and electrospray are similar, with two differences. Firstly, it lies in the concentration of the polymer solution. When the liquid concentration is high, the jet from the Taylor cone is relatively stable, leading to fiber deposition on the receiving device [49]. Conversely, when the liquid concentration is low, the jet from the Taylor cone becomes unstable due to the effect of high electric voltage, resulting in the transformation of larger droplets into smaller ones and ultimately forming particles. Secondly, it depends on the degree of interaction between polymer solutions. If the degree of interaction is high, fibers are formed; if the degree of interaction is low, particles are formed. Therefore, by adjusting the concentration of the spraying solution and the type of polymer, the transition between electrospinning and electrospray can be achieved.

The working fluid for an effective EHDA technique has to be electrospinnable [50]. It can be understood that electrospinning typically produces solid products in the form of nanofibers through the continuous stretching of a viscous polymer fluid jet [51,52]. On the other hand, electrospray coating typically generates solid products in the form of particles through droplet fragmentation and repulsion [53]. Hence, A new type of hybrid material on different scales can be envisioned based on the outcomes of these two EHDA processes, as shown in Figure 1.

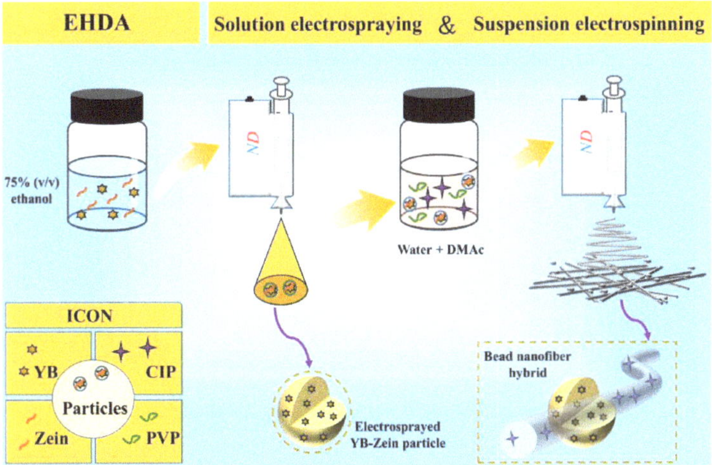

Figure 1. A schematic diagram of the fabrication for bead-on-string nanofibers composed of electrospun nanofibers and electrospray particles.

Electrospraying possesses advantages such as simple preparation, good reproducibility, high drug loading capacity, controllable particle size and surface morphology, and narrow size distribution. Moreover, during the spraying process, the droplets carry like charges, resulting in mutual repulsion and excellent self-dispersion. This property prevents particle aggregation and enables the production of monodisperse drug-loaded nanoparticles. Therefore, it has been widely applied in the field of pharmaceuticals.

The electrospraying process was performed using an electrospinning device. The working Fluid 1 was shown in Figure 2a. As shown in Figure 2b, the entire electrospraying process was successfully carried out, and product E1 was collected on the collector, exhibiting a white area. Figure 2c is an enlarged view of the needle during the electrospraying process, where the Taylor cone is manifested as a typical cone shape. The droplets are affected by Coulomb repulsion and are fragmented into tiny droplets in the air, forming particles that scatter and deposit on the collector as the solvent rapidly evaporates.

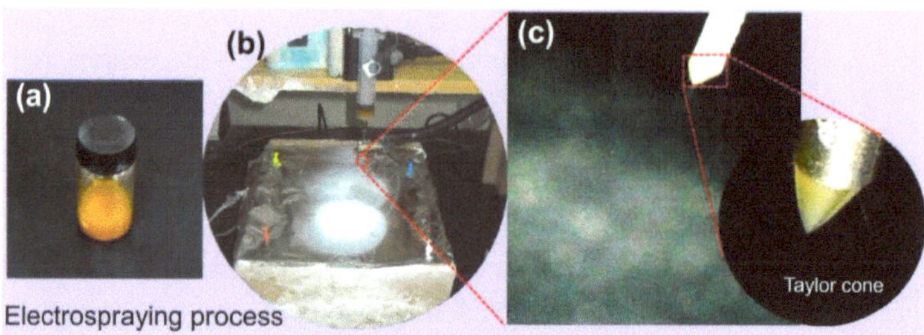

Figure 2. The typical electrospraying process: (**a**) The working Fluid 1; (**b**) Electrospraying equipment operation; (**c**) The electrospraying process and Taylor cone.

The electrospinning process was performed by HHE-1 handheld electrospinning device. Working Fluid 3, depicted in Figure 3a, had a brown color. It was loaded into a 5 mL syringe, connected to a stainless-steel needle, and inserted into the handheld device along with a stainless-steel needle. As shown in Figure 3b, as the device button was continuously pressed, electrospun fibers (E3) were collected on the collector (white area in Figure 3b). In the magnified view of Figure 3b, the fiber-beaded surface morphology can be clearly observed. The liquid flow rate from the stainless-steel needle was carefully monitored, and the electrospinning effect was observed. The electrospinning process was shown in Figure 3c, where a straight jet is ejected from the Taylor cone at the needle tip, which gradually bends and undergoes whipping phenomena as the solvent evaporates, eventually forming fibers [54].

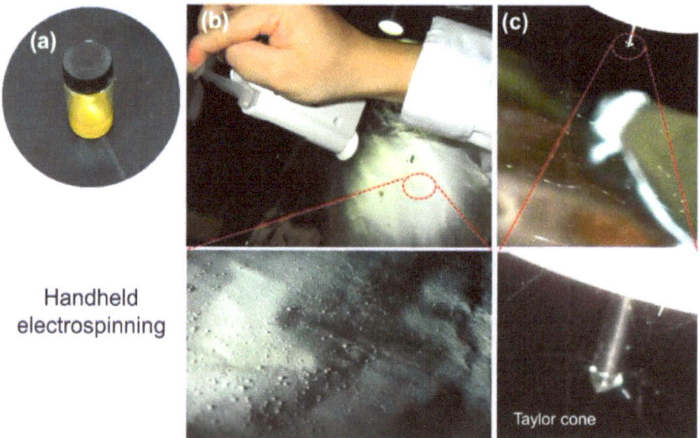

Figure 3. Hand-held electrospinning equipment and typical electrospinning process: (**a**) The working Fluid 3; (**b**) Hand-held electrospinning equipment operation; (**c**) The electrospinning process and Taylor cone.

3.2. Morphology of Product

The morphological changes of different EHDA products are shown in Figure 4. Irregularly arranged YB–Zein particles were obtained on aluminum foil, exhibiting excellent dispersion without significant agglomeration (Figure 4a). Interestingly, there were hardly any "satellites" (submicron particles) observed around the electrosprayed YB–Zein particles (Figure 4a). Due to their small size, these "satellites" may be regarded as a negative aspect for drug release applications because the loaded drug molecules have a short diffusion path to reach the surrounding fluid. As expected, the electrospun CIP-PVP nanofibers (E2) exhibited fine linear morphology. The fibers showed no adhesion, interweaving with each other, and had a certain porosity (Figure 4b). Additionally, the diameter of the E2 electrospun fibers was about 640 ± 130 nm (Figure 4e). PVP is a linear polymer known for its good spinnability. Moreover, PVP has excellent solubility in water, ethanol, methanol, acetone, and other organic solvents [55,56]. Hence, the FDA has approved PVP for biomedical uses, and it is frequently investigated for various dosage forms such as powders, particles, tablets, and capsules. Furthermore, Figure 4b revealed a smooth surface of E2 nanofibers with no drug particles formed due to phase separation during the static process. The electrospun hybrid material YB-CIP beaded nanofiber (E3) prepared from the suspension working Fluid 3 exhibited a mixture of typical beads or spindle bodies along with nanofibers, as shown in Figure 4c,d.

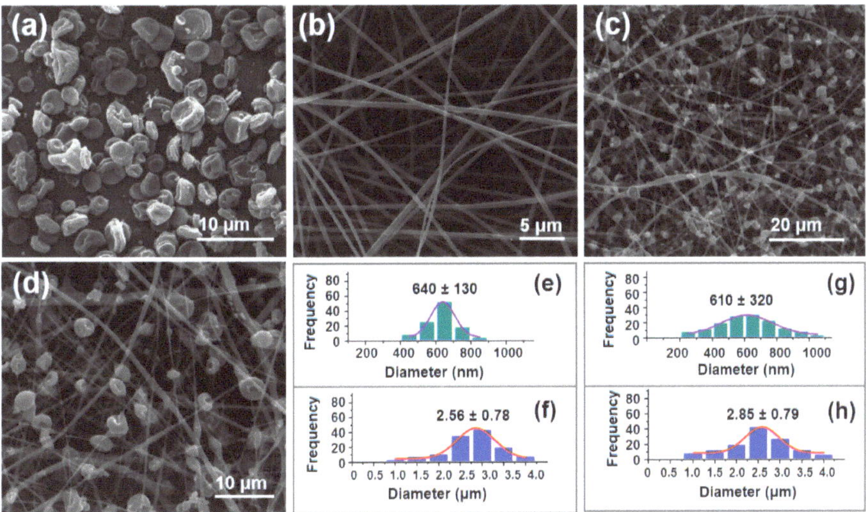

Figure 4. SEM images and the average diameters of EHDA products. (**a**) YB–Zein particles E1; (**b**) CIP-PVP nanofibers E2; (**c**) Hybrid YB-CIP nanofibers E3; (**d**) Magnified image of the hybrid YB-CIP nanofibers E3; (**e**) Average diameter of CIP-PVP nanofibers E2; (**f**) Average diameter of YB–Zein particles E1; (**g**) Average diameter of CIP-PVP nanofibers E3; (**h**) Average diameter of hybrids E3 microparticles.

The size of the EHDA products was evaluated using ImageJ software. The average diameter of CIP-PVP nanofibers (E2) and YB particles (E1) was determined to be 640 ± 130 nm (Figure 4e) and 2.56 ± 0.78 μm (Figure 4f), respectively. The diameter of the PVP fibers and Zein particles in the YB-CIP beaded nanofiber (E3) was measured to be 610 ± 320 nm (Figure 4g) and 2.85 ± 0.79 μm (Figure 4h), respectively. The difference in PVP fiber diameter between E2 and E3 was attributed to the addition of Zein particles. The E1 particles were uniformly distributed within the fibers, and their addition increased the stretching effect on the flowing fluid jet, resulting in an uneven thickness. However, there was no significant difference in the average diameter of Zein particles between E1 and E3, indicating that E1 could be redispersed as a suspension in Working Fluid 3 without altering the particle size.

3.3. Physical State and Compatibility of Components

The crystalline nature of raw materials and compounds were analyzed by X-ray diffraction. As shown in Figure 5, characteristic peaks corresponding to CIP and YB are observed in their respective XRD spectra, indicating the crystalline form of CIP and YB. However, the XRD spectra of PVP, Zein, and hybrid compounds exhibit broad peaks within specific angular ranges, indicating the presence of amorphous phases. The broad peaks suggest a lower degree of crystallinity in the samples. PVP is an amorphous linear polymer known for its excellent electrospinnability, which prevents the crystallization of many drugs. It can be inferred that the crystalline component of CIP loses its crystal state during the electrospinning process and exists in an amorphous form within the hybrid membrane. This result is favorable for the rapid dissolution and release of the active ingredient CIP, as it eliminates the need to overcome lattice energy for dissolution [57,58].

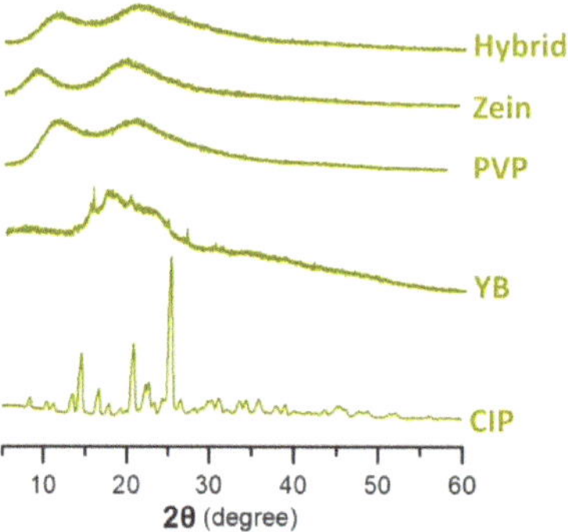

Figure 5. XRD spectra of CIP, YB, PVP, Zein, and Hybrid.

The FTIR spectra of the materials (CIP, YB, Zein, and PVP) and hybrid were shown in Figure 6a. The characteristic peak at 3275 cm^{-1} corresponds to the stretching vibration of O-H in CIP. The characteristic peak at 3054 cm^{-1} is attributed to the stretching vibration of C-H, and the absorption peak at 1986 cm^{-1} represents the vibration of water molecules absorbed by CIP. Additionally, there are several sharp peaks in the fingerprint region of CIP. However, when CIP is encapsulated in compounds containing corn protein, these sharp peaks almost disappear. It could be explained by secondary interactions between the (O-H) and (C-H) of CIP molecules and the Zein carriers (Figure 6b), which typically include hydrogen bonding, hydrophobic contacts, and electrostatic interactions [59]. On the one hand, this facilitates the rapid dissolution of CIP. On the other side, it contributes to the stable transport and storage of the thin film. Moreover, in the hybrid spectrum, the characteristic peak of C=O was found at 1658 cm^{-1}, which exhibits a slight blue shift compared to Zein (1661 cm^{-1}) and a slight red shift compared to PVP (1658 cm^{-1}). Combined with the molecular structure in Figure 6b, this further confirms the interactions between the components and the encapsulation effect of E3 on CIP and YB.

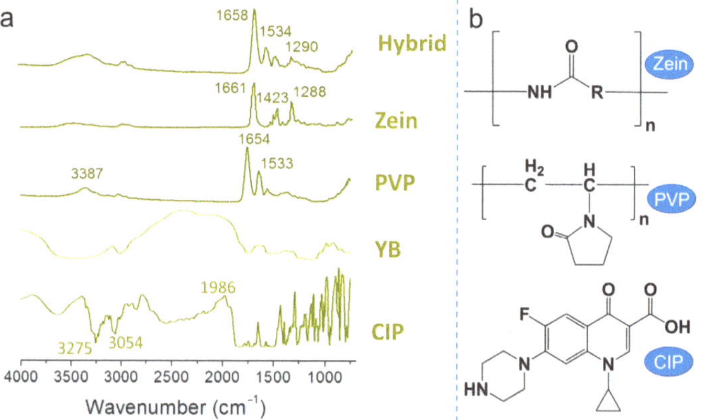

Figure 6. (**a**) FTIR spectra of CIP, YB, PVP, Zein, and Hybrid; (**b**) The molecular formats of the components (Zein, PVP, and CIP).

3.4. Rapid Disintegration of YB-CIP Beaded Nanofiber Membrane

The dissolution experiment was conducted using a glass slide, as shown in Figure 7a. After depositing YB-CIP fibers for 5 min, a drop of water was added to the YB-CIP fibers. A camera was used to photograph the rapid dissolution process. Upon contact with the water droplet, the YB-CIP fibers rapidly disintegrated into a transparent gel, gradually revealing the logo of "USST" The time from "a1" to "a6" was about 4.8 ± 0.4 s. By morphological analysis of the water-dripped place, different images were observed. One typical image is that the water-soluble CIP-PVP nanofibers were swollen and dissolved upon water absorption, while the water-insoluble Zein particles dispersed uniformly on the glass slide (Figure 7b). Another typical image is that the dissolved drug CIP precipitated and re-crystallized into elongated strips, as shown in Figure 7c.

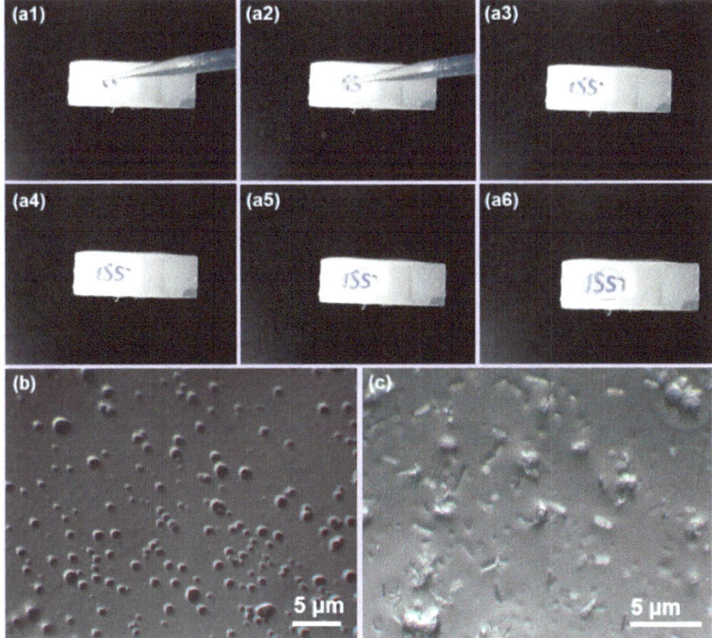

Figure 7. Rapid disintegration of YB-CIP beaded nanofiber membrane. (**a1–a6**) Dissolution process; (**b,c**) Different OM images of the water-dripped place after drying.

Furthermore, as shown in Figure 8a, the fiber membrane was cut into circular pieces with a diameter of 1.2 cm using a puncher, which was used to simulate the rapid disintegration experiment of an artificial tongue. First, the water-soaked filter paper was placed in a Petri dish, and then a circular YB-CIP fiber membrane was placed on the surface of the wet paper. As shown in Figure 8b, the dissolution and passive diffusion process of the YB-CIP fiber membrane transformed into a water-absorbing gelation process, which was captured by a camera. The time from "1" to "6" was about 1.7 ± 0.4 s. The fiber membrane changed from opaque white to semi-transparent. Obviously, the gelation process was promoted by the hygroscopicity of the PVP matrix, the tiny diameter of the nanofibers, and the three-dimensional network structure of the nanofiber membrane. It is worth noting that a faint yellowish color gradually shows in fiber membranes "7" to "8" in Figure 8b. This process had a relatively long duration of approximately 91.7 ± 11.4 s, indicating a passive diffusion process. However, the YB–Zein particles could not be dissolved and removed; thus, a light-yellow trace was left there, as indicated by the red arrow in Figure 8b-8. Of course, when additional stirring was introduced (e.g., mimicking tongue movements or a wound place), the diffusion and transportation process remained very rapid. It should be mentioned that due to the hygroscopic nature of the polymer carrier (i.e., PVP), the membrane needs to be stored in a low-humidity environment, which is a common concern for many conventional dosage forms.

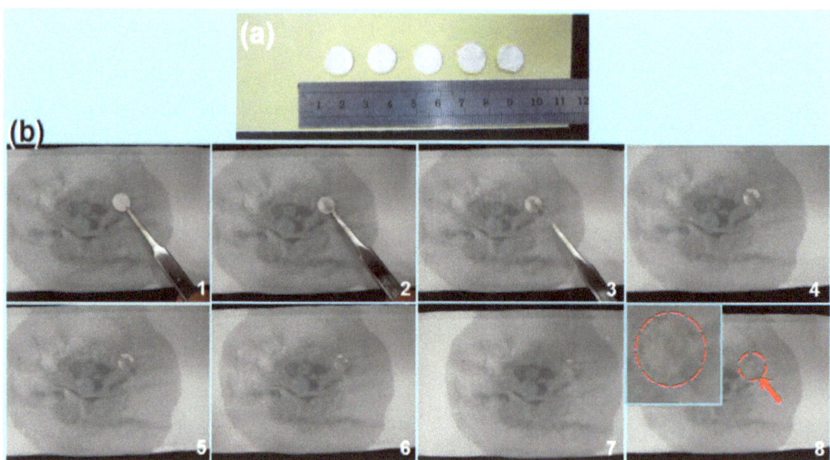

Figure 8. Rapid disintegration experiment of simulated artificial tongue. (**a**) Circular YB-CIP fiber membrane; (**b**) Artificial tongue process, the dissolution process is according to the order from "1" to "8", the up-left inset of "8" is an enlarged image indicted by the red arrow and circle, in which are YB-Zein particles.

3.5. In Vitro Drug Release

The in vitro release profiles of CIP from E2 and E3 were determined using UV–vis spectrophotometry. The wavelength scan of CIP in the range of 170–370 nm was performed, as shown in Figure 9a. The results indicated maximum absorbance of CIP at 276 nm. Subsequently, a series of CIP solutions with different concentrations was formulated to construct a standard curve of $A = 0.1148\,C + 9.8924 \times 10^{-4}$ with a R value of 0.9997 (Figure 9b), where A was the absorbance and C was CIP concentration (1–50 µg/mL).

The CIP release performances of EHDA product E2 and E3 are shown in Figure 9c,d, respectively. These release profiles were constructed based on the accumulative percentage of CIP released over time. Clearly, both the CIP-PVP nanofibers E2 and the beads-on-a-string E3 exhibited a pulsatile release of the loaded CIP within 1 min. These results demonstrated that the E3, on one hand, was able to release the Western medicine as rapidly as the electrospun nanofibers E2, by which it can be expected that the therapeutic effect of CIP can be rapidly initiated for urgent wound treatment requiring fast antimicrobial activity. On the other hand, E3 contained the insoluble YB-Zein microparticles, which could manipulate the sustained release of active ingredients in YB for playing their important roles in wound healing (Figure 9e). YB is famous for its excellent functions such as promoting myogenesis and stopping pain and detumescence. A sustained release profile of YB is better for this functional performance.

It is noteworthy that rapid release of CIP can be attributed to two main factors: (1) the excellent hydrophilic nature of the PVP K60 matrix [55] and (2) the electrospun membrane exhibits characteristics such as a large fiber surface area, small diameter, and high porosity [60–62], which accelerate the dissolution and diffusion of CIP from the PVP matrix. In contrast, the YB released from the Zein particles through a diffusion process due to the insolubility of Zein in water. The micro-scale diameters mean a longer way for the YB diffusion routes and thus a better sustained release effect. Although UV–vis cannot be explored for quantitative analyses of the YB release profiles due to multiple active ingredients, the direct observations of the taupe color by eyes can judge the sustained release effect of YB to 12 h.

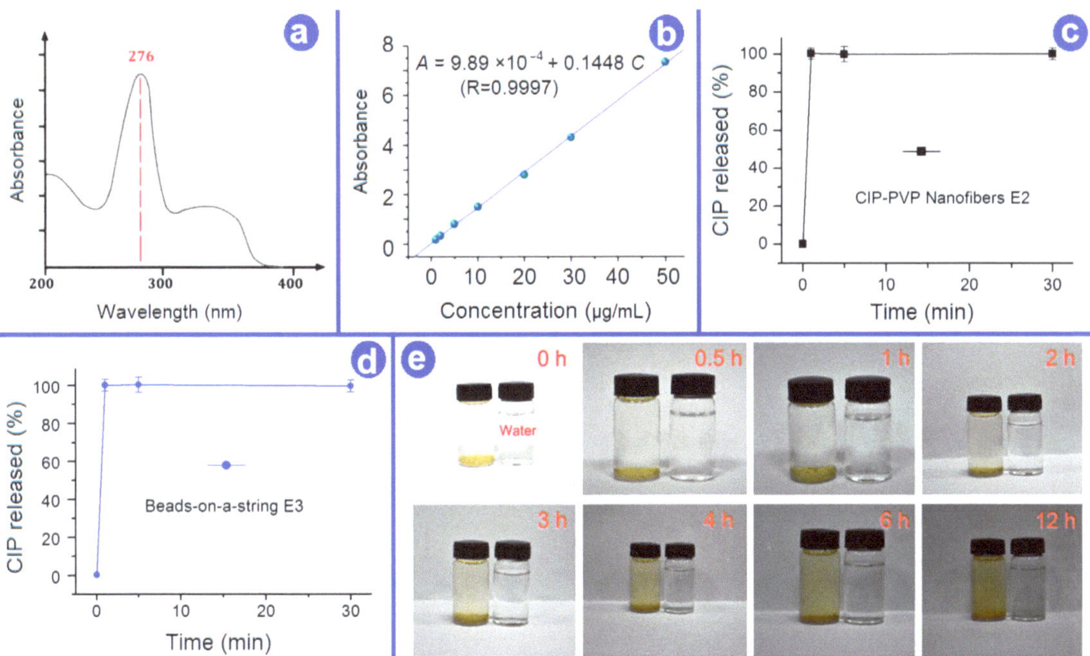

Figure 9. In vitro drug release. (**a**) Absorbance curve of CIP; (**b**) Standard calibration curve of CIP; (**c,d**) In vitro CIP pulsatile release from E2 and E3, respectively; (**e**) An observation of the YB sustained release effect from the electrosprayed insoluble Zein microparticles E1.

Additionally, an interesting phenomenon is the color change of the materials conversion. The raw YB powders show a gray color. In Figure 2a, the YB–Zein solution showed a taupe color, and the YB–Zein particles E1 showed a deeper taupe color. However, the suspensions for handheld electrospinning showed a yellow color in Figure 3a, and the electrospun E3 showed almost a white color (Figure 8a). In the YB sustained-release experiments, the taupe colors were gradually increased as the YB ingredients were gradually freed from the Zein particles to the dissolution media in Figure 9e.

3.6. Analysis of Antibacterial Performances

As a useful wound dressing, the first and foremost thing is its antibacterial performance [63–65]. Bacterial counts and antibacterial rates in the culture medium were determined using the plate counting method at 6, 12, and 24 h of incubation. The *Escherichia coli* and *Staphylococcus aureus* represent Gram-negative and Gram-positive bacteria, respectively. As shown in Table 2, the powders of YB have a certain antibacterial performance. When they were loaded into Zein microparticles through electrospraying, the formed EHDA product E1 still had antibacterial effects. Furthermore, with the increase of incubation time, the antibacterial effects increased significantly, suggesting the sustained release of active ingredients from the Zein particles. For *Wb800* and *Escherichia coli dh5α*, the increases are from 88.3% to 99.9% and from 81.1% to 99.9% after 2 and 12 h, respectively.

As for the little molecule CIP, it has a fine sterilizing effect. After 2 h of incubation, the ABE% could reach a value of around 99.0%, suggesting its fast effect on antibacterial performance. As anticipated, when the YB-loaded microparticles were combined with the electrospun CIP-loaded PVP nanofibers, synergistic antibacterial effects were achieved, having fast initiation and also an extension effect in antibacterial performance. YB is famous for its hemostatic and myogenic effects. Its encapsulation into Zein particles can benefit the myogenic effects for wound healing, whereas the combination with CIP can endow the hybrid EHDA products with a desired antibacterial performance. The strategy of tailoring components, compositions, and organization formats through advanced fabrication methods can promote the development of novel biomedicines [66–68]. The present protocol is a fine example of taking into consideration several factors together to create new kinds of medicated products. Additionally, the handheld electrospinning in situ for treating open wound places should be more convenient than the raw powders of YB.

Table 2. The antibacterial performance of the EHDA products against *Wb800* and *Escherichia coli dh5α* (n = 6) [a].

Bacteria	Samples	Initial CFU	CFU after 2 h	CFU after 6 h	CFU after 12 h
			CFU (ABE%)	CFU (ABE%)	CFU (ABE%)
Wb800	YB	1.5×10^5	2.8×10^4 (88.3%)	1.4×10^4 (98.7%)	6.7×10^3 (99.9%)
	E1	1.5×10^5	4.7×10^4 (80.4%)	2.4×10^4 (97.8%)	8.5×10^3 (99.9%)
	E2	1.5×10^5	1.2×10^3 (99.5%)	2.8×10^2 (>99.9%)	1.9×10^2 (>99.9%)
	E3	1.5×10^5	2.1×10^4 (91.3%)	3.6×10^2 (>99.9%)	2.1×10^2 (>99.9%)
	Blank	1.5×10^5	2.4×10^5	1.1×10^6	7.3×10^6
Escherichia coli dh5α	YB	1.5×10^5	5.1×10^4 (81.1%)	2.2×10^4 (98.1%)	4.7×10^3 (99.9%)
	E1	1.5×10^5	7.4×10^4 (72.6%)	3.5×10^4 (97.1%)	5.3×10^3 (99.9%)
	E2	1.5×10^5	4.2×10^3 (98.4%)	8.9×10^2 (>99.9%)	3.4×10^2 (>99.9%)
	E3	1.5×10^5	7.6×10^3 (97.2%)	1.3×10^3 (>99.9%)	5.7×10^2 (>99.9%)
	Blank	1.5×10^5	2.7×10^5	1.2×10^6	8.1×10^6

[a] Abbreviations: YB, Yunnan Baiyao powders (10 mg, an equal amount loaded in microparticles E1); CFU, colony-forming units; ABE, antibacterial efficacy.

3.7. Release Mechanism

The process of multi-drug release from the hybrid product E3, which is made up of electrospun PVP nanofibers and electrosprayed YB–Zein particles, is rather evident based on the analysis outlined above, as shown in Figure 10. When the hybrid product E3 is immersed in the dissolution medium, the CIP-PVP nanofibers dissolve rapidly. This represents a typical first-stage erosion process for the rapid release of CIP. Then, YB molecules distributed or adsorbed on the surface of YB–Zein particles dissolve into the dissolution medium, gradually opening pathways for water molecules to penetrate the interior cross-sections of the Zein particles. As water molecules permeate from the surface to the core of Zein particles, the loaded YB molecules freely diffuse back into the bulk solution from the Zein particles. Throughout the entire process, the Zein skeletons maintain the diffusion and exchange of water and YB molecules. In theory, the diffusion process continues until achieving a uniform drug concentration distribution throughout the entire bulk solution and reaching dynamic equilibrium in terms of absorbance between the dissolution medium and the solid Zein skeletons. Human health always relies on the establishment of new methods [69–72], with the continuous emergence of "bottom-up" approaches involving molecular reactions to create new materials and "top-down" methods for preparing nanomaterials [73–75]. In this study, the trans-scale combination of nanofibers and microparticles has been utilized to generate functional hybrid materials, enhancing the functional performance of the materials. Meanwhile, the release profile of a drug always relies on the property of the host matrix that is exploited to encapsulate the drug [76–79]. Here, another combination of water-soluble PVP and water-insoluble Zein has been utilized to provide the different release behaviors of different drugs. The above-mentioned two "combinations" are finally aimed to the third combination, i.e., the combination of Chinese Herbs and Western Medicine for an improved wound healing effect.

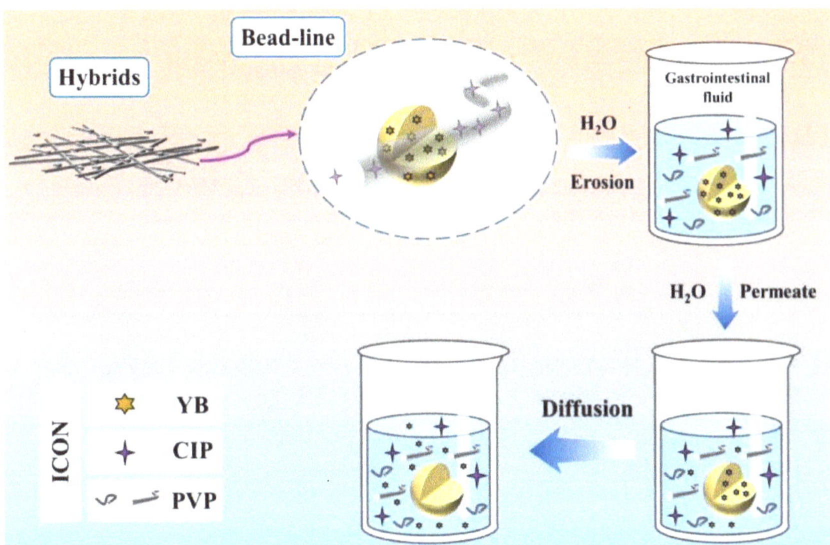

Figure 10. Multi-drug dissolution-controlled release and synergistic mechanism.

4. Conclusions

In this study, a continuous EHDA process was developed to prepare a novel drug mixture, E3. Zein was used as a sustained-release carrier PVP K60 was used as the spinning matrix; and YB and CIP were used as model drugs for hemostasis, analgesia, and antimicrobial purposes, respectively. A handheld electrospinning device was successfully employed to fabricate bead-on-string structured nanofiber blend membranes for rapid antimicrobial activity and long-lasting hemostasis. SEM revealed well-dispersed YB–Zein particles with almost no unfavorable smaller particles on the surface, providing the possibility of sustained release of YB. The CIP–PVP fibers exhibited a good linear morphology with a uniform and smooth surface, free from beads or spindles, enabling rapid drug release for antimicrobial efficacy. The YB–CIP hybrid membrane presented a fiber-beaded structure. In vitro dissolution tests demonstrated the complete release of the antimicrobial drug CIP within 1 min. The prepared E3 exhibited desirable functional performance in facilitating the rapid breakdown and dissolution of CIP, thereby enhancing patient convenience. The antimicrobial analysis experiment demonstrated the excellent antibacterial performance of YB–CIP against *Escherichia coli*. Based on the combination of handheld electrospinning and electrospraying and the combination of nanofibers of soluble polymeric matrix and microparticles of insoluble protein, a new concept about the development of combined medicine from the traditional Chinese herbs and modern Western medicine was demonstrated. The present protocols pioneered a new approach for developing many new sorts of combined biomedicines.

Author Contributions: Conceptualization, J.Z. and L.W.; methodology, J.Z., L.W., W.G., B.W. and Y.Z.; writing—original draft preparation, J.Z. and D.-G.Y.; writing—review and editing, D.-G.Y. and Y.Z.; visualization, J.Z., B.W. and L.W.; supervision, D.-G.Y.; project administration, D.-G.Y.; funding acquisition, D.-G.Y. and Y.Z. All authors have read and agreed to the published version of the manuscript.

Funding: This investigation was financially supported by Medical Engineering Cross Project between University of Shanghai for Science & Technology and Naval Medical University (No. 2020-RZ05) and the Shanghai Natural Science Foundation (No. 20ZR1439000).

Institutional Review Board Statement: Not applicable.

Informed Consent Statement: Not applicable.

Data Availability Statement: The data supporting the findings of this manuscript are available from the corresponding authors upon reasonable.

Conflicts of Interest: The authors declare no conflict of interest.

References

1. Tung, N.-T.; Dong, T.-H.-Y.; Tran, C.-S.; Nguyen, T.-K.-T.; Chi, S.-C.; Dao, D.-S.; Nguyen, D.-H. Integration of lornoxicam nanocrystals into hydroxypropyl methylcellulose-based sustained release matrix to form a novel biphasic release system. *Int. J. Biol. Macromol.* **2022**, *209*, 441–451. [CrossRef] [PubMed]
2. Adala, I.; Ramis, J.; Ntone Moussinga, C.; Janowski, I.; Amer, M.H.; Bennett, A.J.; Alexander, C.; Rose, F.R.A.J. Mixed polymer and bioconjugate core/shell electrospun fibers for biphasic protein release. *J. Mater. Chem. B* **2021**, *9*, 4120–4133. [CrossRef] [PubMed]
3. Conceicao, J.; Adeoye, O.; Cabral-Marques, H.; Concheiro, A.; Alvarez-Lorenzo, C.; Sousa Lobo, J.M. Carbamazepine bilayer tablets combining hydrophilic and hydrophobic cyclodextrins as a quick/slow biphasic release system. *J. Drug Deliv. Sci. Technol.* **2020**, *57*, 101611. [CrossRef]
4. Yang, B.; Dong, Y.; Shen, Y.; Hou, A.; Quan, G.; Pan, X.; Wu, C. Bilayer dissolving microneedle array containing 5-fluorouracil and triamcinolone with biphasic release profile for hypertrophic scar therapy. *Bioact. Mater.* **2021**, *6*, 2400–2411. [CrossRef] [PubMed]
5. Chen, Y.; Yu, W.; Qian, X.; Li, X.; Wang, Y.; Ji, J. Dissolving microneedles with a biphasic release of antibacterial agent and growth factor to promote wound healing. *Biomater. Sci.* **2022**, *10*, 2409–2416. [CrossRef]
6. Tabakoglu, S.; Kolbuk, D.; Sajkiewicz, P. Multifluid electrospinning for multi-drug delivery systems: Pros and cons, challenges, and future directions. *Biomater. Sci.* **2022**, *11*, 37–61. [CrossRef]
7. Hameed, A.; Rehman, T.U.; Rehan, Z.A.; Noreen, R.; Iqbal, S.; Batool, S.; Qayyum, M.A.; Ahmed, T.; Farooq, T. Development of polymeric nanofibers blended with extract of neem (azadirachta indica), for potential biomedical applications. *Front. Mater.* **2022**, *9*, 1042304. [CrossRef]
8. Brimo, N.; Serdaroglu, D.C.; Uysal, B. Comparing antibiotic pastes with electrospun nanofibers as modern drug delivery systems for regenerative endodontics. *Curr. Drug Deliv.* **2022**, *19*, 904–917.
9. Kuang, G.; Zhang, Z.; Liu, S.; Zhou, D.; Lu, X.; Jing, X.; Huang, Y. Biphasic drug release from electrospun polyblend nanofibers for optimized local cancer treatment. *Biomater. Sci.* **2018**, *6*, 324–331. [CrossRef]
10. Wang, M.; Ge, R.L.; Zhang, F.; Yu, D.G.; Liu, Z.P.; Li, X.; Shen, H.; Williams, G.R. Electrospun fibers with blank surface and inner drug gradient for improving sustained release. *Biomater. Adv.* **2023**, *150*, 213404. [CrossRef]
11. Lee, H.; Xu, G.; Kharaghani, D.; Nishino, M.; Song, K.H.; Lee, J.S.; Kim, I.S. Electrospun tri-layered zein/PVP-GO/zein nanofiber mats for providing biphasic drug release profiles. *Int. J. Pharm.* **2017**, *531*, 101–107. [CrossRef]
12. Khalek, M.A.A.; Gaber, S.A.A.; El-Domany, R.A.; El-Kemary, M.A. Photoactive electrospun cellulose acetate/polyethylene oxide/methylene blue and trilayered cellulose acetate/polyethylene oxide/silk fibroin/ciprofloxacin nanofibers for chronic wound healing. *Int. J. Biol. Macromol.* **2021**, *193*, 1752–1766. [CrossRef]
13. Ejeta, F.; Gabriel, T.; Joseph, N.M.; Belete, A. Formulation, optimization and in vitro evaluation of fast disintegrating tablets of salbutamol sulphate using a combination of superdisintegrant and subliming agent. *Curr. Drug Delivery* **2022**, *19*, 129–141.
14. Kose, M.D.; Ungun, N.; Bayraktar, O. Eggshell membrane based turmeric extract loaded orally disintegrating films. *Curr. Drug Deliv.* **2022**, *19*, 547–559.
15. Zhang, G.-F.; Liu, X.; Zhang, S.; Pan, B.; Liu, M.-L. Ciprofloxacin derivatives and their antibacterial activities. *Eur. J. Med. Chem.* **2018**, *146*, 599–612. [CrossRef]
16. Miao, H.; Wang, P.; Cong, Y.; Dong, W.; Li, L. Preparation of ciprofloxacin-based carbon dots with high antibacterial activity. *Int. J. Mol. Sci.* **2023**, *24*, 6814. [CrossRef]
17. Che, C.-T.; Wang, Z.J.; Chow, M.S.S.; Lam, C.W.K. Herb-herb combination for therapeutic enhancement and advancement: Theory, practice and future perspectives. *Molecules* **2013**, *18*, 5125–5141. [CrossRef] [PubMed]
18. Hallmann, E.; Sabała, P. Organic and conventional herbs quality reflected by their antioxidant compounds concentration. *Appl. Sci.* **2020**, *10*, 3468. [CrossRef]
19. Feng, N.; Yang, Z. (Eds.) *Novel Drug Delivery Systems for Chinese Medicines*; Springer: Singapore, 2021.
20. Wang, X.; Feng, C. Chiral fiber supramolecular hydrogels for tissue engineering. *Wires. Nanomed. Nanobi.* **2023**, *15*, e1847. [CrossRef]
21. Zheng, W.; Bai, Z.; Huang, S.; Jiang, K.; Liu, L.; Wang, X. The effect of angiogenesis-based scaffold of mesoporousbioactive glass nanofiber on osteogenesis. *Int. J. Mol. Sci.* **2022**, *23*, 12670. [CrossRef]
22. Guadagno, L.; Raimondo, M.; Vertuccio, L.; Lamparelli, E.P.; Ciardulli, M.C.; Longo, P.; Mariconda, A.; Della Porta, G.; Longo, R. Electrospun membranes designed for burst release of new gold-complexes inducing apoptosis of melanoma cells. *Int. J. Mol. Sci.* **2022**, *23*, 7147. [CrossRef] [PubMed]
23. De la Ossa, J.G.; Danti, S.; Salsano, J.E.; Azimi, B.; Tempesti, V.; Barbani, N.; Digiacomo, M.; Macchia, M.; Uddin, M.J.; Cristallini, C.; et al. Electrospun poly(3-hydroxybutyrate-co-3-hydroxyvalerate)/olive leaf extract fiber mesh as prospective bio-based scaffold for wound healing. *Molecules* **2022**, *27*, 6208. [CrossRef] [PubMed]

24. Li, W.; Zhang, Y.; Huang, P.; Liu, Y.; Zhang, Y.; Wang, Z.; Feng, N. Application of nmp and neusilin US$_2$-integrated liquisolid technique in mini-tablets for improving the physical performances and oral bioavailability of liposoluble supercritical fluid extracts. *J. Drug Deliv. Sci. Technol.* **2023**, *81*, 104205. [CrossRef]
25. Zhang, Y.; Xia, Q.; Wu, T.; He, Z.; Li, Y.; Li, Z.; Hou, X.; He, Y.; Ruan, S.; Wang, Z.; et al. A novel multi-functionalized multicellular nanodelivery system for non-small cell lung cancer photochemotherapy. *J. Nanobiotechnol.* **2021**, *19*, 245. [CrossRef]
26. Jing, Q.; Ruan, H.; Li, J.; Wang, Z.; Pei, L.; Hu, H.; He, Z.; Wu, T.; Ruan, S.; Guo, T.; et al. Keratinocyte membrane-mediated nanodelivery system with dissolving microneedles for targeted therapy of skin diseases. *Biomaterials* **2021**, *278*, 121142. [CrossRef]
27. Bayer, I.S. A review of sustained drug release studies from nanofiber hydrogels. *Biomedicines* **2021**, *9*, 1612. [CrossRef] [PubMed]
28. Malekmohammadi, S.; Sedghi Aminabad, N.; Sabzi, A.; Zarebkohan, A.; Razavi, M.; Vosough, M.; Bodaghi, M.; Maleki, H. Smart and biomimetic 3D and 4D printed composite hydrogels: Opportunities for different biomedical applications. *Biomedicines* **2021**, *9*, 1537. [CrossRef] [PubMed]
29. Javed, B.; Zhao, X.; Cui, D.; Curtin, J.; Tian, F. Enhanced anticancer response of curcumin- and piperine-loaded lignin-g-p (NIPAM-co-DMAEMA) gold nanogels against U-251 MG glioblastoma multiforme. *Biomedicines* **2021**, *9*, 1516. [CrossRef]
30. Huang, J.; Feng, C. Aniline dimers serving as stable and efficient transfer units for intermolecular charge-carrier transmission. *iScience* **2023**, *26*, 105762. [CrossRef] [PubMed]
31. Aleemardani, M.; Zare, P.; Seifalian, A.; Bagher, Z.; Seifalian, A.M. Graphene-based materials prove to be a promising candidate for nerve regeneration following peripheral nerve injury. *Biomedicines* **2022**, *10*, 73. [CrossRef]
32. Wang, M.; Hou, J.; Yu, D.G.; Li, S.; Zhu, J.; Chen, Z. Electrospun tri-layer nanodepots for sustained release of acyclovir. *J. Alloys Compd.* **2020**, *846*, 156471. [CrossRef]
33. Shen, S.-F.; Zhu, L.-F.; Liu, J.; Ali, A.; Zaman, A.; Ahmad, Z.; Chen, X.; Chang, M.-W. Novel core-shell fiber delivery system for synergistic treatment of cervical cancer. *J. Drug Deliv. Sci. Technol.* **2020**, *59*, 101865. [CrossRef]
34. Yildiz, Z.I.; Topuz, F.; Kilic, M.E.; Durgun, E.; Uyar, T. Encapsulation of antioxidant beta-carotene by cyclodextrin complex electrospun nanofibers: Solubilization and stabilization of beta-carotene by cyclodextrins. *Food Chem.* **2023**, *423*, 136284. [PubMed]
35. Shen, Y.; Yu, X.; Cui, J.; Yu, F.; Liu, M.; Chen, Y.; Wu, J.; Sun, B.; Mo, X. Development of biodegradable polymeric stents for the treatment of cardiovascular diseases. *Biomolecules* **2022**, *12*, 1245. [CrossRef]
36. Yao, Z.-C.; Zhang, C.; Ahmad, Z.; Huang, J.; Li, J.-S.; Chang, M.-W. Designer fibers from 2D to 3D-simultaneous and controlled engineering of morphology, shape and size. *Chem. Eng. J.* **2018**, *334*, 89–98. [CrossRef]
37. Zhou, J.; Wang, P.; Yu, D.-G.; Zhu, Y. Biphasic drug release from electrospun structures. *Exp. Opin. Drug Deliv.* **2023**, *20*, 621–640. [CrossRef]
38. Zhou, J.; Dai, Y.; Fu, J.; Yan, C.; Yu, D.-G.; Yi, T. Dual-step controlled release of berberine hydrochloride from the trans-scale hybrids of nanofibers and microparticles. *Biomolecules* **2023**, *13*, 1011. [CrossRef]
39. Mirzaeei, S.; Mansurian, M.; Asare-Addo, K.; Nokhodchi, A. Metronidazole-and amoxicillin-loaded PLGA and PCL nanofibers as potential drug delivery systems for the treatment of periodontitis: In vitro and in vivo evaluations. *Biomedicines* **2021**, *9*, 975. [CrossRef]
40. Wang, Q.; Liu, Q.; Gao, J.; He, J.; Zhang, H.; Ding, J. Stereo coverage and overall stiffness of biomaterial arrays underly parts of topography effects on cell adhesion. *ACS Appl. Mater. Interfaces* **2023**, *15*, 6142–6155. [CrossRef]
41. Zhang, Y.; Liu, X.; Geng, C.; Shen, H.; Zhang, Q.; Miao, Y.; Wu, J.; Ouyang, R.; Zhou, S. Two hawks with one arrow: A review on bifunctional scaffolds for photothermal therapy and bone regeneration. *Nanomaterials* **2023**, *13*, 551. [CrossRef]
42. Qi, Q.; Wang, Q.; Li, Y.; Silva, D.Z.; Ruiz, M.E.L.; Ouyang, R.; Liu, B.; Miao, Y. Recent development of rhenium-based materials in the application of diagnosis and tumor therapy. *Molecules* **2023**, *28*, 2733. [CrossRef] [PubMed]
43. Sivan, M.; Madheswaran, D.; Hauzerova, S.; Novotny, V.; Hedvicakova, V.; Jencova, V.; Kostakova, E.K.; Schindler, M.; Lukas, D. AC electrospinning: Impact of high voltage and solvent on the electrospinnability and productivity of polycaprolactone electrospun nanofibrous scaffolds. *Mater. Today Chem.* **2022**, *26*, 101025. [CrossRef]
44. Wang, L.; Ahmad, Z.; Huang, J.; Li, J.-S.; Chang, M.-W. Multi-compartment centrifugal electrospinning based composite fibers. *Chem. Eng. J.* **2017**, *330*, 541–549. [CrossRef]
45. Lv, H.; Liu, Y.; Bai, Y.; Shi, H.; Zhou, W.; Chen, Y.; Liu, Y.; Yu, D.-G. Recent combinations of electrospinning with photocatalytic technology for treating polluted water. *Catalysts* **2023**, *13*, 758. [CrossRef]
46. Couto, A.F.; Favretto, M.; Paquis, R.; Estevinho, B.N. Co-encapsulation of epigallocatechin-3-Gallate and vitamin B12 in zein microstructures by electrospinning/electrospraying technique. *Molecules* **2023**, *28*, 2544. [CrossRef]
47. Du, Y.; Yang, Z.; Kang, S.; Yu, D.-G.; Chen, X.; Shao, J. A sequential electrospinning of a coaxial and blending process for creating double-layer hybrid films to sense glucose. *Sensors* **2023**, *23*, 3685. [CrossRef]
48. Guler, E.; Nur Hazar-Yavuz, A.; Tatar, E.; Morid Haidari, M.; Sinemcan Ozcan, G.; Duruksu, G.; Graça, M.P.F.; Kalaskar, D.M.; Gunduz, O.; Emin Cam, M. Oral empagliflozin-loaded tri-layer core-sheath fibers fabricated using tri-axial electrospinning: Enhanced *in vitro* and *in vivo* antidiabetic performance. *Int. J. Pharm.* **2023**, *635*, 122716. [CrossRef]
49. Yao, L.; Sun, C.; Lin, H.; Li, G.; Lian, Z.; Song, R.; Zhuang, S.; Zhang, D. Electrospun bi-decorated BixBiyOz/TiO$_2$ flexible carbon nanofibers and their applications on degradating of organic pollutants under solar radiation. *J. Mater. Sci. Technol.* **2023**, *150*, 114–123. [CrossRef]

50. Li, H.; Zhang, Z.; Ren, Z.; Chen, Y.; Huang, J.; Lei, Z.; Qian, X.; Lai, Y.; Zhang, S. A quadruple biomimetic hydrophilic/hydrophobic janus composite material integrating $Cu(OH)_2$ micro-needles and embedded bead-on-string nanofiber membrane for efficient fog harvesting. *Chem. Eng. J.* **2023**, *455*, 140863. [CrossRef]
51. Yu, D.-G.; Xu, L. Impact evaluations of articles in current drug delivery based on web of science. *Curr. Drug Deliv.* **2023**, *20*. [CrossRef]
52. Xu, J.; Zhong, M.; Song, N.; Wang, C.; Lu, X. General synthesis of pt and ni co-doped porous carbon nanofibers to boost HER performance in both acidic and alkaline solutions. *Chin. Chem. Lett.* **2023**, *34*, 107359. [CrossRef]
53. Yao, Z.-C.; Yuan, Q.; Ahmad, Z.; Huang, J.; Li, J.-S.; Chang, M.-W. Controlled morphing of microbubbles to beaded nanofibers via electrically forced thin film stretching. *Polymers* **2017**, *9*, 265. [CrossRef]
54. Zhu, M.; Yu, J.; Li, Z.; Ding, B. Self-healing fibrous membranes. *Angew. Chem. Int. Edit.* **2022**, *61*, e202208949. [CrossRef] [PubMed]
55. Huang, H.; Song, Y.; Zhang, Y.; Li, Y.; Li, J.; Lu, X.; Wang, C. Electrospun nanofibers: Current progress and applications in food systems. *J. Agric. Food. Chem.* **2022**, *70*, 1391–1409. [CrossRef]
56. Wang, Y.; Yu, D.-G.; Liu, Y.; Liu, Y.-N. Progress of electrospun nanofibrous carriers for modifications to drug release profiles. *J. Func. Biomater.* **2022**, *13*, 289. [CrossRef]
57. Wang, H.; Lu, Y.; Yang, H.; Yu, D.-G.; Lu, X. The influence of the ultrasonic treatment of working fluids on electrospun amorphous solid dispersions. *Front. Mol. Biosci.* **2023**, *10*, 1184767. [CrossRef] [PubMed]
58. Kang, S.; Hou, S.; Chen, X.; Yu, D.-G.; Wang, L.; Li, X.; Williams, R.G. Energy-saving electrospinning with a concentric teflon-core rod spinneret to create medicated nanofibers. *Polymers* **2020**, *12*, 2421. [CrossRef] [PubMed]
59. Huang, X.; Jiang, W.; Zhou, J.; Yu, D.-G.; Liu, H. The applications of ferulic-acid-loaded fibrous films for fruit preservation. *Polymers* **2022**, *14*, 4947. [CrossRef]
60. Kumar, R.G.; Rajan, T.P. A review on electrospinning of natural bio herbs blended with polyvinyl alcohol nanofibres for biomedical applications. *J. Nat. Fibers* **2022**, *19*, 11984–12003.
61. Yang, S.; Li, X.; Liu, P.; Zhang, M.; Wang, C.; Zhang, B. Multifunctional chitosan/polycaprolactone nanofiber scaffolds with varied dual-drug release for wound-healing applications. *ACS Biomater. Sci. Eng.* **2020**, *6*, 4666–4676. [CrossRef]
62. Saraiva, M.M.; Campelo, M.D.S.; Câmara Neto, J.F.; Lima, A.B.N.; Silva, G.D.A.; Dias, A.T.D.F.F.; Ricardo, N.M.P.S.; Kaplan, D.L.; Ribeiro, M.E.N.P. Alginate/polyvinyl alcohol films for wound healing: Advantages and challenges. *J. Biomed. Mater. Res.* **2023**, *111*, 220–233. [CrossRef] [PubMed]
63. Ajalli, N.; Pourmadadi, M.; Yazdian, F.; Abdouss, M.; Rashedi, H.; Rahdar, A. PVA Based Nanofiber Containing GO Modified with Cu Nanoparticles and Loaded Curcumin; High Antibacterial Activity with Acceleration Wound Healing. *Curr. Drug Deliv.* **2023**, *20*, 1569–1583. [PubMed]
64. Grosu, O.-M.; Dragostin, O.-M.; Gardikiotis, I.; Chitescu, C.L.; Lisa, E.L.; Zamfir, A.-S.; Confederat, L.; Dragostin, I.; Dragan, M.; Stan, C.D.; et al. Experimentally Induced Burns in Rats Treated with Innovative Polymeric Films Type Therapies. *Biomedicines* **2023**, *11*, 852. [CrossRef]
65. Patel, P.; Thanki, A.; Viradia, D.; Shah, P.A. Honey-based Silver Sulfadiazine Microsponge-Loaded Hydrogel: In vitro and In vivo Evaluation for Burn Wound Healing. *Curr. Drug. Deliv.* **2023**, *20*, 608–628. [CrossRef]
66. Alosaimi, A.M.; Alorabi, R.O.; Katowah, D.F.; Al-Thagafi, Z.T.; Alsolami, E.S.; Hussein, M.A.; Qutob, M.; Rafatullah, M. Recent biomedical applications of coupling nanocomposite polymeric materials reinforced with variable carbon nanofillers. *Biomedicines* **2023**, *11*, 967. [CrossRef] [PubMed]
67. Deineka, V.; Sulaieva, O.; Pernakov, M.; Korniienko, V.; Husak, Y.; Yanovska, A.; Yusupova, A.; Tkachenko, Y.; Kalinkevich, O.; Zlatska, A.; et al. Hemostatic and tissue regeneration performance of novel electrospun chitosan-based materials. *Biomedicines* **2021**, *9*, 588. [CrossRef]
68. Nan, N.; Hu, W.; Wang, J. Lignin-based porous biomaterials for medical and pharmaceutical applications. *Biomedicines* **2022**, *10*, 747. [CrossRef]
69. Yu, D.-G.; Du, Y.; Chen, J.; Song, W.; Zhou, T. A Correlation Analysis between Undergraduate Students' Safety Behaviors in the Laboratory and Their Learning Efficiencies. *Behav. Sci.* **2023**, *13*, 127. [CrossRef]
70. Xie, D.; Zhou, X.; Xiao, B.; Duan, L.; Zhu, Z. Mucus-penetrating silk fibroin-based nanotherapeutics for efficient treatment of ulcerative colitis. *Biomolecules* **2022**, *12*, 1263. [CrossRef]
71. Sivan, M.; Madheswaran, D.; Valtera, J.; Kostakova, E.K.; Lukas, D. Alternating current electrospinning: The impacts of various high-voltage signal shapes and frequencies on the spinnability and productivity of polycaprolactone nanofibers. *Mater. Des.* **2022**, *213*, 110308. [CrossRef]
72. Liu, H.; Dai, Y.; Li, J.; Liu, P.; Zhou, W.; Yu, D.-G.; Ge, R. Fast and convenient delivery of fluidextracts liquorice through electrospun core-shell nanohybrids. *Front. Bioeng. Biotechnol.* **2023**, *11*, 1172133. [CrossRef] [PubMed]
73. Wu, Y.; Li, Y.; Lv, G.; Bu, W. Redox dyshomeostasis strategy for tumor therapy based on nanomaterials chemistry. *Chem. Sci.* **2022**, *13*, 2202–2217. [CrossRef] [PubMed]
74. Zhao, P.; Li, H.; Bu, W. A forward vision for chemodynamic therapy: Issues and opportunities. *Angew. Chem. Int. Ed.* **2023**, *62*, e202210415. [CrossRef] [PubMed]
75. Meng, Y.; Chen, L.; Chen, Y.; Shi, J.; Zhang, Z.; Wang, Y.; Wu, F.; Jiang, X.; Yang, W.; Zhang, L.; et al. Reactive metal boride nanoparticles trap lipopolysaccharide and peptidoglycan for bacteria-infected wound healing. *Nat. Commun.* **2022**, *13*, 7353. [CrossRef]

76. Tan, P.K.; Kuppusamy, U.R.; Chua, K.H.; Arumugam, B. Emerging Strategies to Improve the Stability and Bioavailability of Insulin: An Update on Formulations and Delivery Approaches. *Curr. Drug Deliv.* **2022**, *19*, 1141–1162.
77. Assi, S.; El Hajj, H.; Hayar, B.; Pisano, C.; Saad, W.; Darwiche, N. Development and Challenges of Synthetic Retinoid Formulations in Cancer. *Curr. Drug Deliv.* **2022**, *20*, 1314–1326.
78. Yu, D.G.; Huang, C. Electrospun Biomolecule-Based Drug Delivery Systems. *Biomolecules* **2023**, *13*, 1152. [CrossRef]
79. Yu, D.-G.; Zhao, P. The Key Elements for Biomolecules to Biomaterials and to Bioapplications. *Biomolecules* **2022**, *12*, 1234. [CrossRef]

Article

Development of a Multifunctional Oral Dosage Form via Integration of Solid Dispersion Technology with a Black Seed Oil-Based Self-Nanoemulsifying Drug Delivery System

Abdelrahman Y. Sherif [1,2] **and Ahmad Abdul-Wahhab Shahba** [1,*]

[1] Department of Pharmaceutics, College of Pharmacy, King Saud University, P.O. Box 2457, Riyadh 1145, Saudi Arabia; ashreef@ksu.edu.sa
[2] Kayyali Research Chair for Pharmaceutical Industries, College of Pharmacy, King Saud University, P.O. Box 2457, Riyadh 11451, Saudi Arabia
[*] Correspondence: shahba@ksu.edu.sa

Abstract: Lansoprazole (LZP) is used to treat acid-related gastrointestinal disorders; however, its low aqueous solubility limits its oral absorption. Black seed oil (BSO) has gastroprotective effects, making it a promising addition to gastric treatment regimens. The present study aims to develop a stable multifunctional formulation integrating solid dispersion (SD) technology with a bioactive self-nanoemulsifying drug delivery system (SNEDDS) based on BSO to synergistically enhance LZP delivery and therapeutic effects. The LZP-loaded SNEDDS was prepared using BSO, Transcutol P, and Kolliphor EL. SDs were produced by microwave irradiation and lyophilization using different polymers. The formulations were characterized by particle apparent hydrodynamic radius analysis, zeta potential, SEM, DSC, PXRD, and in vitro dissolution testing. Their chemical and physical stability under accelerated conditions was also examined. Physicochemical characterization revealed that the dispersed systems were in the nanosize range (<500 nm). DSC and PXRD studies revealed that lyophilization more potently disrupted LZP crystallinity versus microwave heating. The SNEDDS effectively solubilized LZP but degraded completely within 1 day. Lyophilized SDs with Pluronic F-127 demonstrated the highest LZP dissolution efficiency (3.5-fold vs. drug) and maintained chemical stability (>97%) for 1 month. SDs combined with the SNEDDS had variable effects suggesting that the synergistic benefits were dependent on the formulation and preparation method. Lyophilized LZP-Pluronic F127 SD enabled effective and stable LZP delivery alongside the bioactive effects of the BSO-based SNEDDS. This multifunctional system is a promising candidate with the potential for optimized gastrointestinal delivery of LZP and bioactive components.

Keywords: lansoprazole; solid dispersion; bioactive SNEDDS; black seed oil; multifunctional drug delivery systems

Citation: Sherif, A.Y.; Shahba, A.A.-W. Development of a Multifunctional Oral Dosage Form via Integration of Solid Dispersion Technology with a Black Seed Oil-Based Self-Nanoemulsifying Drug Delivery System. *Biomedicines* **2023**, *11*, 2733. https://doi.org/10.3390/biomedicines11102733

Academic Editors: Yongtai Zhang and Zhu Jin

Received: 4 September 2023
Revised: 2 October 2023
Accepted: 5 October 2023
Published: 9 October 2023

1. Introduction

Gastrointestinal disorders associated with gastric hyperacidity, such as peptic ulcers and gastroesophageal reflux disease (GERD), are prevalent conditions that can be exacerbated by factors including *Helicobacter pylori* infection, non-steroidal anti-inflammatory drug (NSAID) use, corticosteroid administration, alcohol consumption, stress conditions, and continuous intake of spicy and caffeine-containing products [1–4].

Proton pump inhibitors (PPIs) like lansoprazole (Figure 1) are commonly prescribed to reduce gastric acid production through inhibition of the $H+/K+$-ATPase enzyme with a typical daily dosage of 15–30 mg [5]. However, lansoprazole has low aqueous solubility and is classified as a Biopharmaceutics Classification System Class II drug, resulting in incomplete dissolution from oral dosage forms, and low and erratic oral bioavailability [6,7]. Various formulation strategies were conducted to improve solubility, dissolution, and absorption of LZP. For example, Zhang et al. prepared an SD formulation of LZP using fluid-bed coating technology where 90% of drug was dissolved at the end of experiment [8]. In alignment with such an achievement, various studies showed that SD formulations were able to enhance drug bioavailability compared to pure LZP [9,10]. Moreover, Ubgade et al. prepared a nanosuspension formulation of LZP and in vitro dissolution showed that the prepared formulation was able to enhance initial and total drug dissolution behavior compared to the pure drug [11].

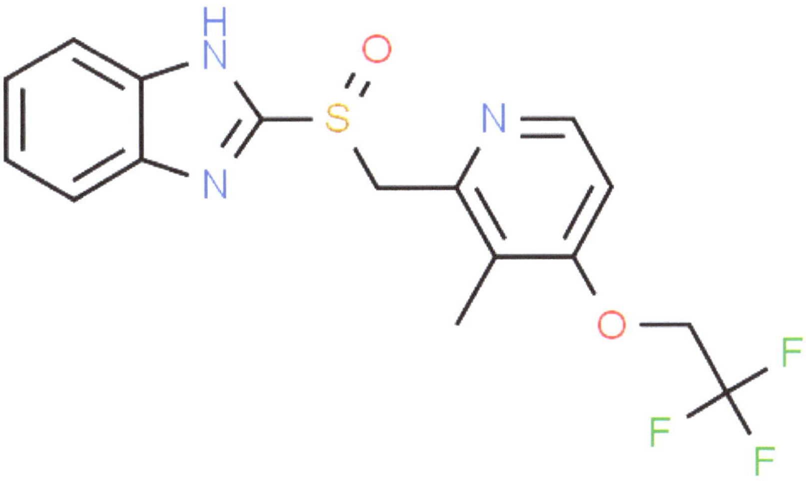

Figure 1. The chemical structure of lansoprazole (obtained from ChemSpider chemical structure database, http://www.chemspider.com/, accessed on 28 September 2023).

Self-nanoemulsifying drug delivery systems (SNEDDSs) incorporate oils, surfactants, and co-surfactants to form nano-scale emulsion droplets upon mild agitation, enhancing the surface area for dissolution and absorption compared to the drug alone [12]. Recent research incorporated naturally derived oils with biological activities as an attempt to potentially augment the therapeutic activity of administered therapeutic molecules [13].

In light of this, several reports revealed that black seed oil (BSO) has a protective and healing effect on gastric ulcers [14,15]. Along with this, various studies showed that the administration of thymoquinone (TMQ), the major constituent in BSO, reduces peptic ulcers produced by NSAIDs and other agents like ethanol [16–18]. In detail, Kanter et al. studied the impact of BSO and TMQ on gastric ulcers in rats. Their study reported that both BSO and TMQ were able to reduce peptic ulcer indices, with a more prominent reduction observed in the BSO-treated group [14]. This suggests that the incorporation of BSO rather than TMQ alone into SNEDDS formulations could combine the benefits of TMQ alongside additional protective components intrinsic to BSO. Furthermore, Radwan et al. prepared a TMQ-loaded SNEDDS to study the impact of increasing drug solubility on the therapeutic effects. Their study reported that the formulated TMQ-SNEDDS decreased the ulcer index by 2-fold compared to free TMQ alone, as indicated in their findings [17]. Despite their reported in vivo activity, the FDA has not yet approved black seed extracts or concentrated thymoquinone for treating any medical conditions like high cholesterol, diabetes, or high blood pressure. However, black seed extracts remain available for purchase over-the-counter as dietary supplements marketed to aid digestion and promote energy levels. Research has consistently shown black seed derivatives to be well-tolerated and safe, as evidenced by their designation as "generally recognized as safe" (GRAS) by the FDA [19].

While the formulation of drugs within SNEDDSs addresses solubility limitations, practical challenges with physical/chemical instability over the shelf-life still restrict clinical translation. Chemical instability might arise from incompatibilities between excipients and the loaded drug, leading to its degradation in the presence of these excipients [20]. Thus, formulated SNEDDSs are often separated from the therapeutic agent to circumvent stability issues [21,22]. As an alternative technology, SDs represent stabilized carrier systems capable of maintaining improved drug performance through nano-dispersed or molecularly dissolved drug domains within hydrophilic polymer matrices [23].

The current research proposes a novel formulation approach combining solubility- and stability-enhancing technologies with a bioactive natural oil into a single optimized product. The specific aims are to prepare and characterize drug-free and drug-loaded BSO-SNEDDSs; fabricate lansoprazole SDs and evaluate their in vitro dissolution; assess the effects of BSO-SNEDDSs on SD dissolution behavior; and conduct stability testing to identify a lead candidate formulation. The findings could provide clinically translatable dosage forms with synergistic delivery of lansoprazole and bioactive BSO components for a potentially more effective treatment for acid-related gastrointestinal disorders.

2. Materials and Methods

2.1. Procurement of Plant Material and Isolation of Bioactive Components

The methodologies implemented for assembling, isolating, and calibrating black seed oil (BSO) have been delineated comprehensively within our prior investigative studies [13,24,25].

2.2. Chemical and Reagents

The proton pump inhibitor lansoprazole was acquired from Mesochem Technology (Beijing, China). The surfactant Kolliphor EL (KrEL) was obtained from BASF (Ludwigshafen, Germany). The triblock copolymer Pluronic-F127 (PF-127) was sourced from Sigma Aldrich (St. Louis, MO, USA). The co-solvent Transcutol® P (TCP) was provided by Gattefossé (Lyon, France). The polymer polyethylene glycol 4000 (PEG 4000) was procured from BDH Chemicals Ltd. (Poole, UK). The gelatin capsules size 0 were supplied by Capsugel (Morristown, NJ, USA). The cellulose derivative hydroxypropyl methylcellulose (HPMC) E3 was acquired from JRS Pharma (Rosenberg, Germany).

2.3. Preparation and Characterization of LZP-SNEDDS

Self-nanoemulsifying drug delivery systems (SNEDDSs) were developed utilizing black seed oil, Transcutol P, and Kolliphor EL at optimized concentration ratios (25/25/50 w/w). For the preparation of a drug-free SNEDDS, the components (2 g) were added to vials in the specified amounts and blended using a vortex mixer. Drug-loaded SNEDDSs were formed by incorporating lansoprazole (30 mg) into the formulation ingredients using a similar procedure. Both drug-free and drug-loaded SNEDDSs, with and without accompanying solid dispersions, were subjected to dilution with deionized water (at a 1:1000 w/w ratio) and then subjected to mixing for 1 min. The resulting solutions were centrifuged and analyzed using a Zetasizer dynamic light scattering instrument (Model ZEN3600, Malvern Instruments Co., Worcestershire, UK) to determine the particle apparent hydrodynamic radius and zeta potential of the dispersed systems. This enabled the characterization of the nanoemulsion properties formed upon dilution of the self-emulsifying formulations [25].

2.4. Preparation of Solid Dispersion (SD) Formulation

Solid dispersions (SDs) of lansoprazole (LZP) were prepared in this work using microwave irradiation (MW) and lyophilization (LP) techniques. The polymers Pluronic F-127 and polyethylene glycol 4000 (PEG4000) were selected for MW preparation due to their relatively low melting points, which enabled uniform heating and mixing during irradiation. In contrast, Pluronic F-127 and hydroxypropyl methylcellulose (HPMC) were used for LP preparation. The compositions of the LZP/polymer ratios and their corresponding preparation methods are outlined in Table 1.

Table 1. Composition of solid dispersion formulations.

Formulation Code	LZP	PF-127	PEG-4000	HPMC	Sodium Hydrogen Carbonate (NaHCO$_3$)	Disodium Carbonate (Na$_2$CO$_3$)
MW-PF-127	1.0	4.0	-	-	-	-
MW-PEG 4000	1.0	-	4.0	-	-	-
LP-PF-127	1.0	4.0	-	-	1.4	12.6
LP-HPMC	1.0	-	-	4.0	1.4	12.6

All numbers in table are expressed as mass ratios (*w/w*). LZP (lansoprazole), PF-127 (Pluronic-F127), PEG-4000 (polyethylene glycol-4000), HPMC (hydroxypropyl methylcellulose), MW-PF-127 (solid dispersion prepared using Pluronic-F127 by microwave method), MW-PEG-4000 (solid dispersion prepared using polyethylene glycol 4000 by microwave method), LP-PF-127 (solid dispersion prepared using Pluronic-F127 by lyophilization method), and LP-HPMC (solid dispersion prepared using hydroxypropyl methylcellulose by lyophilization method).

2.4.1. Microwave Method

Lansoprazole (LZP) was blended with the polymers Pluronic F-127 and polyethylene glycol 4000 (PEG4000) in a 1:4 ratio by weight to produce mixtures for microwave solid dispersions (MW-SDs) [26]. Approximately 1 g of each drug–polymer mixture was thoroughly mixed in a porcelain mortar to achieve a homogeneous preparation. Domestic microwave irradiation (Samsung Model ME0113M1) was then utilized to prepare the MW-SDs [21,27]. The microwave instrument was preheated for around 2 min prior to irradiation of the mixtures. The LZP-polymer mixtures were subjected to microwave radiation at 900 W power for about 2 min and 6 min to obtain the MW-PF-127 and MW-PEG-4000 MW-SD formulations, respectively. The molten dispersions were stirred continuously with a glass rod during irradiation to maintain homogeneity. Upon cooling to room temperature, the solidified dispersions were gently crushed and sieved through a 315 µm screen to achieve uniform fine powders.

2.4.2. Lyophilization Method

A preliminary LZP solubility study was conducted to select the optimum pH for preparing the LZP solution. Among the three tested pH levels (9.2, 10.0, and 10.8), the latter showed the highest LZP solubility (1.882 ± 0.069 mg/mL) and therefore, was selected as the optimum pH for the lyophilization process. LZP was dissolved in the prepared bicarbonate buffer (pH 10.8) using a magnetic stirrer to generate an ~0.74 mg/mL solution. Predetermined quantities of the polymers Pluronic F-127 and hydroxypropyl methylcellulose (HPMC) were added to the LZP solution at a 4:1 ratio by weight and mixed thoroughly to produce formulations for lyophilized solid dispersions (LP-SDs). The prepared drug–polymer solutions were frozen at −60 °C prior to lyophilization. The frozen dispersions were then lyophilized for at least 48 h at −60 °C using a freeze dryer (Al-pha 1-4 LD Plus, Osterode am Harz, Germany) to allow solvent sublimation. This could potentially achieve a porous matrix with the drug molecularly dispersed in the polymer scaffold. The obtained LP-SDs were gently crushed and sieved through a 315 µm screen to achieve uniform fine powders [28].

2.5. Scanning Electron Microscopy (SEM)

The prepared microwave solid dispersions with Pluronic F-127 and PEG4000 (MW-PF-127 and MW-PEG-4000) and the lyophilized solid dispersions with Pluronic F-127 and HPMC (LP-PF-127 and LP-HPMC) were analyzed using scanning electron microscopy (SEM). The samples were mounted on stubs and sputter coated with gold for 60 s at 20 mA using a Q150R sputter coating unit (Quorum Technologies Ltd., East Sussex, UK) under an argon atmosphere. This allowed examination of the surface morphology and topography of the different solid dispersion formulations using SEM imaging (Carl Zeiss EVO LS10, Cambridge, MA, USA) under high vacuum [24,28].

2.6. Differential Scanning Calorimetry (DSC)

Differential scanning calorimetry (DSC) was utilized to characterize the prepared solid dispersion samples using a DSC-60 instrument (Shimadzu, Kyoto, Japan). Approximately 2 mg of each sample was weighed into a non-hermetically sealed aluminum pan. The samples were heated from 25 °C to 250 °C at a 10 °C/minute heating rate to obtain thermographs. The DSC measurements were performed under a nitrogen atmosphere with a 40 mL/min flow rate. Post-analysis, the DSC curves were subjected to baseline manual correction using TA 60 thermal analysis software. This enabled investigation of the thermal behavior and identification of any thermal events such as melting, crystallization, or degradation in the solid dispersion formulations [29].

2.7. Powder X-ray Powder Diffraction (PXRD)

Powder X-ray diffraction analysis was conducted to investigate crystallinity changes after SD preparation. The LZP, polymers, physical mixtures, and SD formulations were subjected to an X-ray diffractometer instrument (Ultima IV, Rigaku Inc. Tokyo, Japan). The obtained PXRD pattern was investigated to assess the crystalline state of LZP within the prepared SD formulations. Each sample was measured in the scanning range of 3–60° with a scanning rate of 1°/min using an X-ray diffractometer. The characteristic peak of each sample was assessed by collecting the data by monochromatic radiation (Cu Kα′ 1, λ = 1.54 Å), operating at a voltage of 40 kV and current of 40 mA. This allowed evaluation of the nature of crystallinity of lansoprazole within the solid dispersion formulations compared to pure drug and physical mixtures [28].

2.8. In Vitro Dissolution Test

Dissolution tests were performed to evaluate and compare the drug release behavior of formulations. A USP Type II dissolution testing apparatus (UDT-814, LOGAN Inst. Corp., Franklin, NJ, USA) was employed for the study. Formulations containing equivalent amounts (15 mg) of lansoprazole were placed in capsules, surrounded by sinkers, and placed into vessels containing 900 mL of pH 6.8 phosphate buffer (prepared according to European pharmacopeia specifications). Paddles were set to rotate at 75 rpm for the duration of the dissolution experiments. Prior to initiating the release studies, buffer media were equilibrated to 37 °C in the jacketed vessels. Samples were manually drawn from vessels at pre-defined time points of 5, 10, 15, 30, 60, and 120 min. An in-line filter assembly was used to withdraw aliquots, which were subsequently analyzed by a validated UPLC method to determine the amount of drug dissolved over time [28]. Formulation performance was compared based on the dissolution efficiency (DE)% [21,30].

2.9. Stability Study

The thermal stability of SD, drug-loaded SNEDDS, and raw LZP drug substances was evaluated under accelerated conditions. All samples were packaged in tightly capped amber glass containers to protect them from moisture and light exposure. These stability test units were then placed in programmable climate-controlled chambers (Binder GmbH, Tuttlingen, Germany) preset to maintain 40 ± 2 °C and 75 ± 5% relative humidity. At time points of 1, 7, and 30 days, samples were retrieved and their drug content was analyzed using the UHPLC analytical method. This experiment was designed to assess and compare the changes in lansoprazole content and overall stability of the various pharmaceutical systems exposed to elevated heat and moisture over time [31].

2.10. Quantification of LZP Using the Developed UPLC-UV Method

Lansoprazole (LZP) was quantified using the ultra-high performance liquid chromatography with ultraviolet detection (UHPLC-UV) method. The analyses were performed on a Dionex UltiMate 3000 UHPLC system equipped with an autosampler and DAD detector (Thermo Scientific, Bedford, MA, USA). An Acquity BEH C18 column (2.1 × 50 mm, 1.7 μm) was used for separation. The column temperature was maintained at 40 ± 0.5 °C. An isocratic mobile phase consisting of 0.1% triethylamine (pH 6.7)/acetonitrile (58:42, *v/v*) achieved separation of LZP. The flow rate was 0.4 mL/min and the detection wavelength was set at 320 nm.

2.11. Software

The data from the current study were analyzed primarily using the Python programming language (version 3.9.13) within a Jupyter Notebook environment. The specific Python packages utilized included NumPy for data manipulation, Pandas for data frames, Matplotlib and Seaborn for visualization, StatAnnotations for statistical validation, and itertools for additional data processing functions. Portions of this article describing the data analysis methods and presenting results were composed with writing assistance from Claude (developed by Anthropic and operated by Poe) and Bing AI chat. Some Python scripts supporting the data analysis were also developed with input from these AI tools. However, the authors maintained overall responsibility for the direction, ideas, content, and finalization of the manuscript.

2.12. Statistical Analysis

The normality of the data was assessed by the Shapiro–Wilk test (Scipy.stats python package) [32]. The homogeneity of the variances was assessed by Levene's test (Scipy.stats python package) and homoscedasticity (pingouin python package). For dependent variables (with a fairly normal distribution and equal variances), the independent T-test was used to test the statistical significance between two independent samples, the dependent T-test (Scipy.stats and statannotations python package) for two paired samples, and one-way ANOVA followed by the Tukey's post hoc test (Scipy.stats and scikit_posthocs python packages) for >2 samples. If the sample contain significant outliers and/or if the normality assumption is significantly violated, Kruskal–Wallis H followed by the Dunn post hoc test with Bonferroni correction (Pingouin and scikit_posthocs python packages) was used for >2 samples [33]. The two-way ANOVA test was carried out to analyze the effect of two independent factors on one dependent variable (OLS and SM from Statsmodels.formula.api and Statsmodels.api python packages, respectively). A *p*-value of ≤ 0.05 was denoted statistically significant in all the statistical analysis tests.

3. Results

3.1. Characterization of Lansoprazole-Loaded Self-Nanoemulsifying Delivery System

The apparent hydrodynamic radius of the drug-free SNEDDS formulation droplets significantly increased from 295 nm to 431.5 nm after incorporating LZP ($p < 0.05$) (Figure 2A). However, there was no significant difference in the apparent hydrodynamic radius when an SD was added to the SNEDDS formulation. Interestingly, the PDI of the drug-loaded SNEDDS showed a significant increase compared to the combination of the drug-free SNEDDS and SD ($p < 0.05$) (Figure 2B). Additionally, all zeta potential values exhibited significant differences ($p < 0.05$), with the drug-free SNEDDS having the highest value (-39 mV), while the combination with SD had the lowest zeta potential value (-23 mV) (Figure 2C).

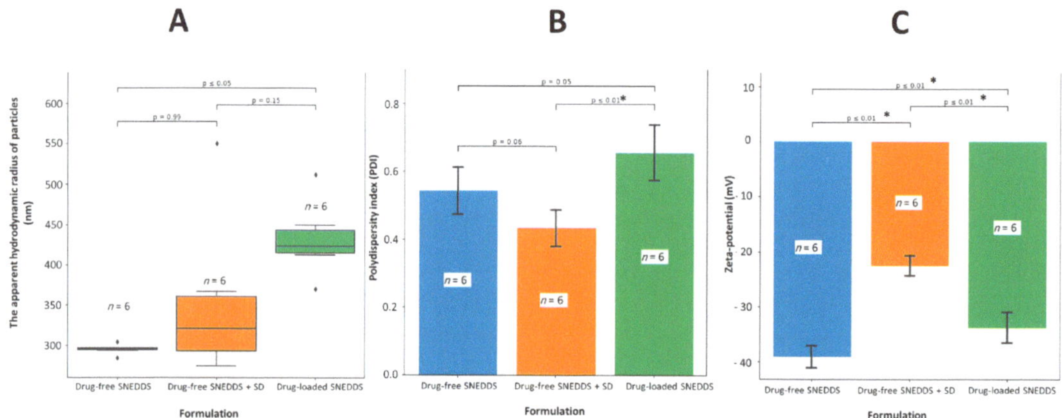

Figure 2. Influence of SNEDDS loading and combination with SD on (**A**) particle apparent hydrodynamic radius, (**B**) PDI (polydispersity index), and (**C**) ZP (zeta potential). SNEDDS (self-emulsifying drug delivery system), and SD (solid dispersion prepared using Pluronic-F127 by lyophilization method). The PS data were statistically analyzed by Kruskal–Wallis H followed by the Dunn post hoc test (Bonferroni correction) while the PDI and ZP data were analyzed by ANOVA followed by Tukey's post hoc test. A significant *p*-value (<0.05) is marked with an asterisk (*). Outliers in the dataset were symbolized using a diamond shape (♦).

3.2. SEM

3.2.1. Microwave Method

Scanning electron micrographs revealed that raw lansoprazole exists as small crystals with well-defined edges, indicating its crystalline nature (Figure 3). In contrast, the MW-SD formulations appeared as larger particles with smoother surfaces. This change in morphology suggests that the drug was dispersed within the carrier matrices rather than remaining in a crystalline state. No evidence of phase separation or incomplete solidification was observed in the solid dispersion systems.

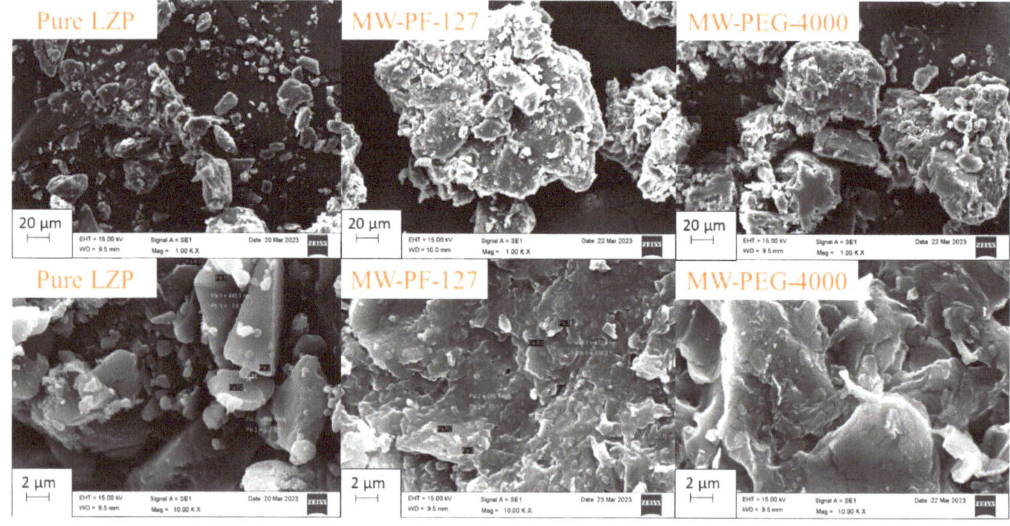

Figure 3. SEM images of pure LZP and MW-SD formulations (MW-PF-127 and MW-PEG-4000).

3.2.2. Lyophilization Method

Comparison of the scanning electron micrographs revealed noticeable differences in particle size between the microwave (MW) and lyophilized (LP) solid dispersions (SDs). The LP-SD particles prepared with Pluronic F-127 appeared remarkably smaller than their MW-SD counterparts (Figure 4). Additionally, images of the LP-SD with hydroxypropyl methylcellulose (LP-HPMC) showed needle-shaped crystalline structures at higher magnification (Figure 4, LP-HPMC). These may be indicative of incomplete amorphization in the LP-HPMC formulation. Similar to the MW-SDs, no evidence of incomplete separation or residual solvents was observed for the LP-SDs.

Figure 4. SEM images of pure LZP and prepared LP-SD formulations (LP-PF-127 and LP-HPMC).

3.3. DSC

3.3.1. Microwave Method

The DSC thermogram of pure lansoprazole (LZP) exhibited a sharp endothermic melting point peak at approximately 178 °C, followed by a decomposition exotherm at around 182 °C (Figure 5). In contrast, the microwave solid dispersion (MW-SD) with Pluronic F-127 and its corresponding physical mixture displayed sharp endothermic peaks at about 58–60 °C. Similarly, the MW-SD with PEG-4000 and its physical mixture showed endotherms at approximately 64 °C. Notably, the characteristic LZP melting and decomposition peaks were completely absent in the thermograms of the MW-SD formulations.

3.3.2. Lyophilization Method

The lyophilized formulations also exhibited changes in thermal behavior compared to the raw materials. The physical mixture of LP-PF-127 showed an endotherm at approximately 60 °C, similar to microwave processed samples (Figure 6). However, the LP-PF-127 solid dispersion itself displayed a broader, lower temperature endotherm centered around 55 °C. Both LP-PF-127 and LP-HPMC formulations presented two diffuse peaks, around 84 °C and 126 °C for the former, and 91 °C and 128 °C for the latter. Notably, the characteristic melting peak for lansoprazole was again absent in the DSC curves of the lyophilized solid dispersions.

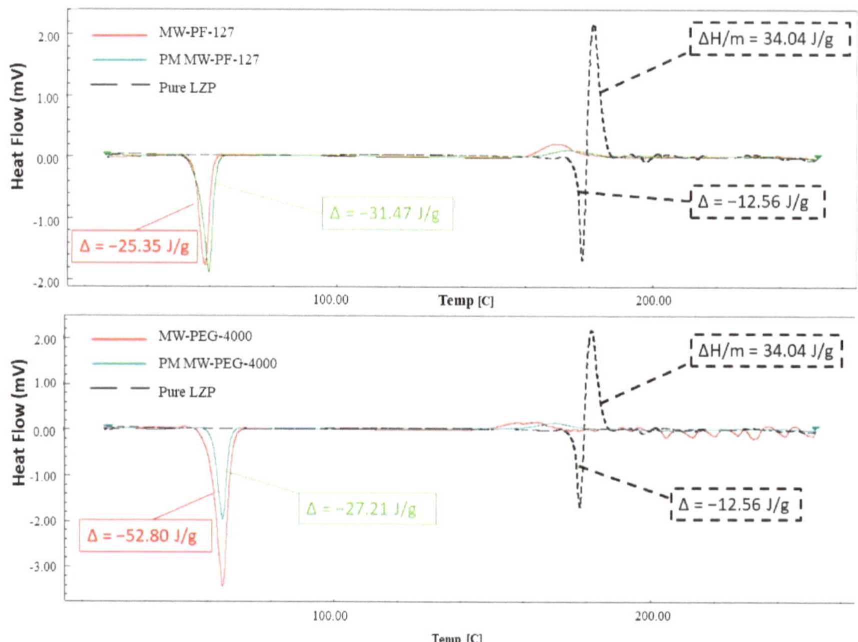

Figure 5. DSC spectra of pure LZP and MW-SD formulations (MW PF-127 and MW-PEG-4000) along with their corresponding physical mixture.

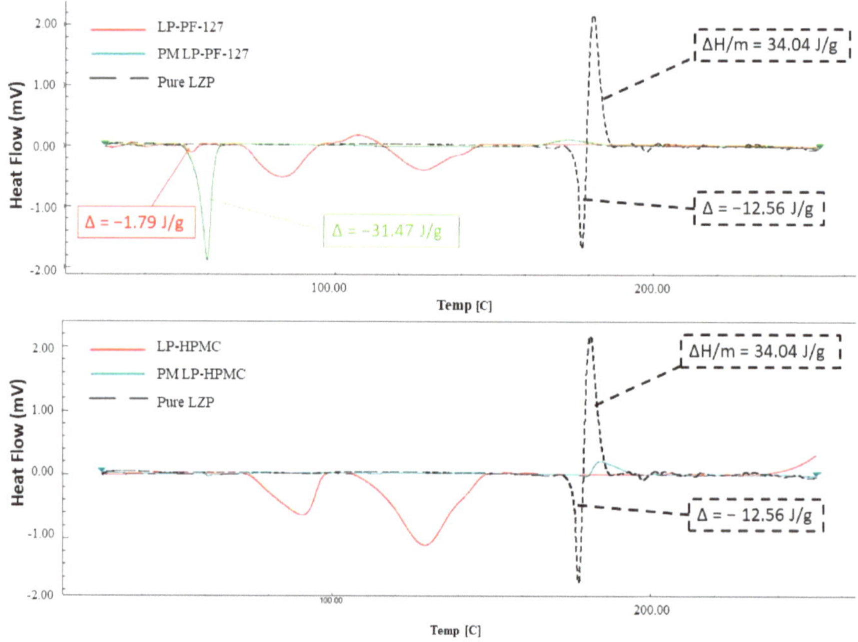

Figure 6. DSC spectra of pure LZP and LP-SD formulations (LP-PF-127 and LP-HPMC) along with their corresponding physical mixture.

3.4. PXRD

The PXRD patterns of lansoprazole (LZP) exhibited characteristic peaks at 5.8, 17.0, 17.6, 22.4°, and 25–26° (Figures 7 and 8). The polymers Pluronic F-127 and PEG-4000 showed peaks near 19° and 23° (Figure 7), while hydroxypropyl methylcellulose (HPMC) displayed peaks at 38.0° and 44.3° (Figure 8). As expected, physical mixtures of LZP with the polymers revealed a combination of the drug and polymer peaks (Figures 7 and 8).

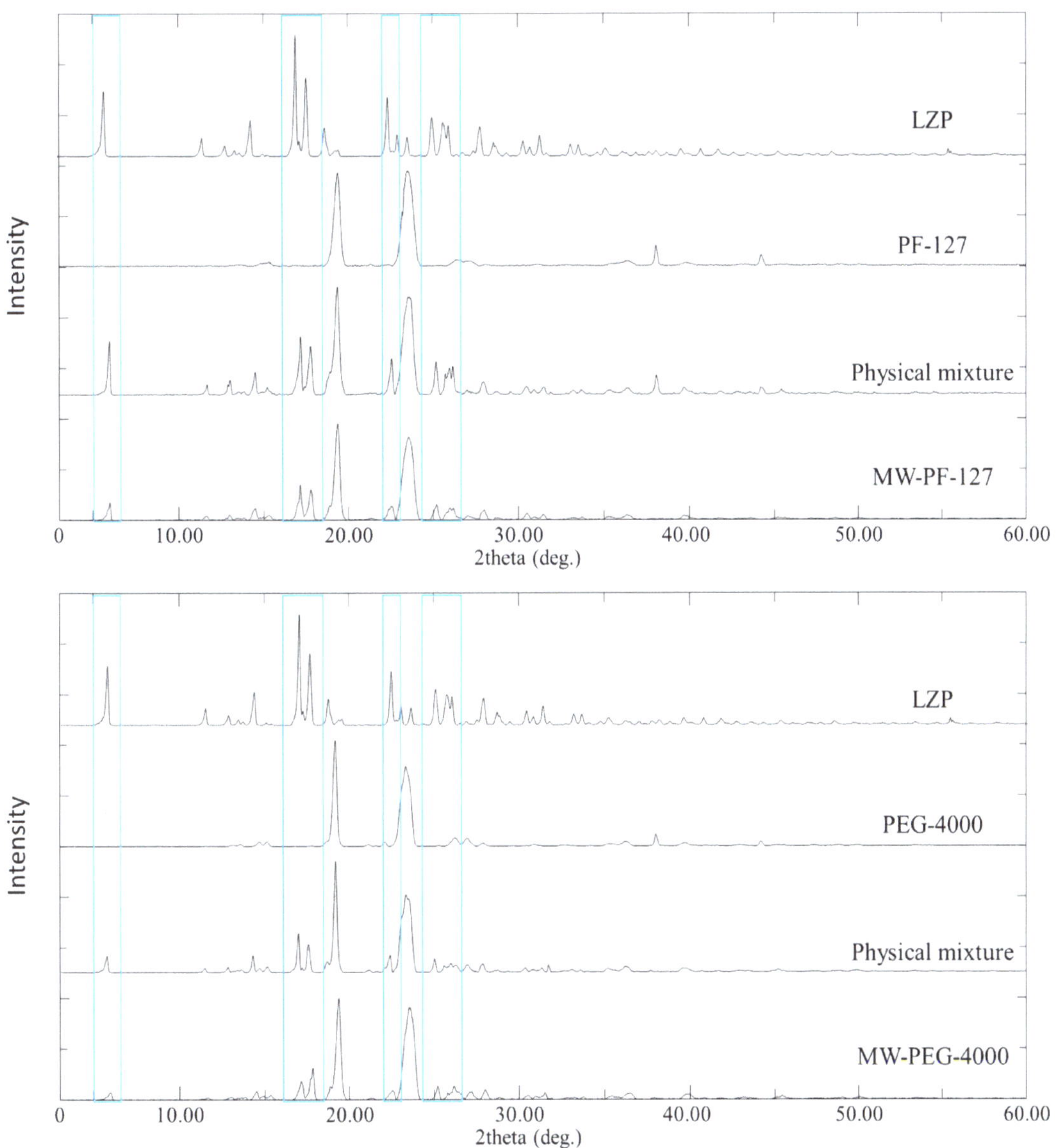

Figure 7. PXRD patterns of LZP, PF-127, PEG 4000, physical mixtures, and the prepared solid dispersion formulations using the microwave method. The Figure employs blue rectangles to visually designate the locations of LNS characteristic peaks of interest.

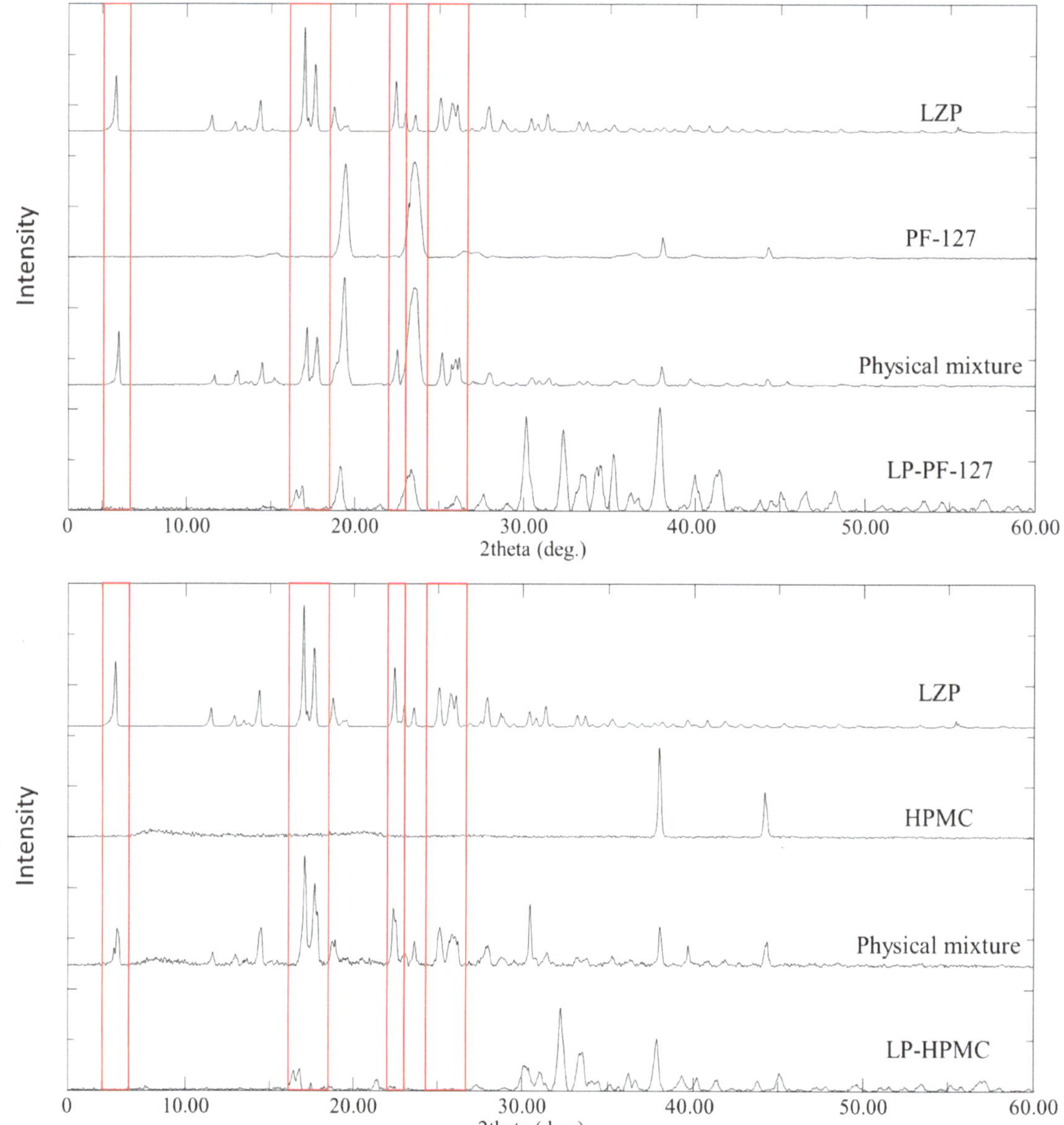

Figure 8. PXRD patterns of LZP, PF-127, HPMC, physical mixtures, and the prepared solid dispersion formulations using the lyophilization method. The Figure employs red rectangles to visually designate the locations of LNS characteristic peaks of interest.

For microwave solid dispersions (MW-SDs), the intensities of the LZP peaks decreased but were still present, indicating some residual crystallinity. In contrast, the lyophilized SDs showed complete disappearance of the LZP peaks at 5.8° and 22.4° and a substantial reduction in the peaks at 17.0° and 17.6°. Additional peaks between 30 and 40° were observed for the lyophilized SDs.

Overall, the results suggest that the lyophilization process was more effective at disrupting the crystallinity of LZP compared to microwave irradiation. However, some residual crystalline drug was detected in both MW-SDs and lyophilized SDs by PXRD.

3.5. In Vitro Dissolution Studies

3.5.1. SNEDDS Formulations

Pure lansoprazole (LZP) exhibited poor dissolution with only 45% drug release and 25% dissolution efficiency (DE) by the end of the experiment (Figure 9A,B). In contrast, the LZP-loaded SNEDDS formulation significantly ($p < 0.05$) enhanced LZP dissolution, increasing the DE over 3-fold compared to the pure drug (Figure 9B). However, the combination of pure LZP and drug-free SNEDDS (in separate capsules) failed to improve LZP dissolution.

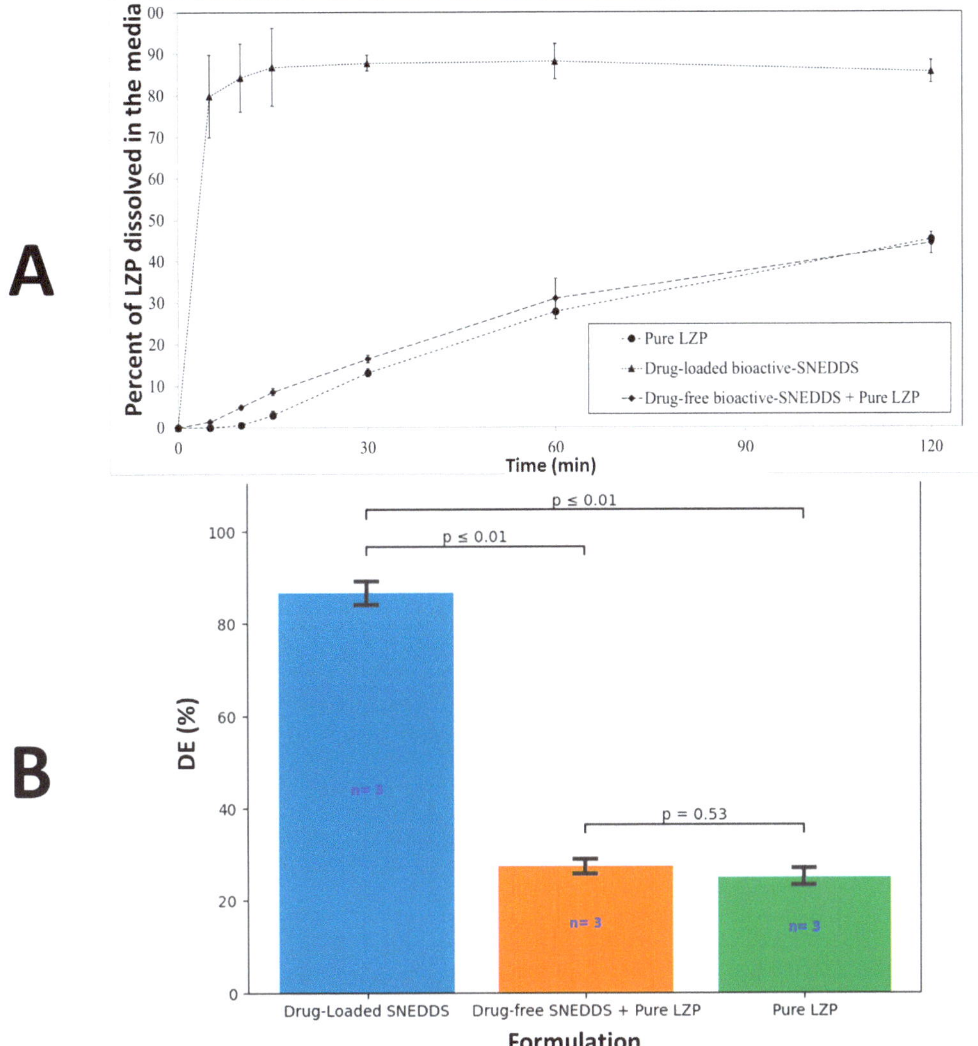

Figure 9. (**A**) In vitro dissolution profile of pure LZP, drug-loaded bioactive SNEDDS, and drug-free bioactive SNEDDS + pure LZP. (**B**) Graphical representation of DE of corresponding formulations.

3.5.2. SD Formulation

All solid dispersions significantly ($p < 0.05$) enhanced LZP release except for MW-PEG-4000 that showed slow drug release and a similar DE as the pure drug (Figure 10A,B). In particular, both LP-SD formulations (prepared using the lyophilization method) showed fast LZP release and a significantly ($p < 0.05$) higher DE compared to the MW-SD formulations. Notably, both lyophilized SDs (LP-SDs) displayed rapid drug release with a 3.5- and 3.3-fold higher DE than pure LZP for LP-PF127 and LP-HPMC, respectively.

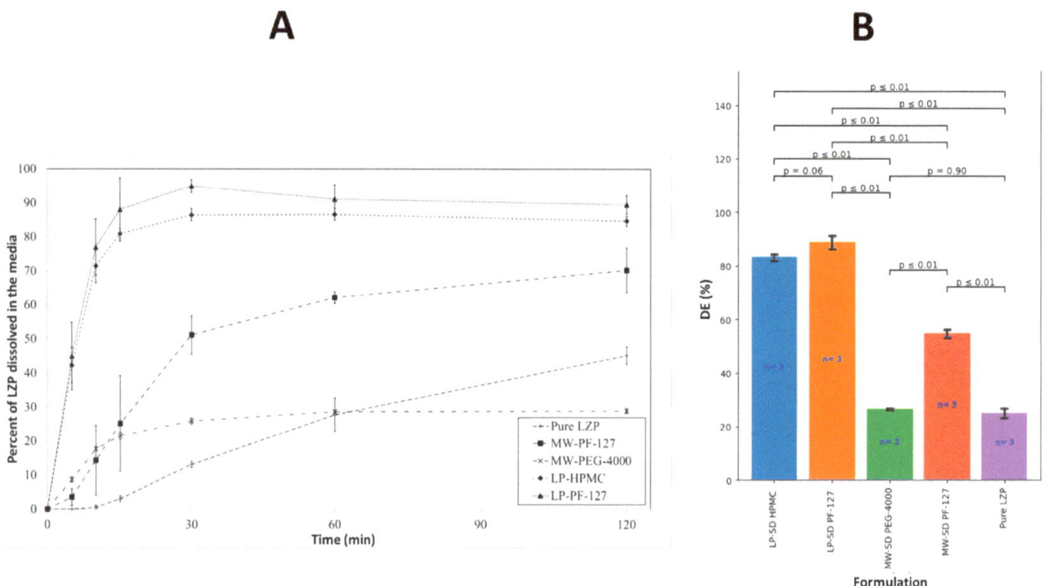

Figure 10. (**A**) In vitro dissolution profile of pure LZP and the prepared SD formulations. (**B**) Graphical representation of DE of corresponding formulations.

3.5.3. SD Formulation + Drug-Free Bioactive SNEDDS

The two-way ANOVA test indicates that combining SDs with drug-free SNEDDS did not significantly impact DE overall ($p = 0.51$). However, the differential analysis of each formulation showed that MW-PEG-4000 and LP-HPMC SDs showed a significantly increased DE upon SNEDDS addition ($p < 0.01$) (Figure 11A,B). Meanwhile, MW-PF-127 and LP-PF-127 SDs exhibited no DE enhancement with the SNEDDS.

3.6. Stability Study

The chemical stability study revealed interesting differences between the formulations. The lansoprazole (LZP)-loaded SNEDDS experienced complete drug degradation after just 1 day of storage under accelerated conditions (Figure 12A). In stark contrast, both pure LZP and the lyophilized solid dispersion LP-PF127 maintained exceptional stability, with >97% of intact drug remaining after 1 month.

Aligning with the chemical stability results, the LZP-loaded SNEDDS exhibited significant discoloration to a deep brown/black color by the end of the storage period. However, pure LZP, LP-PF127, and the drug-free SNEDDS showed no noticeable change in physical appearance after storage.

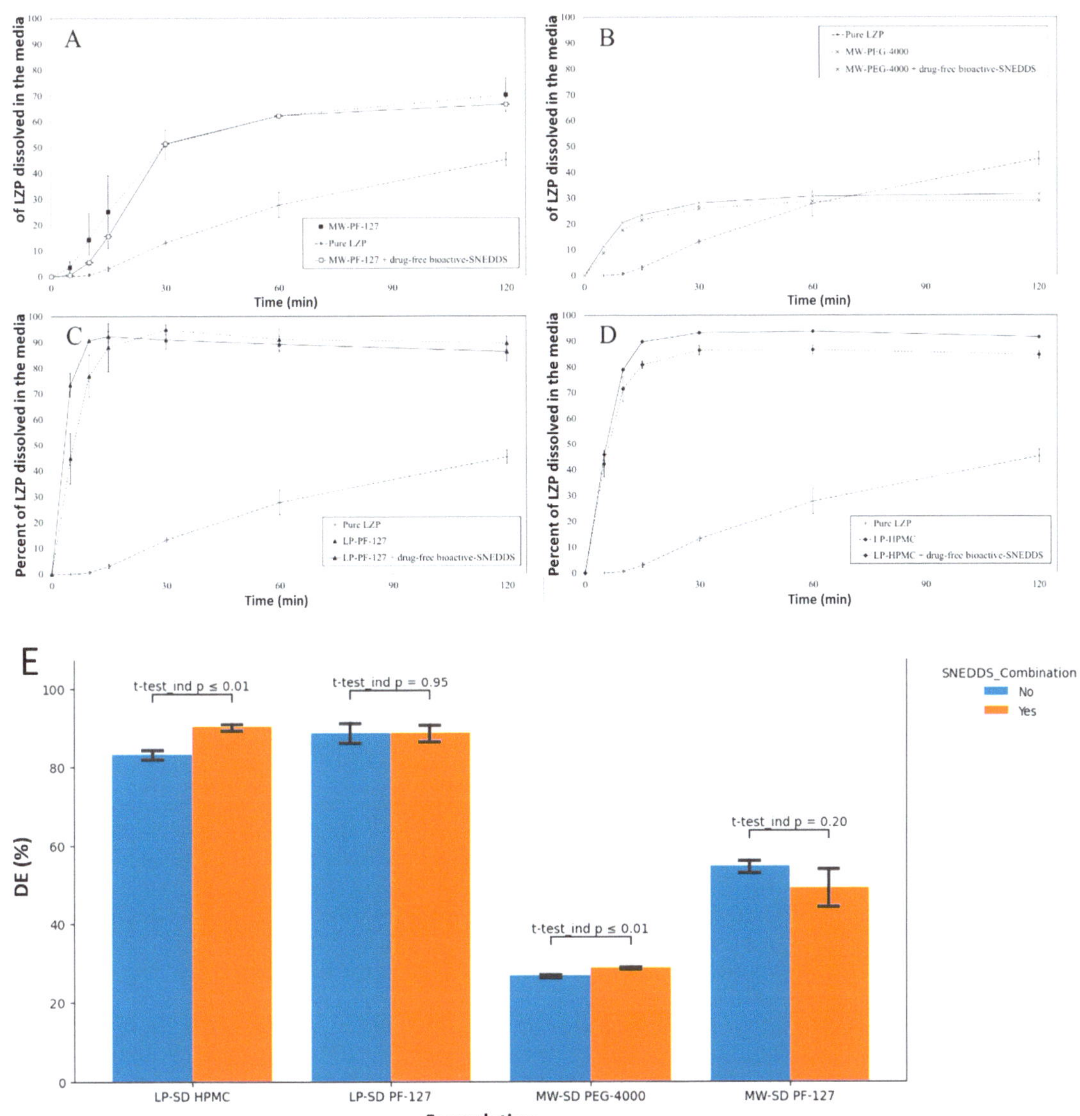

Figure 11. In vitro dissolution profile of pure LZP, SD formulation, and SD formulation + drug-free bioactive SNEDDS (**A**) MW-PF-127, (**B**) MW-PEG-4000, (**C**) LP-PF-127, (**D**) LP-HPMC and (**E**) Graphical representation of DE of corresponding formulations.

Overall, the findings indicate that the SNEDDS system afforded limited protection against drug degradation, while the lyophilized solid dispersion provided excellent protection against degradation under accelerated conditions. The physical discoloration of the SNEDDS correlates with the extensive drug degradation observed.

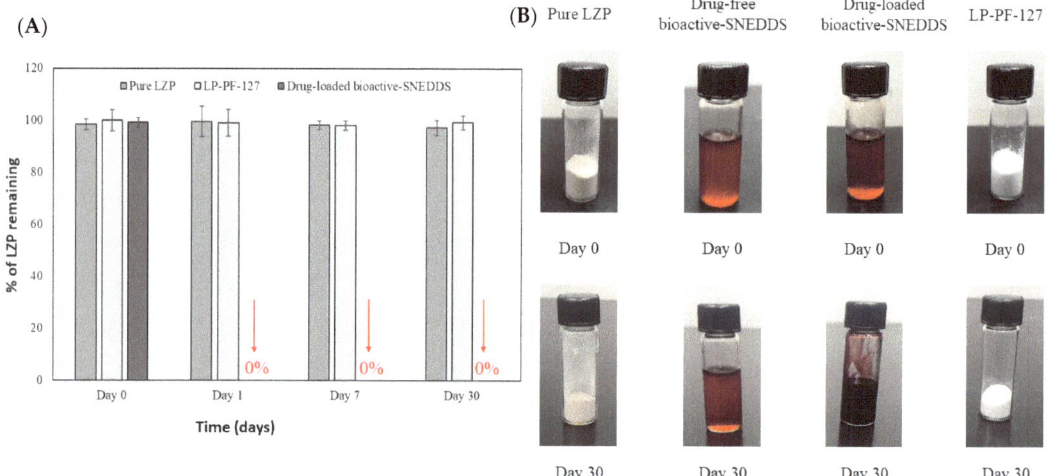

Figure 12. The (**A**) chemical and (**B**) physical stability study of pure LZP, LP-PF-127, and drug-loaded bioactive SNEDDS under accelerated storage conditions for 1 month (left).

4. Discussion

In the present study, an SNEDDS was utilized to enhance the dissolution of LZP based on its previously reported advantages [34,35]. SNEDDS formulations consist of surfactants, cosurfactants, and oils, with the oils able to be derived from natural sources. Recently, there has been interest in incorporating natural compounds with pharmacological activity into SNEDDSs to provide additional therapeutic effects for diseases such as cancer [36,37], hypertension [25], and bacterial infections [38].

Previous studies have shown that thymoquinone (TMQ), a major constituent of black seed oil (BSO), possesses gastroprotective effects and can help treat peptic ulcers more potently when formulated using a SNEDDS-based delivery approach [17]. Additional research found BSO to be more effective than TMQ alone for reducing peptic ulcer indices in animal models [14]. In light of these findings, the current study selected BSO rather than isolated TMQ to develop a "bioactive-SNEDDS" system combining the benefits of the gastro-protective components intrinsic to BSO. It was hypothesized that incorporating BSO as the oil phase may augment the anti-ulcer activity of encapsulated lansoprazole (LZP). Therefore, BSO was employed not only for SNEDDS preparation but also in developing a multifunctional SNEDDS intended to co-deliver LZP and bioactive molecules from BSO which could potentially improve peptic ulcer treatment outcomes.

In the current study, the in vitro dispersion of three systems (Figure 2) were in the nanosize range which could significantly enhance LZP bioavailability following oral administration [39]. Furthermore, all the formulations exhibited negative zeta potential values ranging from -22.5 to -39 mV, which aligns with prior studies [24,40,41]. Specifically, the drug-free SNEDDS produced the highest magnitude potentials, signaling greater colloidal stability likely due to nonionic surfactant inclusion and anionic species binding to droplets. Previous research suggested that hydroxyl ions from water or fatty acid impurities in surfactants could attribute to the observed negative ZP values [42,43]. The increase in zeta potential values after incorporating the drug and solid dispersion (SD) may be linked to residues of lansoprazole and sodium carbonate from the SD, potentially introducing partial positive charges to the surface.

Furthermore, the in vitro dissolution study revealed that the prepared bioactive SNEDDS was able to increase the dissolution efficiency of LZP by 3.5-fold compared to the pure drug. However, the stability study revealed complete drug degradation following just 1 day of incubation in the stability cabinet. This could be attributed to the presence of free fatty acids in the SNEDDS components, which can create an acidic microenvironment promoting LZP degradation during storage, as reported previously [44]. Based on previous experience, the instability issues with SNEDDSs are often addressed by separating the drug from the formulation [21].

Therefore, in the current work, drug-free bioactive SNEDDS and pure LZP were placed in separate capsules and subjected to in vitro dissolution testing. However, this failed to improve the dissolution of pure LZP. Consequently, there is a need to investigate an adjuvant technology to enhance LZP dissolution alongside the drug-free bioactive SNEDDS.

Solid dispersions (SDs) were explored in this study as a means to improve LZP dissolution when co-administered with the drug-free bioactive SNEDDS, as reported previously [21,45]. For MW-SDs, LZP and the polymer (Pluronic F-127 or PEG 4000) were mixed and subjected to microwave irradiation. For the LP-SD, the drug and polymer are typically dissolved in a solvent before freeze-drying. However, the poor aqueous solubility of LZP would require a large volume of solvent for industrial production [46]. Therefore, a bicarbonate buffer was utilized as the solution medium due to its known high solubility for LZP [47] and added advantages regarding stability [48]. Accordingly, the bicarbonate buffer enabled the preparation of an LZP solution for lyophilization using a minimal solvent volume.

The DSC analyses provided vital information corroborating the changes to LZP's physical state induced by SD preparation. The DSC spectrum of raw LZP was consistent with those previously reported in the literature, confirming the purity of the drug substance used [24,49]. In the SDs prepared by lyophilization, two new broad endothermic peaks were observed, which could have resulted from the melting of sodium carbonate and sodium bicarbonate components within the buffer system. The literature indicates that the melting points are approximately 96 °C and 160 °C for these species, respectively [50,51]. Interestingly, both MW- and LP-SDs demonstrated a disappearance of the LZP melting endotherm. This may have been due to either a dilutional effect or the transformation of the drug into an amorphous state within the polymer matrices. Therefore, PXRD analysis was subsequently conducted to further characterize the polymorphism changes in the SD formulations.

The PXRD analysis revealed changes to LZP's crystalline state within the SD matrices prepared by the MW and LP methods. The analysis by PXRD revealed that lansoprazole (LZP) was partially or extensively converted to an amorphous form within the solid dispersion (SD) matrices prepared by the microwave (MW) and lyophilization (LP) methods. In particular, the intensities of LZP's characteristic diffraction peaks were substantially decreased following the LP process, more so than with the MW preparation. This suggests that greater solubilization and amorphization of LZP occurred when it was dissolved within the alkaline carbonate buffer system during lyophilization, compared to the drug potentially maintaining some degree of crystallinity when subjected to microwave irradiation alone. Additionally, new peaks detected in the LP-SDs could be correlated to the presence of bicarbonate buffer components, consistent with previous reports [52,53]. These findings indicate that lyophilization more potently disrupted LZP crystallinity versus microwave heating, likely attributable to enhanced drug solubilization facilitated by the carbonate vehicle during freeze-drying.

In vitro dissolution testing was used to characterize the drug release behavior from the SDs to assess the impact of the preparation method and combination with the drug-free SNEDDS. Statistical analysis revealed that all SDs, except MW-PEG-4000, enhanced lansoprazole (LZP) dissolution relative to the pure drug. This aligns with a prior report showing a slower drug release from microwave-prepared SDs with PEG versus other polymers [54]. Additional studies found that PEG 4000 SDs enabled retardation of drug release compared to the free drug [55] and exhibited a small water absorption and solubility enhancement [56]. The inferior performance of MW-SD PEG-4000 may relate to drug recrystallization during the cooling step, as reported by Hempel et al. [57].

In contrast, the MW-PF127 SD provided a moderate ~2-fold increase in LZP dissolution efficiency, likely attributable to drug amorphization within the matrix as evidenced by PXRD. However, the lyophilized LP-PF127 SD achieved a 3.5-fold enhancement, which can be rationalized by the higher extent of drug amorphization induced during lyophilization along with the alkaline microenvironment generated by the bicarbonate buffer to facilitate wetting and dissolution. A comparable trend was observed with LP-HPMC SDs.

Combining the SDs with the drug-free bioactive SNEDDS showed no significant overall improvement in LZP release. However, MW-PEG-4000 SDs exhibited enhanced dissolution upon SNEDDS addition, potentially due to increased water absorption and wetting enabled by the self-emulsifying system. In contrast, the high water solubility and amphiphilic properties of Pluronic polymers may negate the need for an adjuvant SNEDDS to achieve optimal LZP release from PF127-based SDs [26].

LP-PF-127 exhibited the highest independent dissolution efficiency and stability, suggesting its optimization as a standalone formulation. However, incorporating the drug-free SNEDDS remains justifiable due to potential synergistic anti-ulcer effects from BSO bioactives.

There are several potential avenues for furthering this research. In vivo pharmacokinetic and pharmacodynamic studies evaluating the optimized LP-PF-127 SD-SNEDDS system could provide insights into its translation potential by comparing its performance to commercial products and other test formulations. Additional investigations into the direct gastroprotective effects of the combined drug-free SNEDDS using relevant animal disease models would help validate the hypothesized synergistic benefits from BSO bioactives. Finally, mechanistic studies examining the mode of anti-ulcer action of BSO components and how formulation impacts tissue distribution and bioavailability could provide insights to support platform translation.

5. Conclusions

This work demonstrates the development of an integrated gastroprotective oral delivery system containing lansoprazole solid dispersions and a bioactive black seed oil-based SNEDDS. An SNEDDS incorporating black seed oil provided a 3.5-fold improvement in lansoprazole dissolution; however, stability issues with drug loading prompted separation into a drug-free system. SD characterization revealed varying degrees of crystallinity disruption induced by different preparation techniques and polymers. In vitro dissolution directly correlated to these physicochemical property alterations, with lyophilization generating the optimal amorphous form that dramatically boosted drug release performance. Remarkably, the lyophilized Pluronic F-127 SD formulation demonstrated superior performance, and up to 97% of the drug remained stable under accelerated conditions for a month. Combining lyophilized dispersions with the bioactive SNEDDS offered minimal additional dissolution enhancement but preserved synergistic anti-ulcer potential through its natural oil components. The integrated solid dispersion–SNEDDS formulation is a promising candidate to enhance oral treatment of acid-related gastrointestinal disorders. Future work should evaluate the in vivo performance and continue optimizing the formulation to advance toward an efficacious clinical product.

Author Contributions: Conceptualization, A.Y.S. and A.A.-W.S.; methodology, A.Y.S. and A.A.-W.S.; software, A.A.-W.S.; formal analysis, A.A.-W.S.; investigation, A.Y.S.; resources, A.A.-W.S.; writing—original draft preparation, A.Y.S.; writing—review and editing, A.Y.S. and A.A.-W.S.; visualization, A.Y.S. and A.A.-W.S.; supervision, A.A.-W.S.; project administration, A.A.-W.S. All authors have read and agreed to the published version of the manuscript.

Funding: This research is funded by the Deputyship for Research & Innovation, "Ministry of Education" in Saudi Arabia (IFKSUOR3-428-1).

Institutional Review Board Statement: Not applicable.

Informed Consent Statement: Not applicable.

Data Availability Statement: The data presented in this study are available within the article.

Acknowledgments: The authors extend their appreciation to the Deputyship for Research & Innovation, "Ministry of Education" in Saudi Arabia, for funding this research (IFKSUOR3-428-1).

Conflicts of Interest: The authors declare no conflict of interest.

References

1. Lien, H.-M.; Wang, Y.-Y.; Huang, M.-Z.; Wu, H.-Y.; Huang, C.-L.; Chen, C.-C.; Hung, S.-W.; Chen, C.-C.; Chiu, C.-H.; Lai, C.-H. Gastroprotective Effect of *Anisomeles indica* on Aspirin-Induced Gastric Ulcer in Mice. *Antioxidants* **2022**, *11*, 2327. [CrossRef] [PubMed]
2. Martinsen, T.C.; Fossmark, R.; Waldum, H.L. The phylogeny and biological function of gastric juice—Microbiological consequences of removing gastric acid. *Int. J. Mol. Sci.* **2019**, *20*, 6031. [CrossRef] [PubMed]
3. Tolbert, K.; Gould, E. Gastritis and gastric ulceration in dogs and cats. In *Clinical Small Animal Internal Medicine*; John Wiley & Sons, Inc.: Hoboken, NJ, USA, 2020; pp. 547–555.
4. Vemulapalli, R. Diet and lifestyle modifications in the management of gastroesophageal reflux disease. *Nutr. Clin. Pract.* **2008**, *23*, 293–298. [CrossRef] [PubMed]
5. Kinoshita, Y.; Miwa, H.; Sanada, K.; Miyata, K.; Haruma, K. Clinical characteristics and effectiveness of lansoprazole in Japanese patients with gastroesophageal reflux disease and dyspepsia. *J. Gastroenterol.* **2014**, *49*, 628–637. [CrossRef]
6. Srebro, J.; Brniak, W.; Mendyk, A. Formulation of dosage forms with proton pump inhibitors: State of the art, challenges and future perspectives. *Pharmaceutics* **2022**, *14*, 2043. [CrossRef]
7. Lu, Y.; Guo, T.; Qi, J.; Zhang, J.; Wu, W. Enhanced dissolution and stability of lansoprazole by cyclodextrin inclusion complexation: Preparation, characterization, and molecular modeling. *AAPS PharmSciTech* **2012**, *13*, 1222–1229. [CrossRef]
8. Zhang, X.; Sun, N.; Wu, B.; Lu, Y.; Guan, T.; Wu, W. Physical characterization of lansoprazole/PVP solid dispersion prepared by fluid-bed coating technique. *Powder Technol.* **2008**, *182*, 480–485. [CrossRef]
9. Fang, Y.; Wang, G.; Zhang, R.; Liu, Z.; Liu, Z.; Wu, X.; Cao, D. Eudragit L/HPMCAS blend enteric-coated lansoprazole pellets: Enhanced drug stability and oral bioavailability. *AAPS PharmSciTech* **2014**, *15*, 513–521. [CrossRef]
10. Alsulays, B.B.; Kulkarni, V.; Alshehri, S.M.; Almutairy, B.K.; Ashour, E.A.; Morott, J.T.; Alshetaili, A.S.; Park, J.-B.; Tiwari, R.V.; Repka, M.A. Preparation and evaluation of enteric coated tablets of hot-melt extruded lansoprazole. *Drug Dev. Ind. Pharm.* **2017**, *43*, 789–796. [CrossRef]
11. Ubgade, S.; Bapat, A.; Kilor, V. Effect of various stabilizers on the stability of lansoprazole nanosuspension prepared using high shear homogenization: Preliminary investigation. *J. Appl. Pharm. Sci.* **2021**, *11*, 085–092.
12. Morakul, B. Self-nanoemulsifying drug delivery systems (SNEDDS): An advancement technology for oral drug delivery. *Pharm. Sci. Asia* **2020**, *47*, 205–220. [CrossRef]
13. Kazi, M.; Shahba, A.A.; Alrashoud, S.; Alwadei, M.; Sherif, A.Y.; Alanazi, F.K. Bioactive self-nanoemulsifying drug delivery systems (Bio-SNEDDS) for combined oral delivery of curcumin and piperine. *Molecules* **2020**, *25*, 1703. [CrossRef] [PubMed]
14. Kanter, M.; Coskun, O.; Uysal, H. The antioxidative and antihistaminic effect of *Nigella sativa* and its major constituent, thymoquinone on ethanol-induced gastric mucosal damage. *Arch. Toxicol.* **2006**, *80*, 217–224. [CrossRef]
15. Jarmakiewicz-Czaja, S.; Zielińska, M.; Helma, K.; Sokal, A.; Filip, R. Effect of *Nigella sativa* on Selected Gastrointestinal Diseases. *Curr. Issues Mol. Biol.* **2023**, *45*, 3016–3034. [CrossRef] [PubMed]
16. Zeren, S.; Bayhan, Z.; Kocak, F.E.; Kocak, C.; Akcılar, R.; Bayat, Z.; Simsek, H.; Duzgun, S.A. Gastroprotective effects of sulforaphane and thymoquinone against acetylsalicylic acid–induced gastric ulcer in rats. *J. Surg. Res.* **2016**, *203*, 348–359. [CrossRef] [PubMed]
17. Radwan, M.F.; El-Moselhy, M.A.; Alarif, W.M.; Orif, M.; Alruwaili, N.K.; Alhakamy, N.A. Optimization of Thymoquinone-Loaded Self-Nanoemulsion for Management of Indomethacin-Induced Ulcer. *Dose-Response* **2021**, *19*, 15593258211013655. [CrossRef]
18. Paseban, M.; Niazmand, S.; Soukhtanloo, M.; Meibodi, N.T.; Abbasnezhad, A.; Mousavi, S.M.; Niazmand, M.J. The therapeutic effect of *Nigella sativa* seed on indomethacin-induced gastric ulcer in Rats. *Curr. Nutr. Food Sci.* **2020**, *16*, 276–283. [CrossRef]
19. Bethesda (MD): National Institute of Diabetes and Digestive and Kidney Diseases; 2012-. LiverTox: Clinical and Research Information on Drug-Induced Liver Injury [Internet]. Black Cumin Seed. [Updated 2023 Apr 27]. Available online: https://www.ncbi.nlm.nih.gov/books/NBK591552/ (accessed on 1 October 2023).
20. Ashfaq, M.; Shah, S.; Rasul, A.; Hanif, M.; Khan, H.U.; Khames, A.; Abdelgawad, M.A.; Ghoneim, M.M.; Ali, M.Y.; Abourehab, M.A. Enhancement of the solubility and bioavailability of pitavastatin through a self-nanoemulsifying drug delivery system (SNEDDS). *Pharmaceutics* **2022**, *14*, 482. [CrossRef]
21. Shahba, A.A.; Tashish, A.Y.; Alanazi, F.K.; Kazi, M. Combined Self-Nanoemulsifying and Solid Dispersion Systems Showed Enhanced Cinnarizine Release in Hypochlorhydria/Achlorhydria Dissolution Model. *Pharmaceutics* **2021**, *13*, 627. [CrossRef]
22. Shahba, A.A.-W.; Ahmed, A.R.; Alanazi, F.K.; Mohsin, K.; Abdel-Rahman, S.I. Multi-layer self-nanoemulsifying pellets: An innovative drug delivery system for the poorly water-soluble drug cinnarizine. *AAPS PharmSciTech* **2018**, *19*, 2087–2102. [CrossRef]
23. Nair, A.R.; Lakshman, Y.D.; Anand, V.S.K.; Sree, K.N.; Bhat, K.; Dengale, S.J. Overview of extensively employed polymeric carriers in solid dispersion technology. *AAPS PharmSciTech* **2020**, *21*, 309. [CrossRef] [PubMed]

24. Alshadidi, A.; Shahba, A.A.; Sales, I.; Rashid, M.A.; Kazi, M. Combined Curcumin and Lansoprazole-Loaded Bioactive Solid Self-Nanoemulsifying Drug Delivery Systems (Bio-SSNEDDS). *Pharmaceutics* **2021**, *14*, 2. [CrossRef] [PubMed]
25. Shahba, A.A.-W.; Sherif, A.Y.; Elzayat, E.M.; Kazi, M. Combined Ramipril and Black Seed Oil Dosage Forms Using Bioactive Self-Nanoemulsifying Drug Delivery Systems (BIO-SNEDDSs). *Pharmaceuticals* **2022**, *15*, 1120. [CrossRef] [PubMed]
26. Aboutaleb, A.E.; Abdel-Rahman, S.I.; Ahmed, M.O.; Younis, M.A. Improvement of domperidone solubility and dissolution rate by dispersion in various hydrophilic carriers. *J. Appl. Pharm. Sci.* **2016**, *6*, 133–139. [CrossRef]
27. Alshehri, S.; Shakeel, F.; Ibrahim, M.; Elzayat, E.; Altamimi, M.; Shazly, G.; Mohsin, K.; Alkholief, M.; Alsulays, B.; Alshetaili, A.; et al. Influence of the microwave technology on solid dispersions of mefenamic acid and flufenamic acid. *PLoS ONE* **2017**, *12*, e0182011. [CrossRef] [PubMed]
28. Tashish, A.Y.; Shahba, A.A.-W.; Alanazi, F.K.; Kazi, M. Adsorbent Precoating by Lyophilization: A Novel Green Solvent Technique to Enhance Cinnarizine Release from Solid Self-Nanoemulsifying Drug Delivery Systems (S-SNEDDS). *Pharmaceutics* **2023**, *15*, 134. [CrossRef]
29. Tian, Z.; Yi, Y.; Yuan, H.; Han, J.; Zhang, X.; Xie, Y.; Lu, Y.; Qi, J.; Wu, W. Solidification of nanostructured lipid carriers (NLCs) onto pellets by fluid-bed coating: Preparation, in vitro characterization and bioavailability in dogs. *Powder Technol.* **2013**, *247*, 120–127. [CrossRef]
30. El Maghraby, G.M.; Elzayat, E.M.; Alanazi, F.K. Development of modified in situ gelling oral liquid sustained release formulation of dextromethorphan. *Drug Dev. Ind. Pharm.* **2012**, *38*, 971–978. [CrossRef]
31. Shahba, A.A.-W.; Alanazi, F.K.; Abdel-Rahman, S.I. Stabilization benefits of single and multi-layer self-nanoemulsifying pellets: A poorly-water soluble model drug with hydrolytic susceptibility. *PLoS ONE* **2018**, *13*, e0198469. [CrossRef]
32. Mishra, P.; Pandey, C.M.; Singh, U.; Gupta, A.; Sahu, C.; Keshri, A. Descriptive statistics and normality tests for statistical data. *Ann. Card. Anaesth.* **2019**, *22*, 67–72. [CrossRef]
33. Nahm, F.S. Nonparametric statistical tests for the continuous data: The basic concept and the practical use. *Korean J Anesth.* **2016**, *69*, 8–14. [CrossRef] [PubMed]
34. Rehman, F.U.; Shah, K.U.; Shah, S.U.; Khan, I.U.; Khan, G.M.; Khan, A. From nanoemulsions to self-nanoemulsions, with recent advances in self-nanoemulsifying drug delivery systems (SNEDDS). *Expert Opin. Drug Deliv.* **2017**, *14*, 1325–1340. [CrossRef] [PubMed]
35. Krstić, M.; Medarević, Đ.; Đuriš, J.; Ibrić, S. Self-nanoemulsifying drug delivery systems (SNEDDS) and self-microemulsifying drug delivery systems (SMEDDS) as lipid nanocarriers for improving dissolution rate and bioavailability of poorly soluble drugs. In *Lipid Nanocarriers for Drug Targeting*; Elsevier: Amsterdam, The Netherlands, 2018; pp. 473–508.
36. Kazi, M.; Nasr, F.A.; Noman, O.; Alharbi, A.; Alqahtani, M.S.; Alanazi, F.K. Development, characterization optimization, and assessment of curcumin-loaded bioactive self-nanoemulsifying formulations and their inhibitory effects on human breast cancer MCF-7 cells. *Pharmaceutics* **2020**, *12*, 1107. [CrossRef] [PubMed]
37. Alwadei, M.; Kazi, M.; Alanazi, F.K. Novel oral dosage regimen based on self-nanoemulsifying drug delivery systems for codelivery of phytochemicals–curcumin and thymoquinone. *Saudi Pharm. J.* **2019**, *27*, 866–876. [CrossRef] [PubMed]
38. Kazi, M.; Alhajri, A.; Alshehri, S.M.; Elzayat, E.M.; Al Meanazel, O.T.; Shakeel, F.; Noman, O.; Altamimi, M.A.; Alanazi, F.K. Enhancing oral bioavailability of apigenin using a bioactive self-nanoemulsifying drug delivery system (Bio-SNEDDS): In vitro, in vivo and stability evaluations. *Pharmaceutics* **2020**, *12*, 749. [CrossRef]
39. Pathak, K.; Raghuvanshi, S. Oral bioavailability: Issues and solutions via nanoformulations. *Clin. Pharmacokinet.* **2015**, *54*, 325–357. [CrossRef]
40. Shakeel, F.; Haq, N.; El-Badry, M.; Alanazi, F.K.; Alsarra, I.A. Ultra fine super self-nanoemulsifying drug delivery system (SNEDDS) enhanced solubility and dissolution of indomethacin. *J. Mol. Liq.* **2013**, *180*, 89–94. [CrossRef]
41. Zhao, Y.; Wang, C.; Chow, A.H.; Ren, K.; Gong, T.; Zhang, Z.; Zheng, Y. Self-nanoemulsifying drug delivery system (SNEDDS) for oral delivery of Zedoary essential oil: Formulation and bioavailability studies. *Int. J. Pharm.* **2010**, *383*, 170–177. [CrossRef]
42. Silva, H.D.; Cerqueira, M.A.; Vicente, A.A. Influence of surfactant and processing conditions in the stability of oil-in-water nanoemulsions. *J. Food Eng.* **2015**, *167*, 89–98. [CrossRef]
43. Tian, Y.; Chen, L.; Zhang, W. Influence of Ionic Surfactants on the Properties of Nanoemulsions Emulsified by Nonionic Surfactants Span 80/Tween 80. *J. Dispers. Sci. Technol.* **2016**, *37*, 1511–1517. [CrossRef]
44. He, W.; Yang, M.; Fan, J.H.; Feng, C.X.; Zhang, S.J.; Wang, J.X.; Guan, P.P.; Wu, W. Influences of sodium carbonate on physico-chemical properties of lansoprazole in designed multiple coating pellets. *AAPS PharmSciTech* **2010**, *11*, 1287–1293. [CrossRef] [PubMed]
45. Nora, G.-I.; Venkatasubramanian, R.; Strindberg, S.; Siqueira-Jørgensen, S.D.; Pagano, L.; Romanski, F.S.; Swarnakar, N.K.; Rades, T.; Müllertz, A. Combining lipid based drug delivery and amorphous solid dispersions for improved oral drug absorption of a poorly water-soluble drug. *J. Control. Release* **2022**, *349*, 206–212. [CrossRef] [PubMed]
46. Fang, Z.; Bhandari, B. Spray drying, freeze drying and related processes for food ingredient and nutraceutical encapsulation. In *Encapsulation Technologies and Delivery Systems for Food Ingredients and Nutraceuticals*; Elsevier: Amsterdam, The Netherlands, 2012; pp. 73–109.
47. Tabatar, T.; Makino, T.; Kashihara, T.; Hirai, S.; Kitamori, N.; Toguchi, H. Stabilization of a new antiulcer drug (lansoprazole) in the solid dosage forms. *Drug Dev. Ind. Pharm.* **1992**, *18*, 1437–1447. [CrossRef]
48. Pasic, M. Study to Design Stable Lansoprazole Pellets. Ph.D. Thesis, University of Basel, Basel, Switzerland, 2008.

49. Yu, M.; Sun, L.; Li, W.; Lan, Z.; Li, B.; Tan, L.; Li, M.; Yang, X. Investigation of structure and dissolution properties of a solid dispersion of lansoprazole in polyvinylpyrrolidone. *J. Mol. Struct.* **2011**, *1005*, 70–77. [CrossRef]
50. Harabor, A.; Rotaru, P.; Harabor, N. Two phases in a commercial anhydrous sodium carbonate by air contact. *Phys. AUC* **2013**, *23*, 79–88.
51. Leemsuthep, A.; Mohd Nayan, N.A.; Zakaria, Z.; Uy Lan, D.N. Effect of sodium bicarbonate in fabrication of carbon black-filled epoxy porous for conductive application. In Proceedings of the Macromolecular Symposia—Special Issue: 4th Federation of Asian Polymer Societies International Polymer Congress, Kuala Lumpur, Malaysia, 5–8 October 2015; John Wiley & Sons, Inc.: Hoboken, NJ, USA, 2017; pp. 44–49.
52. Uysal, D.; Dogan, Ö.M.; Uysal, B.Z. Kinetics of absorption of carbon dioxide into sodium metaborate solution. *Int. J. Chem. Kinet.* **2017**, *49*, 377–386. [CrossRef]
53. Reddy, V.V.; Rao, H.S.; Jayaveera, K. Influence of strong alkaline substances (sodium carbonate and sodium bicarbonate) in mixing water on strength and setting properties of concrete. *Indian J. Eng. Mater. Sci.* **2006**, *13*, 123–128.
54. Alshehri, S.; Alanazi, A.; Elzayat, E.M.; Altamimi, M.A.; Imam, S.S.; Hussain, A.; Alqahtani, F.; Shakeel, F. Formulation, in vitro and in vivo evaluation of gefitinib solid dispersions prepared using different techniques. *Processes* **2021**, *9*, 1210. [CrossRef]
55. Moneghini, M.; De Zordi, N.; Grassi, M.; Zingone, G. Sustained-release solid dispersions of ibuprofen prepared by microwave irradiation. *J. Drug Deliv. Sci. Technol.* **2008**, *18*, 327–333. [CrossRef]
56. Tafu, N.N.; Jideani, V.A. Proximate, elemental, and functional properties of novel solid dispersions of *Moringa oleifera* leaf powder. *Molecules* **2022**, *27*, 4935. [CrossRef]
57. Hempel, N.-J.; Knopp, M.M.; Zeitler, J.A.; Berthelsen, R.; Löbmann, K. Microwave-induced in situ drug amorphization using a mixture of polyethylene glycol and polyvinylpyrrolidone. *J. Pharm. Sci.* **2021**, *110*, 3221–3229. [CrossRef] [PubMed]

Article

Transferrin-Modified Triptolide Liposome Targeting Enhances Anti-Hepatocellular Carcinoma Effects

Xiaoli Zhao [1,2,†], Yifan Yang [1,2,†], Xuerong Su [1,2], Ying Xie [1,2], Yiyao Liang [1,2], Tong Zhou [1,2], Yangqian Wu [1,2] and Liuqing Di [1,2,*]

1. College of Pharmacy, Nanjing University of Chinese Medicine, Nanjing 210023, China; leah_zhao@njucm.edu.cn (X.Z.); Yifan_Yang0801@163.com (Y.Y.); suxuer69@163.com (X.S.); xeeyeah@njucm.edu.cn (Y.X.); lyy541@126.com (Y.L.); 20210906@njucm.edu.cn (T.Z.); 20220988@njucm.edu.cn (Y.W.)
2. Jiangsu Provincial TCM Engineering Technology Research Center of High Efficient Drug Delivery System (DDS), Nanjing 210023, China
* Correspondence: diliuqing@njucm.edu.cn
† These authors contributed equally to this work.

Abstract: Triptolide (TP) is an epoxy diterpene lactone compound isolated and purified from the traditional Chinese medicinal plant *Tripterygium wilfordii* Hook. f., which has been shown to inhibit the proliferation of hepatocellular carcinoma. However, due to problems with solubility, bioavailability, and adverse effects, the use and effectiveness of the drug are limited. In this study, a transferrin-modified TP liposome (TF-TP@LIP) was constructed for the delivery of TP. The thin-film hydration method was used to prepare TF-TP@LIP. The physicochemical properties, drug loading, particle size, polydispersity coefficient, and zeta potential of the liposomes were examined. The inhibitory effects of TF-TP@LIP on tumor cells in vitro were assessed using the HepG2 cell line. The biodistribution of TF-TP@LIP and its anti-tumor effects were investigated in tumor-bearing nude mice. The results showed that TF-TP@LIP was spherical, had a particle size of 130.33 ± 1.89 nm and zeta potential of -23.20 ± 0.90 mV, and was electronegative. Encapsulation and drug loading were $85.33 \pm 0.41\%$ and $9.96 \pm 0.21\%$, respectively. The preparation was stable in serum over 24 h and showed biocompatibility and slow release of the drug. Flow cytometry and fluorescence microscopy showed that uptake of TF-TP@LIP was significantly higher than that of TP@LIP ($p < 0.05$), while MTT assays indicated mean median inhibition concentrations (IC_{50}) of TP, TP@LIP, and TF-TP@ of 90.6 nM, 56.1 nM, and 42.3 nM, respectively, in HepG2 cell treated for 48 h. Real-time fluorescence imaging indicated a significant accumulation of DiR-labeled TF-TP@LIPs at tumor sites in nude mice, in contrast to DiR-only or DiR-labeled, indicating that modification with transferrin enhanced drug targeting to the tumor tissues. Compared with the TP and TP@LIP groups, the TF-TP@LIP group had a significant inhibitory effect on tumor growth. H&E staining results showed that TF-TP@LIP inhibited tumor growth and did not induce any significant pathological changes in the heart, liver, spleen, and kidneys of nude mice, with all liver and kidney indices within the normal range, with no significant differences compared with the control group, indicating the safety of the preparation. The findings indicated that modification by transferrin significantly enhanced the tumor-targeting ability of the liposomes and improved their anti-tumor effects in vivo. Reducing its distribution in normal tissues and decreasing its toxic effects suggest that the potential of TF-TP@LIP warrants further investigation for its clinical application.

Keywords: targeted therapy; lipidosome; transferrin; anti-tumor; hepatocellular carcinoma; fluorescence imaging

Citation: Zhao, X.; Yang, Y.; Su, X.; Xie, Y.; Liang, Y.; Zhou, T.; Wu, Y.; Di, L. Transferrin-Modified Triptolide Liposome Targeting Enhances Anti-Hepatocellular Carcinoma Effects. *Biomedicines* **2023**, *11*, 2869. https://doi.org/10.3390/biomedicines11102869

Academic Editors: Yongtai Zhang and Zhu Jin

Received: 1 September 2023
Revised: 10 October 2023
Accepted: 16 October 2023
Published: 23 October 2023

1. Introduction

Hepatocellular carcinoma (HCC) is not only the most common form of primary liver cancer, but also the leading cause of cancer-related deaths worldwide. Its pathogenesis is closely related to chronic infection by the hepatitis B and C viruses, as well as alcoholic cirrhosis, and treatment options currently include surgical resection, liver transplantation, liver-directed therapy, and systemic chemotherapy. However, patients with liver cancer are usually diagnosed at an advanced stage, and as such, they cannot undergo surgical resection and liver transplantation [1]. In most countries, mortality from HCC is also almost equal to morbidity, largely due to poor response to both conventional chemotherapy and targeted therapeutic drugs [2–4]. Hence, in addition to early detection, diagnosis, and surgery, it is necessary to develop new treatment strategies to improve HCC treatment [5]. In this context, Chinese herbal medicine, one of the integrated treatment modalities, has gained increasing attention for its anti-inflammatory and anti-cancer properties.

Triptolide (TP), first isolated by Kup-chan et al. from *Tripterygium wilfordii* Hook. f. (TWHF) in 1972, has significant broad-spectrum anti-tumor and anti-autoimmune effects [6]. As the major active ingredient of TWHF, TP exhibits almost all these therapeutic effects, as well as, unfortunately, toxicity [7]. Recent studies have shown that TP displays significant anti-tumor activity and other potential therapeutic effects against various cancers, such as liver, lung, and pancreatic cancers [8,9], prompting extensive exploration of these properties [10,11]. In particular, TP has been reported to inhibit the proliferation and invasion of tumor cells with activities comparable to or even better than some traditional anti-tumor drugs, including adriamycin, mitomycin, paclitaxel, and cisplatin, to name a few [12]. The anti-tumor activity of TP is reflected in its strong cytotoxicity, which can eliminate drug resistance, while inhibiting neovascularization and tumor metastasis. Early studies have also shown that the anticancer properties of TP mostly involve its induction of apoptosis, thus preventing the development of many tumors [13,14]. However, the mechanism by which TP induces cell death varies depending on the cell type. In addition to apoptosis, TP can also affect the metabolism of tumor cells by reducing cell viability and growth through cell cycle arrest. However, as the clinical use of TP increased, several studies and clinical reports reported that TP also had serious adverse effects, including multi-organ toxicity (hepatotoxicity [15,16], nephrotoxicity [15], cardiotoxicity [17], and reproductive toxicity [18]). This, together with its poor water solubility, resulted in reduced clinical application of TP [19]. Consequently, it would be advantageous to design a drug delivery system that includes modification of the molecular structure, as well as providing effective delivery of TP to targeted sites, while reducing the accumulation of free drugs in other tissues and organs to lessen the incidence of toxic and adverse effects and improve therapeutic efficacy.

Considering the highly potent activity of TP against various types of malignant cells, the current study aimed to improve the treatment of HCC through the targeted delivery of TP to reduce its toxic side effects.

The phospholipid bilayers of liposomes increase the solubility of insoluble drugs, reduce the rate of blood clearance, and prolong the half-life of the drug in the body, thereby increasing bioavailability [20]. The advantages of using liposomes include increased circulation time, improved drug stability, lower toxicities, and enhanced targeting capability. However, it is worth noting that despite their efficacy in treating various cancers, liposomes lack the ability for specific targeting [21]. By introducing targeting molecules (e.g., peptides, monosaccharides, polysaccharides, folic acid, antibodies, or antibody fragments) on the liposome surfaces that can specifically bind to tumor cells, the retention of nano drugs within tumor tissues can be promoted, thus enhancing the efficiency of endocytosis and their enrichment within tumor cells [22]. Human transferrin receptors (TFRs) are single-chain transmembrane glycoproteins composed of 700 amino acids with two disulfide subunits that are involved in the transport of ferric ions [23,24]. Transferrin (TF) acts as a carrier of ferric ions and enters the cell via the TFR. To maintain their rapid proliferation, cancer cells require more iron than normal cells [25], and thus many tumor cells, such as HepG2 and MDA-MB 231 cells, overexpress the TFR. The elevated expression of the TFR can be used to enhance the targeting of therapeutic cargo-loaded nanoparticles, such as liposomes. Through receptor-mediated endocytosis (RME), the TFR-targeted nanocarriers can improve the specificity of the drug cargo towards cancer cells [26,27]. Many studies have used PEGylated liposomes combined with TF and PEG to achieve targetability and longevity for drug delivery to solid tumors [28,29]. To date, there have been several reports of the use of nano agents, including liposomes [30], cubosomes [31], polymer vesicles, and polymer nanoparticles [9,32] for targeting liver cancer.

In this study, a TF-mediated liposomal drug delivery system was designed to slow down its release in vivo, as well as increase its targeting ability in vivo and in vitro, thereby reducing the drug's toxicity, while enhancing its effects against HCC.

2. Materials and Methods

2.1. Materials

Hydrogenated soy lecithin (HSPC), stearoyl phosphatidylethanolamine-polyethylene glycol 2000 (DSPE-PEG2000), and cholesterol (CHO) were purchased from A.V.T. Pharmaceutical Co., Ltd. (Shanghai, China). DSPE-PEG2000-MAL was obtained from the Xi'an Ruixi Biological Technology Co., Ltd. (Xi'an, China), while phosphate buffer (pH = 7.2–7.4) was obtained from Solarbio Science & Technology Co., Ltd. (Beijing, China). Holo-transferrin was obtained from Best Biological Technology Co., Ltd. (Nanjing, China), thiazolyl blue tetrazolium bromide (MTT) was obtained from Yuanye Bio-Technology Co., Ltd. (Shanghai, China), and fetal bovine serum (FBS) and Dulbecco's modified Eagle medium (DMEM) were obtained from SenBeiJia Biological Technology Co., Ltd. (Nanjing, China). Triptolide (TP) was purchased from Nantong Feiyu Biological Technology Co., Ltd. (Nantong, China), and Traut's reagent was purchased from Nanjing JinYibai Biological Technology Co., Ltd. (Nanjing, China), but obtained from the Labgic Technology Co., Ltd. (Beijing, China) together with 4-diamidino-2-phenylindole (DAPI, MW 350.2). Finally, coumarin-6 was purchased from Beyotime Biotechnology Co., Ltd. (Shanghai, China).

2.2. Cell Culture

The human hepatocellular carcinoma cell line (HepG2) was obtained from the Center for Excellence in Molecular Cell Science (Shanghai, China).

2.3. Animals

BALB/c nude mice (female, aged 6–8 weeks, 18–20 g) were purchased from Qinglongshan Laboratory Animal Company Limited (Nanjing, China). The Animal Care and Use Committee of Nanjing University of Chinese Medicine approved the animal experiments (approval number 202107A034). The animals were housed under pathogen-free conditions and provided with sterile food and water at Nanjing University of Chinese Medicine's Laboratory Animal Center.

2.4. Preparation of Triptolide Liposomes (TP@LIP) and Triptolide Liposomes Modified by Transferrin (TF-TP@LIP)

TF-TP@LIP liposomes were prepared by membrane hydration. Using the encapsulation rate and drug loading as the main evaluation indices, we used a star design-response surface optimization method to optimize and verify the best prescription of TP@LIP. Briefly, this involved dissolving TF in PBS at a concentration of 10 mg/mL prior to thiolation with Traut's reagent. The molar ratio of Traut's reagent to TF was 20:1, and the reaction was allowed to proceed for 1 h on a shaker protected from light. DSPE-PEG2000-MAL was then dissolved in 2.5 mL of PBS (pH 6.5), and the mixture was allowed to react with the thiolated TF in darkness overnight to yield DSPE-PEG2000-TF containing different ratios of each substance. In addition, the lipid components of HSPC/Chol/DSPE-PEG2000/TP (69:9:12:12, mass ratio) and HSPC/Chol/DSPE-PEG2000/TP (58:14:4:22, molar ratio) were separately used for several steps, including the preparation of blank and TP liposomes (TP@LIP). To obtain a thin film, HSPC, Chol, DSPE-PEG2000, and TP were dissolved in absolute ethanol in a pear-shaped bottle before drying on a rotary evaporator at 37 °C. An appropriate amount of isotonic buffer was then added, mixed well, and hydrated at atmospheric pressure for 30 min before ultrasonication (SM-1000D, SHUNMATECH, Co., Ltd. Shanghai, China). The resulting mixture was then passed through 0.2 μm and 0.1 μm polycarbonate fibro-lipid membranes six times to obtain the liposome solution. The TF-modified tretinoin liposomes (TF-TP@LIPs) were obtained by incubating the above-mentioned tretinoin liposomes with an appropriate amount of synthetically resoluble TF-PEG2000-DSPE lyophilized powder solution at 60 °C for 1 h. Liposomes loaded with Coumarin 6 (TF-C6@LIP, C6@LIP) were also prepared using the above procedure. To evaluate the stability of the TF-TP@LIP and TP@LIP, the samples were placed in a penicillin bottle and stored at 4 °C. The particle size and PDI (polydispersity index) were determined after 0, 1, 2, 3, 4, 5, 6, and 7 days.

2.5. Characterization of Liposomes

Dynamic light scattering with a Zetasizer Nano ZS90 (Malvern, UK) was used to determine the particle size, polydispersity index, and zeta potential for all liposomes, while transmission electron microscopy (TEM) was used to observe their morphologies (HT7800, Hitachi, Tokyo, Japan). Furthermore, high-performance liquid chromatography (HPLC) was used for measuring the TP content (Alliance HPLC E2695, Waters Corporation, Milford, MA, USA). The unencapsulated drug was separated from the liposome solution by centrifugal ultrafiltration (4500 rpm, 5 min) (TG16-WS, Changsha, China), and the encapsulation rate was determined by HPLC. Specifically, the method involved the addition of 20 volumes of methanol to the liposome solution, followed by demulsification using ultrasound for 20 min and measurement of the total TP amount (Wtotal). In addition, 0.2 mL liposome solution was measured precisely into an ultrafiltration centrifuge tube (MWCO 30,000 Da), diluted to 1 mL with PBS, and centrifuged (4500 rpm, 5 min) to obtain the unwrapped TP. The unwrapped drug content ($W_{Unwrapped\ drug}$) was determined, and the encapsulation efficiency (EE) was calculated as:

$$EE\% = 1 - (W_{Unwrapped\ drug}/W_{total}) \times 100\%,$$

where $W_{Unwrapped}$ and W_{total} represent the unencapsulated amount and the total amount of the drug, respectively. The drug loading (DL%) for TP-loaded liposomes was then calculated using the following equation:

$$DL\% = (WTP - W_{Unwrapped\ drug}/W_{Total\ amount\ of\ lipids}) \times 100\%.$$

2.6. In Vitro Drug Release from LIPs

The drug-releasing properties of TP@Lip and TF-TP@Lip were evaluated in vitro using dialysis, as previously reported [33]. For this purpose, equal volumes of TP@Lip, TF-TP@Lip, and TP solution (5 mL) were placed in dialysis bags (MWCO 8–14 kDa), which were then placed in 200 mL of PBS (pH 7.2~7.4) containing 1% Tween 80. The bags were gently shaken at 100 rpm (37 °C), and at predetermined intervals (0.25, 0.5, 1, 2, 4, 6, 8, 10, and 12 h), 2 mL of the surrounding buffer was collected and replaced with fresh buffer. The samples were analyzed by HPLC to determine the TP content. The cumulative release rate was then calculated and plotted for each time point ($n = 3$).

2.7. Hemolytic Examination of TF-TP@LIP

One milliliter of mouse blood was collected into an anticoagulation tube and centrifuged. After removal of the plasma layer, an appropriate amount of physiological saline was added to the tube and mixed. This was followed by centrifugation at 2000 rpm for 5 min, and after removing the supernatant, saline was again added. The above procedure was repeated two to three times, with the resulting erythrocytes made up to a 2% (*v/v*) suspension with saline and set aside. Corresponding solutions were subsequently added in the order shown in Table 1 with proper mixing prior to a 2 h incubation at 37 °C. Hemolysis and coagulation reactions were observed for each tube (tubes No. 1 to 3 represented different concentrations of the test samples, No. 4 was the negative control, and No. 5 was the positive control). Finally, after the incubation, the tubes were centrifuged (500 rpm, 5 min), with the resulting supernatant collected to determine the absorbance (OD) of each well at 540 nm. Based on the results, the hemolysis rate was determined according to the following equation:

$$\text{Hemolysis rate (\%)} = OD_{Samples}OD_{Negative}/OD_{Positive} - OD_{Negative}$$

Table 1. Hemolysis experiment design table.

Test Tube	1	2	3	4	5
2% Red Blood Cells Suspension (μL)	500	500	500	500	500
Saline solution (μL)	450	400	300	500	
Distilled water (μL)					500
Sample (μL)	50	100	200		

2.8. Cellular Uptake

Fluorescent-labeled Coumarin 6 liposomes (C6@LIP) and TF-modified Coumarin 6 liposomes (TF-C6@LIP) with different coupling densities were investigated by flow cytometry of HepG2 cells. The cells were cultured (5×10^5 cells/well) in 12-well plates for 24 h to allow attachment. C6@LIP and TF-C6@LIP, with different coupling densities, were added to the cultures, with blank liposomes acting as blank controls. The final concentration of Coumarin 6, in this case, was 50 ng/mL. Cells were collected when in the logarithmic growth phase and treated with TF-modified Coumarin-6 liposomes before incubation for 6 h. The cells were then washed with PBS, trypsinized, and centrifuged at 1000 rpm for 5 min. The cells were resuspended in PBS and the fluorescence intensities measured using flow cytometry (Gallios, Beckman Coulter, Brea, CA, USA).

Fluorescence microscopy was also used to verify the uptake of C6@LIP or TT-C6@LIP by HepG2 cells. C6, at a final concentration of 50 ng/mL, was incubated with HepG2 cells together with C6@LIP and TF-C6-LIP, with blank liposomes as a blank control. After incubation for 6 h, the cells were washed with PBS, fixed with 4% paraformaldehyde for 15 min at room temperature (RT), and stained with DAPI. A fluorescence microscope (Axio Vert A1, Zeiss, Germany) was then used for imaging the HepG2 cells.

To verify whether HepG2 cells could take up TF-TP@LIP through TFR-mediated endocytosis, a 20-fold excess of free TF was preincubated with the cells before the competitive inhibition experiment.

2.9. In Vitro Cytotoxicity

To evaluate the cytotoxicity of TF-TP@LIP in vitro, different concentrations of TF-TP@LIP were added to cells in 96-well flat-bottomed plates (8×10^3 cells/well) and grown overnight. After the addition of 5 mg/mL of MTT, the cells were incubated for a further 4 h, after which the culture supernatants were discarded, and 100 μL of DMSO was added. Absorbances were measured at 490 nm with an EnVision Multimode Plate Reader (PerkinElmer, Waltham, MA, USA) to assess cell viability. Comparative studies were also conducted using TP@LIP and the free drug [34].

2.10. Analysis of Apoptosis and the Cell Cycle by Flow Cytometry

To observe cellular apoptosis, HepG2 cells were first incubated with TF-TP@LIP, TP@LIP, and TP alone (TP was administered at a final concentration of 80 nM) for 24 h. The cells were then washed twice with cold PBS, and after being collected using EDTA-free trypsin, were resuspended in binding buffer and stained with Annexin-V-FITC and PI solution (BD, China) for 30 min at room temperature, before analysis by flow cytometry to assess apoptosis.

HepG2 cells were cultured in 6-well plates at a density of 3×10^5 cells/well for 24 h. TP, TP@LIP, and TF-TP@LIP (TP was administered at a final concentration of 80 nM) were then added, and after another 24 h incubation, the cells were resuspended in 1 mL of RNase A solution, according to the instructions of the cell cycle assay kit (MULTI SCIENCE, Hangzhou, China). This was followed by incubation for 30 min in a water bath at 37 °C before measurement of the cell cycle distribution by flow cytometry.

2.11. Animal Tumor Models

Based on previous methods, BALB/c nude mice were used to establish in vivo tumor models. The mice were injected with a 200 μL suspension of HepG2 cells (approximately 6×10^6 cells suspended in PBS) on each side (near the hind limb) prior to fluorescence imaging of the tumors. The growth of the tumors was monitored, and the volumes were calculated as follows:

$$\text{tumor volume (mm}^3) = A \times B2/2,$$

where "A" and "B" represent the longest and shortest diameters of the tumor, respectively.

2.12. Live Fluorescence Imaging

Nude mice were injected with TF-DiR@LIP, DiR@LIP, and DiR via the tail vein at a dose of 0.1 mg/kg ($n = 3$). After 1, 2, 4, 6, 8, 12, and 24 h, the animals were anesthetized by isoflurane inhalation, and the distribution of the nano-formulation in vivo was visualized using an IVIS Spectrum Imaging System (Axio Vert A1, Zeiss, Germany). After 24 h, the mice were euthanized, and tissues (heart, liver, spleen, lung, kidney, and tumor tissues) were harvested, dissected, and imaged.

2.13. Safety Evaluation

At the end of treatment, blood was collected from the mice, centrifuged to obtain the serum, and used to measure the liver index (ALT/AST/TP/ALB) and kidney index (BUN/CRE/UA). Heart, liver, spleen, lung, and kidney tissues were also dissected, and after being weighed, were fixed in paraformaldehyde, paraffin embedded, and cut into 30 μm thick sections. The sections were then stained with hematoxylin–eosin (H&E) and evaluated for morphological changes, signs of inflammation, necrosis, or cell damage.

2.14. Statistical Analysis

All experimental data are presented as the mean ± SD. Statistical analysis was performed using a one-way ANOVA for multiple samples, as well as a Student's *t*-test for paired sample sets. For all analyses, *p*-values less than 0.05 were considered statistically significant, while those less than 0.01 were considered extremely significant. For all analyses, GraphPadPrism 6.0 software was used.

3. Results and Discussion

3.1. Characterization of the Liposomes

The physical properties of liposomes normally affect the amount of drugs that can be loaded, as well as their injectability, biodistribution, in vivo clearance, shelf life, and dose prediction and the pharmacokinetic profile of the drug administered [35]. In this study, thiolated TF was allowed to react with DSPE-PEG2000-MAL to yield DSPE-PEG2000-TF, a functional phospholipid with targeting potential. A thin-film dispersion method was also applied to prepare TF-TP@LIP, and a diluted solution of the latter was observed under TEM. In this case, results showed that TF-TP@LIP was spherical and uniformly distributed (Figure 1). Table 2 provides further information on the physical properties of the liposomes, including the particle size, polydispersity index (PDI), zeta potential, encapsulation efficiency, and drug loading capacity.

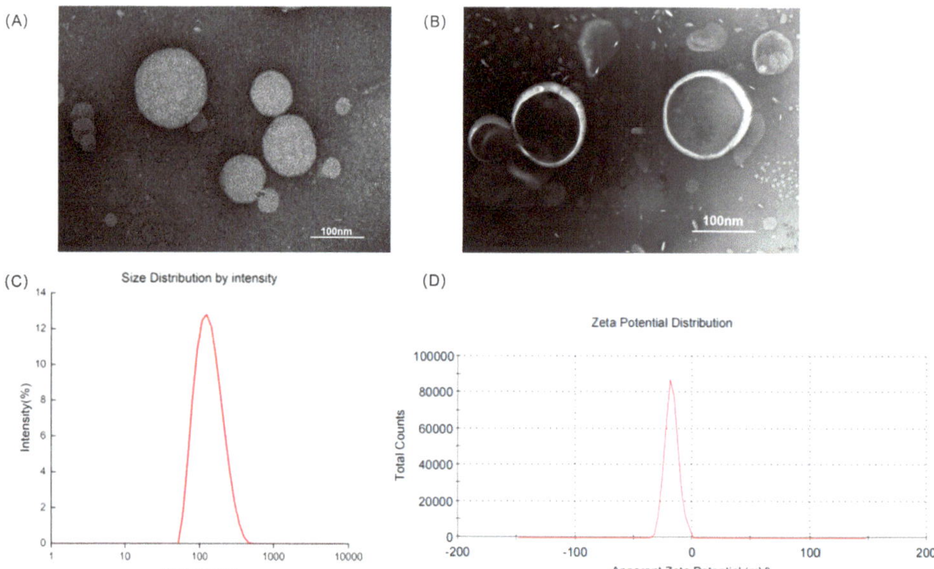

Figure 1. Characteristics of the liposomes. (**A**) Transmission electron microscopy images of TP@LIP; (**B**) Transmission electron microscopy images of TF-TP@LIP; (**C**) Particle size distribution of TF-TP@LIP; (**D**) Zeta potentials of TF-TP@LIP.

Table 2. Physicochemical properties of liposomes. Particle size, zeta potential, encapsulation efficiency, and drug loading of liposomes (unmodified and modified with transferrin).

Liposomes	Size (nm)	PDI	Zeta Potential (mV)	EE%	DL%
TF-Blank@LIP	125.02 ± 2.20	0.19 ± 0.05	−23.90 ± 2.30		
TP@LIP	121.10 ± 3.40	0.15 ± 0.03	−17.50 ± 0.20	88.23 ± 0.14	10.26 ± 0.02
TF-TP@LIP	130.33 ± 1.89	0.20 ± 0.04	−23.20 ± 0.90	85.33 ± 0.41	9.96 ± 0.21

Based on the USP pharmacopeial forum (USP-PF), the mean globular diameter of particles for IV administration should be less than 500 nm, as determined by DLS, irrespective of the lipid concentration [36]. TP@LIP was found to have a mean particle size of 121.10 ± 3.40 nm, a polydispersity index of 0.15 ± 0.03, and a zeta potential of −17.50 ± 0.20 mV, while the values for TF-TP@LIP were 130.33 ± 1.89 nm, 0.20 ± 0.04, and −23.20 ± 0.90 mV, respectively. Therefore, if the particle size of a long-circulating liposome is less than 150 nm, it would be able to penetrate the blood vessels in the tumor area effectively through the enhanced permeability and retention effect (EPR), leading to enrichment within the tumor region. At the same time, this would improve the distribution of the drug in the body, while reducing its toxicity. Interestingly, the particle size of the liposomes prepared in this study was about 130 nm, which is conducive to their passive targeting to reach the tumor tissue [37]. Incubation of TP@LIP with TF-PEG2000-DSPE resulted in an increase in particle size with no significant change in the PDI, indicating that the LIP had a homogeneous particle distribution and that its reproducibility, obtained by the thin-film dispersion method, was high (Figure 1 and Table 2).

The encapsulation rates (EE%) for the TP@LIP and TF-TP@LIP liposomes were $88.23 \pm 0.14\%$ and $85.33 \pm 0.41\%$, respectively (Table 2). In our previous study, we investigated three methods for determining the encapsulation of fat-soluble drugs by liposomes, namely, low-speed centrifugation, dextran gel columns, and ultrafiltration centrifugation. The ultrafiltration centrifugation method does not require much equipment and saves time. This method can also assess drug encapsulation more accurately, and thus the ultrafiltration centrifugation method was selected for determining the encapsulation rate of triptolide liposomes (Supplementary Section S2). Since the ultrafiltration tube can adsorb some of the drug, pre-saturation treatment is required before the ultrafiltration centrifugation operation. In addition, due to the "concentration polarization" effect, the membrane permeability of free drugs tends to be low, so the liposome solution to be tested should be diluted during the experiment to avoid this phenomenon [32].

These results indicated that TF-TP@LIP had a uniform particle size distribution and good electronegative properties. It was also within the nanoscale size and had good physicochemical characteristics, which could potentially regulate the pharmacokinetic and pharmacodynamic characteristics of drugs, thereby improving their therapeutic indices [38].

To investigate the serum stability of TF-TP@LIP, 10% FBS was prepared as the medium, and the particle size distribution, as well as PDI of TF-TP@LIP, were quantified with DLS. It was found that the single peak of the particle size distribution of TF-modified ryanodine liposome was still present after mixing with 10% fetal bovine serum. As shown in Figure 2A, the particle size distribution and PDI of TF-TP@LIP did not change significantly within 24 h after mixing with the serum, thereby indicating good serum stability for TF-TP@LIP. In addition, the results of the stability experiment showed that the particle size did not increase after 7 days in TP@LIP and TF-TP@LIP. This demonstrates good stability for TP@LIP and TF-TP@LIP (Figure 2B,C).

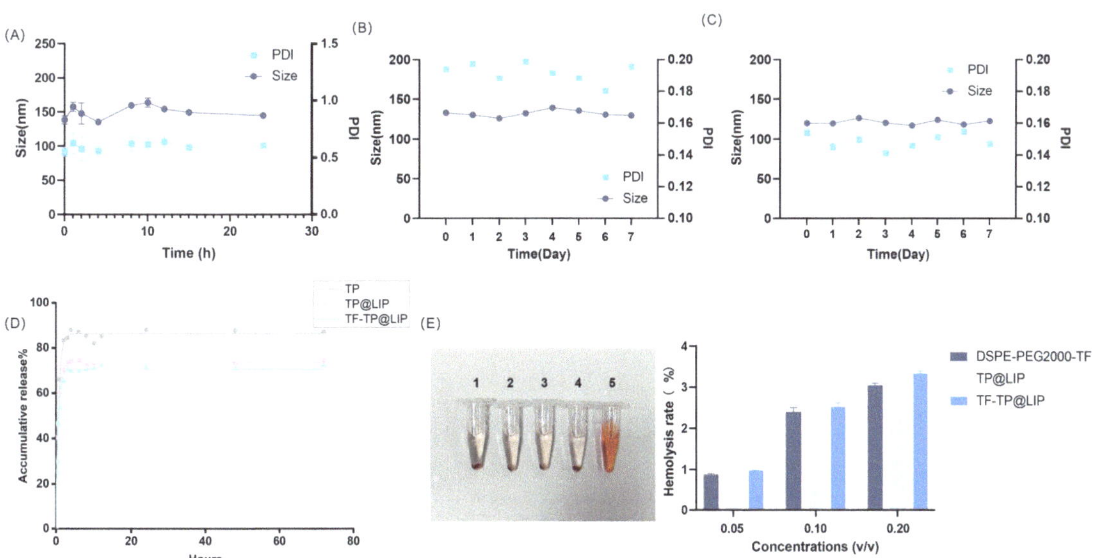

Figure 2. Liposome stability, drug release, and induction of hemolysis. (**A**) Serum stability of TF-TP@LIP; (**B**) Stability of TF-TP@LIP; (**C**) Stability of TP@LIP; (**D**) Release profiles of TP, TP@LIP, TF-TP@LIP; (**E**) Induction of hemolysis. (1. DSPE-PEG2000-TF; 2. TP@LIP; 3. TF-TP@LIP; 4. Negative control group; 5. Positive control group).

3.2. In Vitro Release of TF-TP@LIP

The in vitro release of free TP, TP@LIP, and TF-TP@LIP was investigated using dialysis, and the results are shown in Figure 2D. Due to the rapid release of the TP on the surface of liposomes, there was an initial release of the drug, but this changed to a slow-release phase after 1 h. To gain insight into the release mechanism, three mathematical models, including the first-order, Korsmeyer–Peppas, and Weibull models, were utilized to predict the release behavior of TP; the results are shown in the in Supplementary Materials (Section S5). From the kinetic parameters (Section S5), it appears that the experimental data are best described by the first-order model (Figure 2D). The release experiments further showed that both TP@LIP and TF-TP@Lip had similar drug-release behavior, indicating that the combination with TF had no significant effects on the release behavior of the TP liposomes. The slow-release characteristic at this phase could be attributed to the fact that TP was held by the lipid, and therefore, TP was released gradually from the lipid matrices mainly through dissolution and diffusion. These results showed that TP could be released from the two lipids. Such slow release in the body is not only beneficial to reducing leakage of the drug in the blood circulation before reaching the tumor tissue, but also effectively enhances the stability of the drug.

3.3. Hemolytic Activity of Liposomes

Liposomes are commonly administered intravenously. However, many factors within the bloodstream can disrupt liposomes. For example, phospholipases in the circulatory system hydrolyze phospholipids, high-density lipoproteins disrupt phospholipid membranes, and various modulators of the complement system, such as antibodies and complement, bind to liposomes, thus disrupting the hydrophilic channels in the phospholipid membrane. This leads to leakage of the encapsulated drugs and the entry of water and electrolytes, while accelerating liposome clearance. Therefore, in this study, liposome stability was evaluated by examining liposome–serum interactions [30]. The synthesis of functional phospholipids (DSPE-PEG2000-TF) involves various reactions and reaction materials. To investigate the biosafety of this synthetic material, the hemolysis rate was quantified using a multimode plate reader. It was clear, from the results of the quantitative analysis, that the hemolysis rate was less than 5%, indicating that the synthesis and preparation were safe. The results are shown in Figure 2E.

3.4. Cellular Uptake and Targeting Effects In Vitro

The results of the MTT assay showed that the anti-tumor activity of TF-TP@LIP was dependent on the TF coupling density. When the TF/lipid ratio was 2:1, TF-TP@LIP showed the greatest inhibitory effect on cells (Figure 3A). This may have been due to the high affinity of TF-TP@LIP in binding to TFR to reach saturation. Based on existing studies describing the mechanism of TP endocytosis, C6 dye was used instead of TP to determine the intracellular uptake and distribution of liposomal vectors [39]. TFR was strongly expressed on the surfaces of the human hepatoma HepG2 cells. These cells were used to assess the in vitro targeting of TF-TP@LIP. Targeted fluorescent liposomes (TF-C6@LIP) and non-targeted fluorescent liposomes (C6@LIP) with different coupling densities were also incubated with HepG2 cells for 4 h. The flow cytometry results (Figure 3B) showed that the fluorescence intensity of the targeted liposomes in the cells was significantly higher than that of non-targeted liposomes, thus confirming the fluorescence microscopy findings (Figure 3C). The observations also suggested that the TF modification enhanced the uptake of liposomes by HepG2 cells. In addition, the uptake of TF-C6@LIP by the cells increased with the degree of TF modification (Figure 3B).

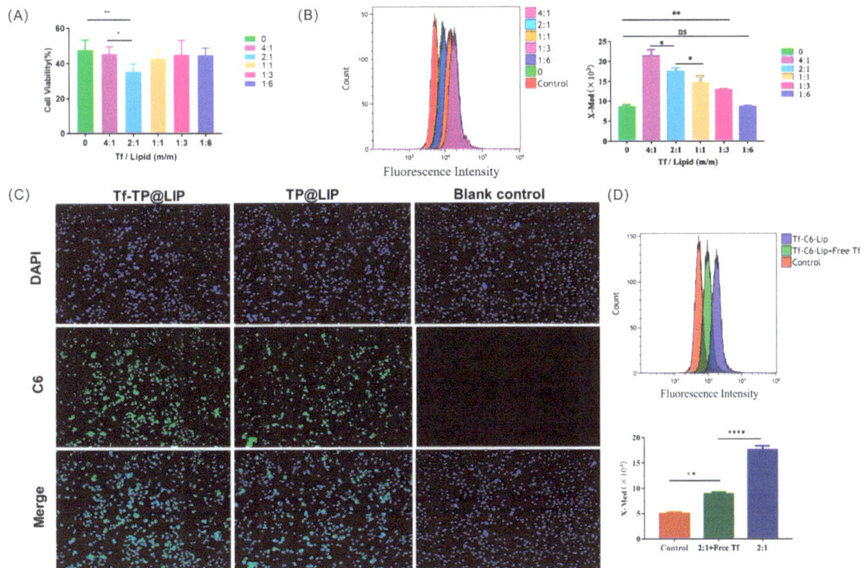

Figure 3. Uptake of liposomes by cells. (**A**) Anti−tumor activity of TF−TP@LIP with different coupling densities; Uptake of TF−TP@LIP investigated by flow cytometry (**B**) and fluorescence microscopy (**C**) (scale bar: 100 μm). (**D**) Uptake of TF−TP@LIP by HepG2 cells. (**** $p < 0.0001$, ** $p < 0.01$, * $p \leq 0.05$, ns, $p > 0.05$).

To verify whether HepG2 cells take up TF-TP@LIP through TFR-mediated endocytosis, a 20-fold excess of free TF was preincubated with the cells in a competitive inhibition experiment. The results (Figure 3D) of this experiment indicated that uptake of TF-C6@LIP by the HepG2 cells was significantly decreased, thereby suggesting that TF-C6@LIP targeting of the TFR enhanced liposome targeting of the cells. Studies [30,40,41] have confirmed that TF-modified liposomes can even enter tumor cells through receptor-mediated endocytosis, indicating that the targeting by TF-TP@LIP could be related to TFR-mediated endocytosis. This would lead to larger amounts of TP entering the tumor tissue due to the strong affinity of the receptor-mediated interaction and its ability to transport the TF-modified nano-targeted liposomes into hepatoma cells.

3.5. Anti-tumor Effects of TF-TP@LIP In Vitro

To investigate the inhibitory effects of TP@LIP on hepatocellular carcinoma cells, TP, TP@LIP, and TF-TP@LIP groups with different mass concentrations were set up to compare the effects of drug administration. As shown in Figure 4A, the IC_{50} values of HepG2 cells in the free TP, TP@LIP, and TF-TP@LIP groups were 90.6 nM, 56.1 nM, and 42.3 nM, respectively. The results further demonstrated that the anti-tumor effects of TP were effectively improved after encapsulation with TF-modified liposomes, and indicated that TF did not undergo structural changes and lose its biological function by the modification, which was the basis for TF-TP@LIP to target tumor tissues. Notably, MTT assays showed that inhibition of HepG2 cell proliferation by TF-TP@LIP was affected by the density of the TF modification. When the TF-to-lipid mass ratio was 2:1, TF-TP@LIP LIP significantly reduced cell viability compared to TP, TP@LIP, and other densities of TF-TP@LIP.

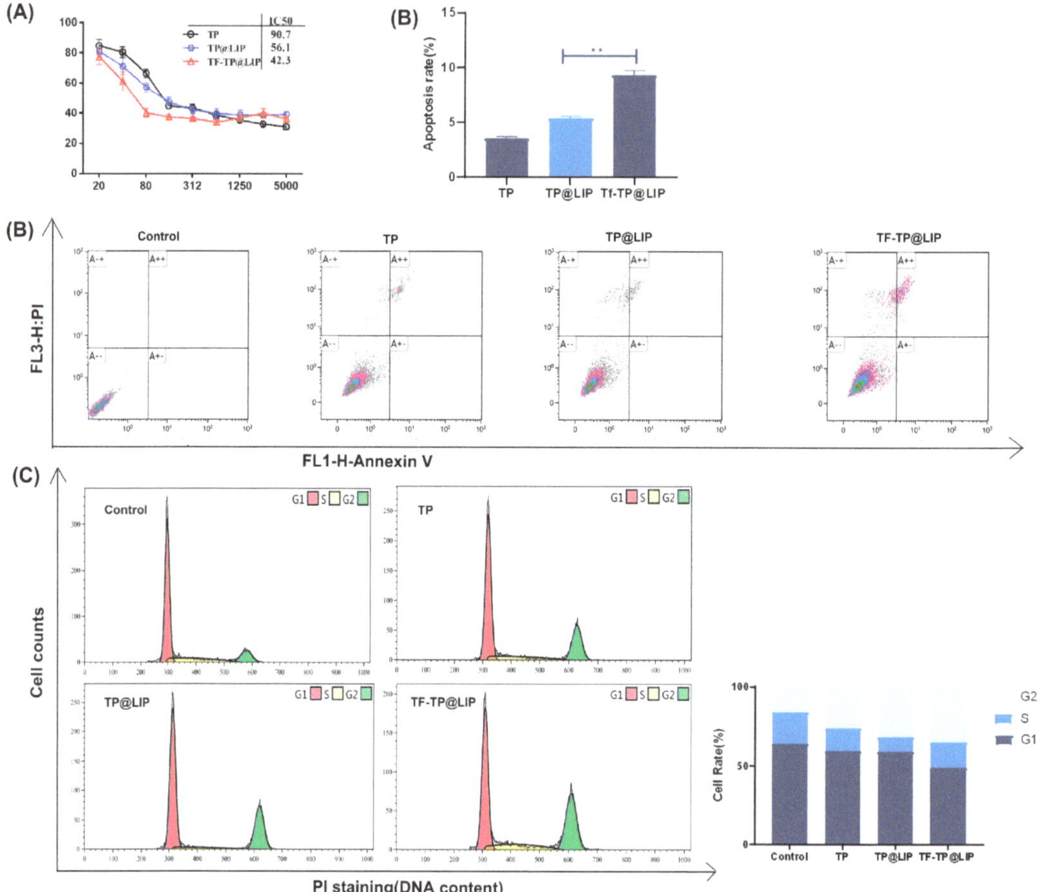

Figure 4. The effects of TF-TP@LIP on the proliferation, apoptosis, and cell cycle of HepG2 cells. (**A**) Cytotoxicity of TP, TP@LIP, and TF-TP@LIP in HepG2 cells after 48 h; (**B**) Flow cytometric analysis and quantification of apoptosis in HepG2 cells following treatment with Control, Free TP, TP@LIP, and TF-TP@LIP for 24 h at an equivalent TP concentration; (**C**) Flow cytometric analysis and quantification of cell cycle distribution of HepG2 cells following treatment with Control, free TP, TP@LIP, and TF-TP@LIP for 24 h at equivalent TP concentrations. (** $p < 0.01$).

Flow cytometry showed that TF-TP@LIP significantly increased the percentage of apoptotic cells in comparison with the free TP and TP@LIP groups (** $p < 0.01$) (Figure 4B). In addition, a comparison of the distribution of cells in the phases of the cell cycle (G0/G1, S, and G2-M phases) among the different treatment groups indicated that treatment with TF-TP@LIP increased the percentage of cells in the G2-M phase and significantly decreased those in the G1 phase, compared with the free TP and TP@LIP groups. This indicated that TF-TP@LIP increased cell cycle arrest and thus inhibited hepatocellular carcinoma cell HepG2 proliferation (Figure 4C). These results confirmed that TF was stably bound to the surface of liposomes and effectively enhanced their uptake by TFR-expressing cancer cells. Studies [42,43] have shown that if DNA damage cannot be repaired during the cell cycle, cells will initiate an apoptotic program, causing the death of cells with DNA damage, as suggested here. Compared with the control group, both the TF-TP@LIP groups showed significantly enhanced cycle arrest and promotion of apoptosis, which could be related to the fact that TF modification enhanced the uptake of the liposomes by hepatoma cells.

3.6. In Vivo Real-Time Imaging

In vivo imaging enables non-invasive real-time monitoring and quantitative analysis of tumor growth and metastasis in living animals. It can also be used to observe preparations targeting tumor organs through in vivo imaging [43]. The present study observed in vivo NIRF imaging of TF-DiR@Lip in tumor-bearing mice to determine its bio-distribution. In this study, DiR, a fluorescence probe, was loaded into liposomes to mimic TP. The intensity of the fluorescence signals represents the accumulation of the drug in the mice. As shown in Figure 5, there was rapid accumulation of the DIR-labeled TF-DIR@LIP over time, with higher fluorescence intensity visible in the tumor tissues 4 h after injection. However, no obvious fluorescence accumulation was seen in the tumor tissues of the DIR-labeled DIR@LIP (non-targeting control) and free DIR mouse groups. Tumor tissue fluorescence was only seen in the DIR@LIP group, demonstrating that TF-TP@LIP was effective for tumor targeting.

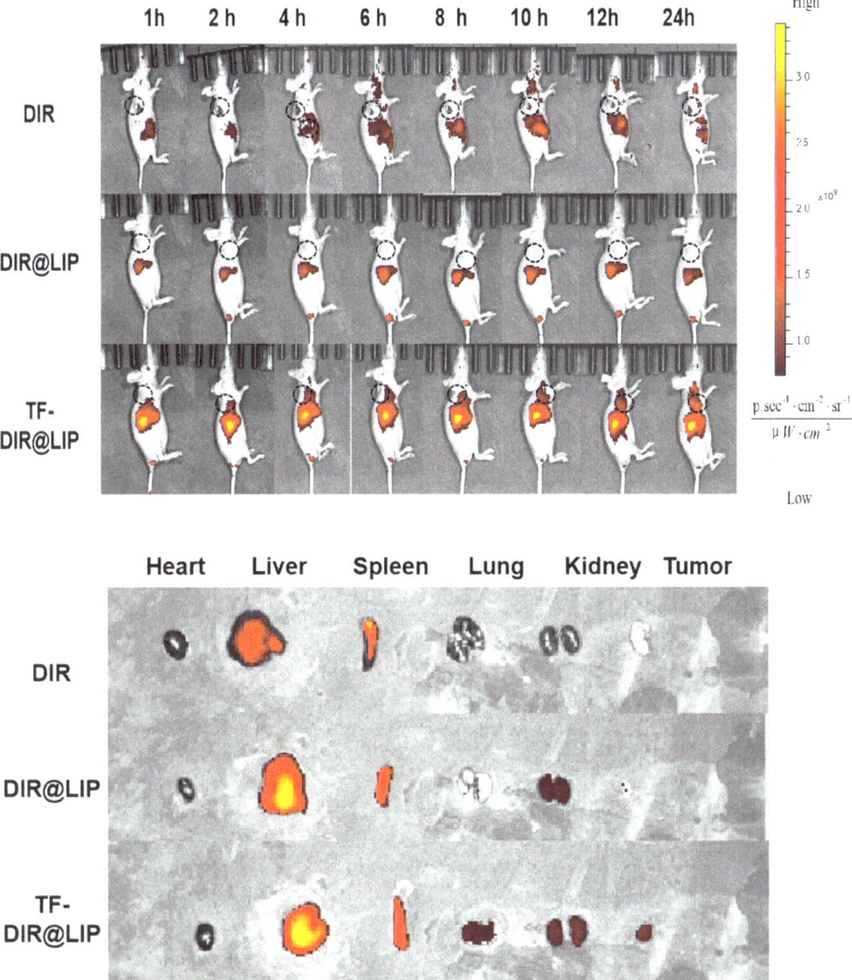

Figure 5. Results of in vivo imaging after tail vein injection of DIR, DIR@LIP, and TF-DIR@LIP in tumor-bearing nude mice. Tissue distribution 24 h after injection in the different treatment groups.

3.7. In Vivo Anti-Hepatoma Efficacy

When the tumor volumes reached 120–150 mm^3, the mice were randomly divided into four groups according to the principle of even distribution of tumor size, namely, the control, TP, TP@LIP, and TF-TP@LIP groups. The drug was administered through the tail vein at a dose of 0.2 mg/kg TP in a volume of 200 µL. A total of seven doses was administered. Tumor suppression was monitored for 18 days, and the body weights, as well as tumor volumes, in each group were measured and recorded every other day from the first day of administration. Compared with other groups, there was significant inhibition of tumor growth in the TF-TP@ LIP group. At the end of the treatment, the tumor volumes and tumor weights of the TF-TP@ LIP group were visibly smaller than those of the other groups (Figure 6). After drug administration, the main organs of nude mice, namely, the heart, liver, spleen, lung, and kidney, were collected for evaluation using H&E staining of tissue sections. The morphological changes of the organ tissues of each group are shown in Figure 7. The organ tissues of the free TP group showed obvious changes, such as the infiltration of inflammatory cells and disordered tissue, indicating toxicity of the free drug. Compared with the control group, no abnormal changes were found in the major organs of the mice in the TF-TP@ LIP treatment group. In addition, the values of the main liver and kidney parameters in the TF-TP@ LIP group were within the normal range and did not differ significantly from those in the control group, demonstrating the safety of the formulation.

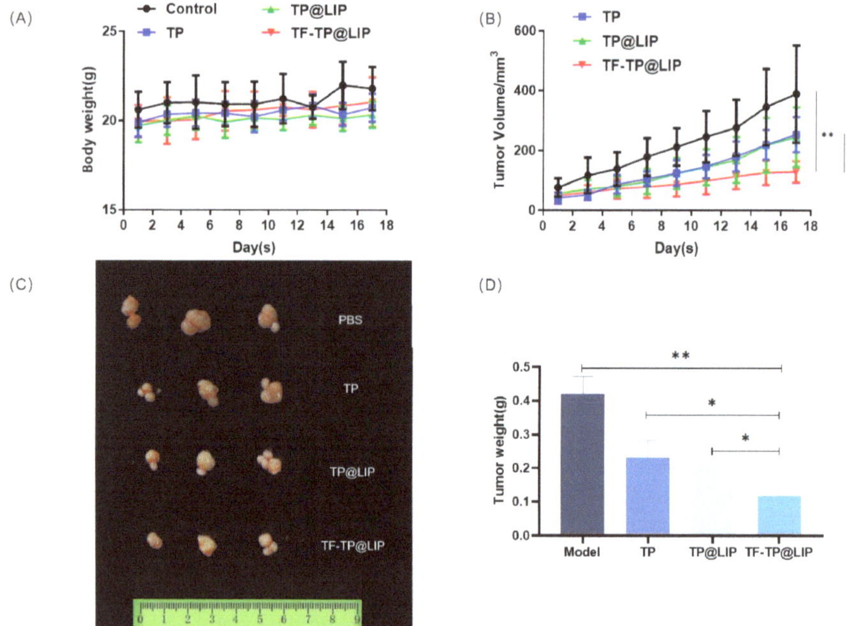

Figure 6. Growth inhibition of HepG2-cell xenograft tumors in the different treatment groups. (**A**) Body weight; (**B**) Tumor volume; (**C,D**) Morphology tumor tissue after treatment in different treatment groups (** $p < 0.01$, * $p < 0.05$).

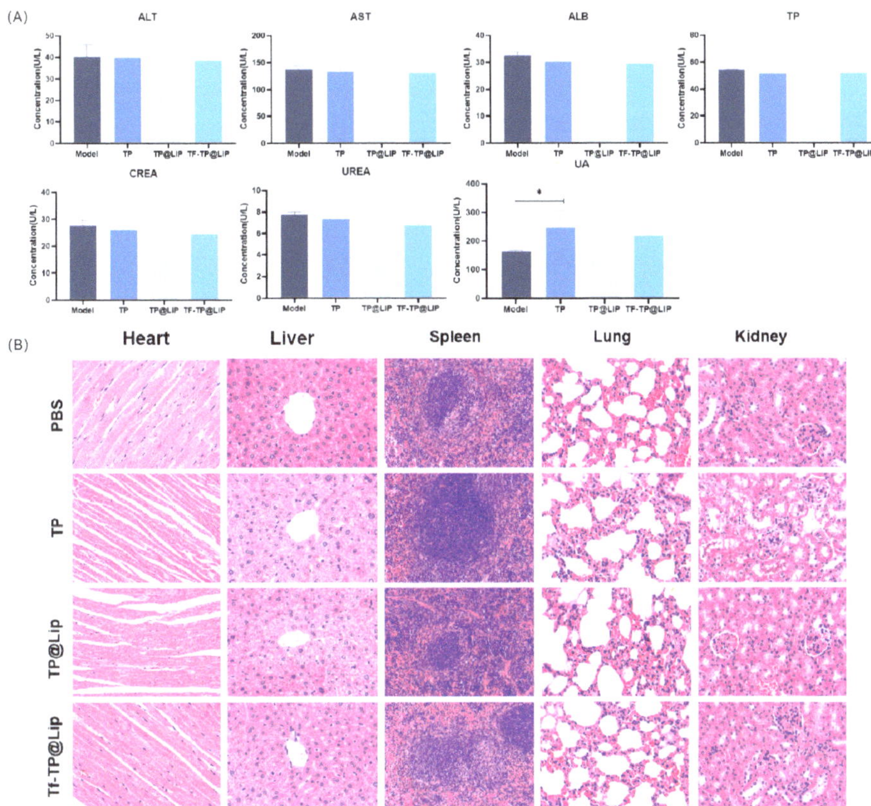

Figure 7. Biochemical indices and histological sections. (**A**) Blood biochemical tests; (**B**) H&E-stained sections of heart, liver, spleen, lung, and kidney tissues from mice in the different treatment groups (400×). (* $p < 0.05$).

4. Conclusions

Drug delivery systems using transferrin-mediated liposome targeting show promise for cancer therapy. In this study, a TF-modified liposome, targeting the TFR on the surface of hepatocellular carcinoma cells, was designed for the delivery of TP. This active targeting delivery system had good biocompatibility and low toxicity. Compared with free TP and non-targeted TP liposomes, the addition of TF significantly enhanced hepatoma cell cytotoxicity, as well as showed enhanced uptake of the TF-TP@LIP in vitro. The inclusion of TF also increased cell cycle arrest and promoted apoptosis in hepatoma cells. The results of in vivo experiments further showed that TF-TP@LIP had stronger anti-tumor effects in vivo compared with the free TP TP@LIP groups, and the results of in vivo and in vitro studies were consistent. In addition, the toxicity of TF-TP@LIP was evaluated by changes in tissue morphology, observed by H&E staining of pathological sections. Compared with the control group, pathological changes were observed in sections of heart, liver, lung, and kidney tissues in the TP group due to toxicity and side effects, while the tissue structures in the TF-TP@LIP group were normal, indicating the relative safety of the preparation and that the toxicity of TF-TP@LIP was less than that of free TP. Therefore, the use of transferrin for targeting TP-containing liposomes may prove to be an effective nanomedicine for the treatment of liver cancers.

Supplementary Materials: The following supporting information can be downloaded at: https://www.mdpi.com/article/10.3390/biomedicines11102869/s1, Figure S1: Size histogram of TEM images of TP@LIP; Figure S2: Size histogram of TEM images of TF-TP@LIP; Figure S3: Elution Curve of TP(A) and TP@LIP(B); Table S1: The choice of centrifugal strength; Table S2: The choice of centrifugal time; Table S3: Ultrafiltration recovery rate of free TP.

Author Contributions: X.Z. and Y.Y.: Conceptualization, Methodology, Software, Roles/writing-original draft, Visualization, Formal analysis. X.S. and Y.X.: Conceptualization, Visualization, Investigation. Y.L.: Data curation, Validation, Writing—review and editing. T.Z.: Funding acquisition, Data Curation, Investigation. Y.W.: Resources, Data Curation. L.D.: Supervision, Project administration, Writing—Review and Editing. All authors have read and agreed to the published version of the manuscript.

Funding: This research was funded by the National Natural Science Foundation of China, grant number 81102825.

Institutional Review Board Statement: The animal study protocol was approved by the Ethics Committee of Nanjing University of Chinese Medicine (No. 202107A034).

Informed Consent Statement: Not applicable.

Data Availability Statement: The data are not publicly available due to the contents of the graduation thesis are subject to confidentiality.

Acknowledgments: We sincerely thank Tingming Fu's research group for their help and guidance in our experiment. We would like to thank the instrument platform of Nanjing University of Chinese Medicine for the experimental equipment support.

Conflicts of Interest: The authors declare no conflict of interest.

References

1. Ladju, R.B.; Ulhaq, Z.S.; Soraya, G.V. Nanotheranostics: A powerful next-generation solution to tackle hepatocellular carcinoma. *World J. Gastroenterol.* **2022**, *28*, 176–187. [CrossRef]
2. Baffy, G. The impact of network medicine in gastroenterology and hepatology. *Clin. Gastroenterol. Hepatol.* **2013**, *11*, 1240–1244. [CrossRef]
3. Maluccio, M.; Covey, A. Recent progress in understanding, diagnosing, and treating hepatocellular carcinoma. *CA Cancer J. Clin.* **2012**, *62*, 394–399. [CrossRef]
4. Zhao, C.; Wu, X.; Chen, J.; Qian, G. The therapeutic effect of IL-21 combined with IFN-γ inducing CD4(+)CXCR5(+)CD57(+)T cells differentiation on hepatocellular carcinoma. *J. Adv. Res.* **2022**, *36*, 89–99. [CrossRef]
5. Li, Z.; Yang, G.; Han, L.; Wang, R.; Gong, C.; Yuan, Y. Sorafenib and triptolide loaded cancer cell-platelet hybrid membrane-camouflaged liquid crystalline lipid nanoparticles for the treatment of hepatocellular carcinoma. *J. Nanobiotechnol.* **2021**, *19*, 360. [CrossRef]
6. Kupchan, S.M.; Court, W.A.; Dailey, R.G., Jr.; Gilmore, C.J.; Bryan, R.F. Triptolide and tripdiolide, novel antileukemic diterpenoid triepoxides from Tripterygium wilfordii. *J. Am. Chem. Soc.* **1972**, *94*, 7194–7195. [CrossRef]
7. Hou, W.; Liu, B.; Xu, H. Triptolide: Medicinal chemistry, chemical biology and clinical progress. *Eur. J. Med. Chem.* **2019**, *176*, 378–392. [CrossRef]
8. Kim, S.T.; Kim, S.Y.; Lee, J.; Kim, K.; Park, S.H.; Park, Y.S.; Lim, H.Y.; Kang, W.K.; Park, J.O. Triptolide as a novel agent in pancreatic cancer: The validation using patient derived pancreatic tumor cell line. *BMC Cancer* **2018**, *18*, 1103. [CrossRef]
9. Zhang, Y.Q.; Shen, Y.; Liao, M.M.; Mao, X.; Mi, G.J.; You, C.; Guo, Q.Y.; Li, W.J.; Wang, X.Y.; Lin, N.; et al. Galactosylated chitosan triptolide nanoparticles for overcoming hepatocellular carcinoma: Enhanced therapeutic efficacy, low toxicity, and validated network regulatory mechanisms. *Nanomedicine* **2019**, *15*, 86–97. [CrossRef]
10. Meng, C.; Zhu, H.; Song, H.; Wang, Z.; Huang, G.; Li, D.; Ma, Z.; Ma, J.; Qin, Q.; Sun, X.; et al. Targets and molecular mechanisms of triptolide in cancer therapy. *Chin. J. Cancer Res.* **2014**, *26*, 622–626. [CrossRef]
11. Park, S.W.; Kim, Y.I. Triptolide induces apoptosis of PMA-treated THP-1 cells through activation of caspases, inhibition of NF-κB and activation of MAPKs. *Int. J. Oncol.* **2013**, *43*, 1169–1175. [CrossRef]
12. Liu, H.J.; Wang, M.; Hu, X.; Shi, S.; Xu, P. Enhanced Photothermal Therapy through the In Situ Activation of a Temperature and Redox Dual-Sensitive Nanoreservoir of Triptolide. *Small* **2020**, *16*, e2003398. [CrossRef]
13. Pefanis, E.; Wang, J.; Rothschild, G.; Lim, J.; Kazadi, D.; Sun, J.; Federation, A.; Chao, J.; Elliott, O.; Liu, Z.P.; et al. RNA exosome-regulated long non-coding RNA transcription controls super-enhancer activity. *Cell* **2015**, *161*, 774–789. [CrossRef]
14. Qin, G.; Li, P.; Xue, Z. Triptolide induces protective autophagy and apoptosis in human cervical cancer cells by downregulating Akt/mTOR activation. *Oncol. Lett.* **2018**, *16*, 3929–3934. [CrossRef]

15. Fu, Q.; Huang, X.; Shu, B.; Xue, M.; Zhang, P.; Wang, T.; Liu, L.; Jiang, Z.; Zhang, L. Inhibition of mitochondrial respiratory chain is involved in triptolide-induced liver injury. *Fitoterapia* **2011**, *82*, 1241–1248. [CrossRef]

16. Cao, L.J.; Hou, Z.Y.; Li, H.D.; Zhang, B.K.; Fang, P.F.; Xiang, D.X.; Li, Z.H.; Gong, H.; Deng, Y.; Ma, Y.X.; et al. The Ethanol Extract of Licorice (*Glycyrrhiza uralensis*) Protects against Triptolide-Induced Oxidative Stress through Activation of Nrf2. *Evid. Based Complement. Altern. Med.* **2017**, *2017*, 2752389. [CrossRef]

17. Wang, S.R.; Chen, X.; Ling, S.; Ni, R.Z.; Guo, H.; Xu, J.W. MicroRNA expression, targeting, release dynamics and early-warning biomarkers in acute cardiotoxicity induced by triptolide in rats. *Biomed. Pharmacother.* **2019**, *111*, 1467–1477. [CrossRef]

18. Ma, B.; Qi, H.; Li, J.; Xu, H.; Chi, B.; Zhu, J.; Yu, L.; An, G.; Zhang, Q. Triptolide disrupts fatty acids and peroxisome proliferator-activated receptor (PPAR) levels in male mice testes followed by testicular injury: A GC-MS based metabolomics study. *Toxicology* **2015**, *336*, 84–95. [CrossRef]

19. Sun, R.; Dai, J.; Ling, M.; Yu, L.; Yu, Z.; Tang, L. Delivery of triptolide: A combination of traditional Chinese medicine and nanomedicine. *J. Nanobiotechnol.* **2022**, *20*, 194. [CrossRef]

20. Yu, L.; Wang, Z.; Mo, Z.; Zou, B.; Yang, Y.; Sun, R.; Ma, W.; Yu, M.; Zhang, S.; Yu, Z. Synergetic delivery of triptolide and Ce6 with light-activatable liposomes for efficient hepatocellular carcinoma therapy. *Acta Pharm. Sin. B* **2021**, *11*, 2004–2015. [CrossRef]

21. Fulton, M.D.; Najahi-Missaoui, W. Liposomes in Cancer Therapy: How Did We Start and Where Are We Now. *Int. J. Mol. Sci.* **2023**, *24*, 6615. [CrossRef]

22. Wei, Y.; Gu, X.; Cheng, L.; Meng, F.; Storm, G.; Zhong, Z. Low-toxicity transferrin-guided polymersomal doxorubicin for potent chemotherapy of orthotopic hepatocellular carcinoma in vivo. *Acta Biomater.* **2019**, *92*, 196–204. [CrossRef]

23. Kakudo, T.; Chaki, S.; Futaki, S.; Nakase, I.; Akaji, K.; Kawakami, T.; Maruyama, K.; Kamiya, H.; Harashima, H. Transferrin-modified liposomes equipped with a pH-sensitive fusogenic peptide: An artificial viral-like delivery system. *Biochemistry* **2004**, *43*, 5618–5628. [CrossRef]

24. Kobayashi, T.; Ishida, T.; Okada, Y.; Ise, S.; Harashima, H.; Kiwada, H. Effect of transferrin receptor-targeted liposomal doxorubicin in P-glycoprotein-mediated drug resistant tumor cells. *Int. J. Pharm.* **2007**, *329*, 94–102. [CrossRef]

25. Daniels, T.R.; Bernabeu, E.; Rodríguez, J.A.; Patel, S.; Kozman, M.; Chiappetta, D.A.; Holler, E.; Ljubimova, J.Y.; Helguera, G.; Penichet, M.L. The transferrin receptor and the targeted delivery of therapeutic agents against cancer. *Biochim. Biophys. Acta* **2012**, *1820*, 291–317. [CrossRef]

26. Choi, C.H.; Alabi, C.A.; Webster, P.; Davis, M.E. Mechanism of active targeting in solid tumors with transferrin-containing gold nanoparticles. *Proc. Natl. Acad. Sci. USA* **2010**, *107*, 1235–1240. [CrossRef]

27. Koshkaryev, A.; Piroyan, A.; Torchilin, V.P. Increased apoptosis in cancer cells in vitro and in vivo by ceramides in transferrin-modified liposomes. *Cancer Biol. Ther.* **2012**, *13*, 50–60. [CrossRef]

28. Torchilin, V.P. Recent approaches to intracellular delivery of drugs and DNA and organelle targeting. *Annu. Rev. Biomed. Eng.* **2006**, *8*, 343–375. [CrossRef]

29. Li, X.; Ding, L.; Xu, Y.; Wang, Y.; Ping, Q. Targeted delivery of doxorubicin using stealth liposomes modified with transferrin. *Int. J. Pharm.* **2009**, *373*, 116–123. [CrossRef]

30. Sakpakdeejaroen, I.; Somani, S.; Laskar, P.; Mullin, M.; Dufès, C. Regression of Melanoma Following Intravenous Injection of Plumbagin Entrapped in Transferrin-Conjugated, Lipid-Polymer Hybrid Nanoparticles. *Int. J. Nanomed.* **2021**, *16*, 2615–2631. [CrossRef]

31. Gowda, B.H.J.; Ahmed, M.G.; Alshehri, S.A.; Wahab, S.; Vora, L.K.; Singh Thakur, R.R.; Kesharwani, P. The cubosome-based nanoplatforms in cancer therapy: Seeking new paradigms for cancer theranostics. *Environ. Res.* **2023**, *273*, 116894. [CrossRef]

32. Wang, K.; Shang, F.; Chen, D.; Cao, T.; Wang, X.; Jiao, J.; He, S.; Liang, X. Protein liposomes-mediated targeted acetylcholinesterase gene delivery for effective liver cancer therapy. *J. Nanobiotechnol.* **2021**, *19*, 31. [CrossRef]

33. Yang, Y.; Liu, B.; Su, X.; Qiao, P.; Di, L.; Zhao, X. Preparation of transferrin modified triptolide liposome and in vitro evaluation. *Chin. Tradit. Herb. Drugs* **2022**, *53*, 687–695. (In Chinese)

34. Lin, Y.; Wan, Y.; Du, X.; Li, J.; Wei, J.; Li, T.; Li, C.; Liu, Z.; Zhou, M.; Zhong, Z. TAT-modified serum albumin nanoparticles for sustained-release of tetramethylpyrazine and improved targeting to spinal cord injury. *J. Nanobiotechnol.* **2021**, *19*, 28. [CrossRef]

35. Maritim, S.; Boulas, P.; Lin, Y. Comprehensive analysis of liposome formulation parameters and their influence on encapsulation, stability and drug release in glibenclamide liposomes. *Int. J. Pharm.* **2021**, *592*, 120051. [CrossRef]

36. Driscoll, D.F. Globule-size distribution in injectable 20% lipid emulsions: Compliance with USP requirements. *Am. J. Health Syst. Pharm.* **2007**, *64*, 2032–2036. [CrossRef]

37. Kuai, R.; Yuan, W.; Qin, Y.; Chen, H.; Tang, J.; Yuan, M.; Zhang, Z.; He, Q. Efficient delivery of payload into tumor cells in a controlled manner by TAT and thiolytic cleavable PEG co-modified liposomes. *Mol. Pharm.* **2010**, *7*, 1816–1826. [CrossRef]

38. Duncan, R.; Gaspar, R. Nanomedicine(s) under the microscope. *Mol. Pharm.* **2011**, *8*, 2101–2141. [CrossRef]

39. Zheng, Y.; Kong, F.; Liu, S.; Liu, X.; Pei, D.; Miao, X. Membrane protein-chimeric liposome-mediated delivery of triptolide for targeted hepatocellular carcinoma therapy. *Drug Deliv.* **2021**, *28*, 2033–2043. [CrossRef]

40. Sakpakdeejaroen, I.; Somani, S.; Laskar, P.; Mullin, M.; Dufès, C. Transferrin-bearing liposomes entrapping plumbagin for targeted cancer therapy. *J. Interdiscip. Nanomed.* **2019**, *4*, 54–71. [CrossRef]

41. Liu, X.; Dong, S.; Dong, M.; Li, Y.; Sun, Z.; Zhang, X.; Wang, Y.; Teng, L.; Wang, D. Transferrin-conjugated liposomes loaded with carnosic acid inhibit liver cancer growth by inducing mitochondria-mediated apoptosis. *Int. J. Pharm.* **2021**, *607*, 121034. [CrossRef]

42. Sutar, Y.B.; Mali, J.K.; Telvekar, V.N.; Rajmani, R.S.; Singh, A. Transferrin conjugates of antitubercular drug isoniazid: Synthesis and in vitro efficacy. *Eur. J. Med. Chem.* **2019**, *183*, 111713. [CrossRef]
43. Weissleder, R.; Pittet, M.J. Imaging in the era of molecular oncology. *Nature* **2008**, *452*, 580–589. [CrossRef]

Article

Application of I-Optimal Design for Modeling and Optimizing the Operational Parameters of Ibuprofen Granules in Continuous Twin-Screw Wet Granulation

Jie Zhao, Geng Tian and Haibin Qu *

Pharmaceutical Informatics Institute, College of Pharmaceutical Sciences, Zhejiang University, Hangzhou 310058, China; zhaojie_1021@zju.edu.cn (J.Z.); iamtiangeng@zju.edu.cn (G.T.)
* Correspondence: quhb@zju.edu.cn; Tel./Fax: +86-571-88208428

Abstract: The continuous twin-screw wet granulation (TSWG) process was investigated and optimized with prediction-oriented I-optimal designs. The I-optimal designs can not only obtain a precise estimation of the parameters that describe the effect of five input process parameters, including the screw speed, liquid-to-solid (L/S) ratio, TSWG feed rate, and numbers of the 30° and 60° mixing elements, on the granule quality in a TSWG process, but it can also provide a prediction of the response to determine the optimum operating conditions. Based on the constraints of the desired granule properties, a design space for the TSWG was determined, and the ranges of the operating parameters were defined. An acceptable degree of prediction was confirmed through validation experiments, demonstrating the reliability and effectiveness of using the I-optimal design method to study the TSWG process. The I-optimal design method can accelerate the screening and optimization of the TSWG process.

Keywords: twin-screw wet granulation (TSWG); I-optimal design; continuous manufacturing; process understanding; design space

Citation: Zhao, J.; Tian, G.; Qu, H. Application of I-Optimal Design for Modeling and Optimizing the Operational Parameters of Ibuprofen Granules in Continuous Twin-Screw Wet Granulation. *Biomedicines* **2023**, *11*, 2030. https://doi.org/10.3390/biomedicines11072030

Academic Editors: Ali Nokhodchi and Concettina La Motta

Received: 8 June 2023
Revised: 6 July 2023
Accepted: 14 July 2023
Published: 19 July 2023

1. Introduction

The pharmaceutical industry is currently undergoing a paradigm shift from traditional batch production to continuous manufacturing (CM) [1,2]. Twin-screw wet granulation (TSWG) is a typical method of continuous granulation; it has the advantages of a high processing volume, higher production efficiency, short residence time, and better mixing and controlling processes compared with traditional batch manufacturing [3]. In addition, TSWG can be readily integrated into the CM of pharmaceutical dosage forms and provides easier process scale-up, better quality assurance, low production costs, and material waste [4].

During the TSWG process, the powder and liquid binders are added at the entrance of the granulator through powder feeders and nozzles, respectively, and then the low-strength wetted agglomerates are formed and blended inside according to the conveying and mixing profile of the system. These primary agglomerates are broken apart into smaller rounded granules, allowing for growth through layering as the primary powder adheres to the surface. Uniform granules are produced by breakage and layering of the primary agglomerates under the two main rate-controlling processes [5]. In this granulation process, the critical quality attributes (CQAs) of the granules produced by TSWG are affected by many parameters. The present literature shows that the materials and binder properties, the process parameters, such as the liquid-to-solid (L/S) ratio, the rotation speed of the screw, the material feed rates, and the configuration of the screw elements (conveying elements, kneading elements, distributive elements, etc.) are the key process variables of the TSWG process [6].

An accurate model for the TSWG process would provide a thorough understanding of the process dynamics and can be used to optimize the operating conditions of the TSWG process [1]. Therefore, it is important for the process understanding to clarify the influence mechanism of the process conditions on the particle properties. Experimental studies using the design of experiment (DoE) method are typically performed to analyze the mechanism involved during the granulation process and have been extensively used to develop models to quantify the relationship between critical process parameters (CPPs) and CQAs of particles. Different DoE methods have been performed to evaluate the influence of material properties and process variables on the characteristics of granules [6–9]. Seem et al. summarized the comprehensive review of the experimental twin-screw granulator literature, indicating the complex interactions between the role of the screw element type, screw configuration, feed formulation, and liquid flow rates on the granules [10]. Liu et al. analyzed the effects of throughput, screw speed, and screw components on the properties of granules and tablets in the TSWG process using a Box–Behnken experimental design, and the design space for TSWG was defined and validated to demonstrate the robustness of the optimal operating conditions [7]. Kumar et al. used a full-factor experimental design to study the effects of process parameters (feed rate, screw speed, and L/S ratio) and equipment parameters (number of kneading elements and staggered angles) on residence time distribution, solid–liquid mixing, and final particle size distribution in the TSWG process. The results showed that it is necessary to strive for a balance between material throughput and screw speed to achieve a specific granulation time and solid–liquid mixing to achieve a high granulation yield [11]. Meng et al. studied the relationship between liquid content, throughput, and rotational speed with key performance indicators such as particle size, porosity, flowability, and particle morphology in the acetaminophen formula TSWG process using a face-centered cubic experimental design method. The results showed that the screw configuration should be fully utilized to achieve different particle characteristics [12]. It can be noticed that the previous research mainly focused on certain aspects of material properties, process parameters, or screw configurations, but it must be added that, in the case of process optimization, all aspects of the critical process parameters (CPPs) need to be considered at the same time, making it a complex process for the traditional approach that can be managed to balance the screening design and prediction optimization. In addition, the process parameters of TSWG include both discontinuous factors (e.g., screw elements configuration) and continuous factors (e.g., screw speed), and it is difficult for the common DoE method to handle.

Compared to other DoE methods, the optimal design approach can handle different types of models and experimental factors, such as continuous factors, categorical factors, and mixture factors [13,14]. It can help to obtain an improved process understanding and characterization and make predictions by the models at the same time, which provides a new and effective tool for the study of TSWG. Willecke et al. used D-optimal designs as a non-standard experimental design to investigate the impact of fillers properties (three principal components derived from eight selected pharmaceutical fillers), binder type, and binder concentration in granules together on the properties of the granule and tablet. The results showed that the filler properties mainly affected the granule characteristics, such as particle size, friability, and specific surface area, and the binder type and concentration had a relevant influence on granule flowability, friability, and compactibility [15]. Stauffer et al. used D-optimal designs to investigate the impact of raw material variability upon the granule size distribution, density, and flowability of granules produced via TSWG. Three principal components from raw material variability together with screw speed and L/S ratio were used as factors, and then the significant factors obtained from analysis were used to determine the design space of the TSWG process to reduce the active pharmaceutical ingredient (API) batch-to-batch variability [16]. Meng et al. developed an interaction model between input and output variables in the continuous TSWG of anhydrous caffeine particles using a D-optimal design and stepwise regression. Response surface design was used to study the dependence of key quality attributes of particles and tablets (D_{10},

D_{50}, D_{90}, loose density, compacted density, and Hausner ratio) on selected key process parameters (L/S, barrel temperature, and screw speed) and screw configuration. The results showed that the impact of throughput and barrel temperature was relatively less than the L/S ratio. Higher liquid saturation leads to narrower particle size distribution, smaller porosity, enhanced flowability, and decreased tablet tensile strength, but slower drug release [17]. The usefulness of prediction-oriented optimal design criteria, such as the I-optimality criterion and the G-optimality criterion in the response surface of TSWG, is more reasonable. The I-optimality criterion can determine regions in the design space where the response falls within an acceptable range by minimizing the average variance in prediction over the design space. To build an accurate response surface in predicting the response and determining optimum operating conditions, the I-optimal design is more suitable [14]. However, the use of an I-optimal design for the process optimization of TSWG has not been reported.

This work proposed a systematic understanding of the TSWG process by the DoE method to investigate the effect of process parameters on granules properties. Ibuprofen was selected as the model drug due to its high market demand. Five process variables of screw speed, L/S ratio, powder feed rate, and screw element number ($30°$ kneading and $60°$ kneading) of a twin-screw granulator were investigated using the I-optimal experimental design method. The analysis of variance was used to quantify the process response to variation in the parameters. The predictability of the developed models was validated within and without the defined design space. The I-optimal design was used for the first time to acquire better insight into the TSWG process.

2. Materials and Methods

2.1. Materials

The model formulation used in this study comprised 10% ibuprofen (Lot No. A2107096, Aladdin Biochemical Technology Co., Ltd., Shanghai, China) as API with a melting point of 75~78 °C and a density of 1.03 g/cm^3; 40% lactose (Lot No. 1320020470) and 40% microcrystalline cellulose (MCC, Lot No. 20200719) were used as fillers (Infinitus Company Ltd., Guangzhou, China). Additionally, 10% polyvinylpyrrolidone (PVP K-30, ISP Technologies, Inc., Wayne, NJ, USA) was used as a dry binder and premixed with the raw materials using a V-mixer (Chenli powder equipment Co., Ltd., Wuxi, China). Distilled water was added as the granulation liquid.

2.2. Continuous Twin-Screw Wet Granulation

TSWG experiments were performed on a Pharma 11 twin-screw granulator (Thermo Fisher Scientific Inc., Waltham, MA, USA). The granulator comprised two co-rotating screws with a diameter of 11 mm and a length-to-diameter (L/D) ratio of 40:1. The Pharma 11 granulator employed the types of screw elements including conveying, kneading, and chopping elements. The pre-blend raw materials were fed into the barrel of the granulator by a single-screw feeder which was controlled by a digital governor. The relationship between the opening value (x, %) and mass flow (y, g/min) followed a linear relationship $y = 0.6560x - 5.1585$ ($R^2 = 0.9990$). Granulation liquid was transferred to the liquid feed nozzle by a peristaltic pump (WCL Fluid Technology Co., Ltd., Changzhou, China) through silicon tubing (2.4 mm × 5.6 mm, 19 #). The relationship between the rotating speed (x, r/min) and mass flow (y, g/min) followed a linear relationship $y = 0.5822x - 0.5756$ ($R^2 = 0.9959$).

The TSWG setup is illustrated in Figure 1, which shows the Pharma 11 co-rotating parallel twin-screw granulator used for manufacturing granules. The powder inlet port was located on top of the conveying zone, and the liquid nozzle was adjacent to it.

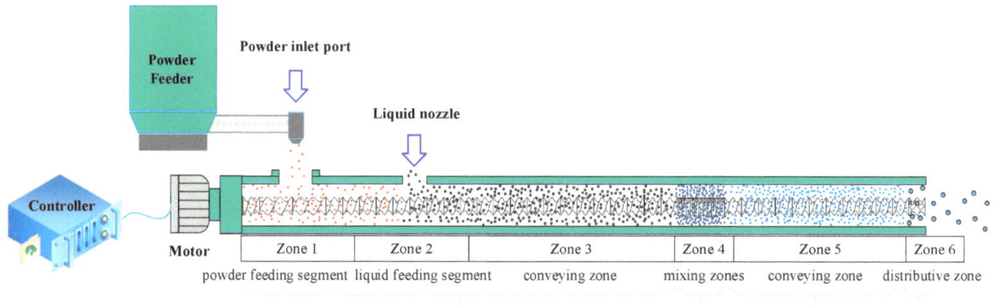

Figure 1. Setup and screw configuration of TSWG.

The screw configuration is a simple system composed of conveying zone, kneading zone, and discharge zone. As the screw configuration shows in Figure 1, the conveying zone comprised two types of feed screw elements, i.e., a long-hellx feed screw (2 L/D) and a feed screw (1 L/D); the kneading zone was a combination of mixing elements in 90° and 0° with mixing elements in forward stagger angles of 30° and 60° and one distributive feed screw element at the end of the granulator barrel in order to reduce the number of oversized agglomerates.

In the granulation process, the temperature of the jacket around the granulator barrel was kept constant at 25 °C using an active cooling system, and other parameters were set according to the designed experiments.

2.3. Experimental Design

I-optimal design was structured based on an iterative search algorithm and provided lower average prediction variance across the region of experimentation, which is desirable for response surface methods (RSMs) as prediction is important. In the current study, five process variables, screw speed (X_1), L/S ratio (X_2), powder feed rate (X_3), number of the 60° mixing elements (X_4), and number of the 30° mixing elements (X_5), were systematically investigated to understand their effects on granule properties and optimize the granulation process. The common scale was utilized to describe each variable, whereby the highest coded value was equal to +1, the middle coded value was equal to 0, and the lowest was assigned a value of −1. Table 1 shows the independent variables and their levels in the DoE. The CQAs of TSWG were the moisture content, D_{50}, span, and yield, and they were identified as Y_1–Y_4, respectively.

Table 1. Independent variables and their levels in I-optimal design.

Independent Variables		Levels		
		Minimum (−1)	Intermediate (0)	Maximum (+1)
Continuous	X_1: screw speed (r/min)	200	250	300
	X_2: L/S ratio	0.35	0.4	0.45
	X_3: powder feed rate (%)	20	25	30
Discrete numeric	X_4: number of the 60° mixing elements (pcs)	4	7	10
	X_5: number of the 30° mixing elements (pcs)	3	6	9

Considering that the independent variables included both continuous numeric variables (X_1, X_2, X_3) and discrete numeric variables (X_4, X_5), an I-optimal design which contained 27 runs was performed, as shown in Table 2. The I-optimal design determined important factors and fit a quadratic polynomial model to the response. The results from the I-optimal design were adopted to build the nonlinear quadratic mathematical model, as follows:

$$y = b_0 + b_1 X_1 + b_2 X_2 + \cdots + b_n X_n + \sum b_{ik} X_i X_k + \sum b_{ii} X_i^2 \tag{1}$$

Table 2. Independent variable sets in I-optimal design and responses.

Run Order	Pattern	Independent Variables				
		X_1	X_2	X_3	X_4	X_5
1	0--++	250	0.35	20	10	9
2	0----	250	0.35	20	4	3
3	+-0-+	300	0.35	25	4	9
4	0+-+0	250	0.45	20	10	6
5	-++00	200	0.45	30	7	6
6	0-00-	250	0.35	25	7	3
7	+++-0	300	0.45	30	4	6
8	--0+-	200	0.35	25	10	3
9	-0-0-	200	0.4	20	7	3
10	-00++	200	0.4	25	10	9
11	+0--+	300	0.4	20	4	9
12	+-+0-	300	0.35	30	7	3
13	00+-+	250	0.4	30	4	9
14	--+0+	200	0.35	30	7	9
15	-+--+	200	0.45	20	4	9
16	00000	250	0.4	25	7	6
17	+-++0	300	0.35	30	10	6
18	00000	250	0.4	25	7	6
19	+0-+-	300	0.4	20	10	3
20	00000	250	0.4	25	7	6
21	++0++	300	0.45	25	10	9
22	0+++-	250	0.45	30	10	3
23	-0+--	200	0.4	30	4	3
24	0--00	250	0.35	20	7	6
25	--0-0	200	0.35	25	4	6
26	+---0	300	0.35	20	4	6
27	0+0--	250	0.45	25	4	3

In the equation, the measured response is denoted as y; b_0 is a constant. The effect of each calculated term is described by the regression coefficients b_1 to b_i. Independent variable X_i is coded in terms of factors. The interaction and quadratic terms are denoted as $X_i X_k$ and X_i^2, respectively.

2.4. Characterization of Particle Properties

The granules exiting the system were collected and analyzed using different techniques to quantify the attributes. After drying for 24 h in a 60 °C oven, the samples were analyzed for particle size distribution, span of the size distribution (span), and granule yield. All measurements were performed in triplicate, and the test methods followed a previously published paper [18].

2.5. Method Validation

Validation studies were conducted to evaluate the process models using the percentage of prediction error, which was calculated with Equation (2).

$$e = \frac{|y - \hat{y}|}{y}\% \tag{2}$$

2.6. Data Analysis and Modeling

Design-Expert 12 software (Stat-Ease Inc., Minneapolis, MN, USA) was utilized for conducting the five-factor, three-level I-optimal design in this study. Additionally, this software was employed for generating response contour plots and performing the relevant analyses.

3. Results and Discussion

3.1. Fitting Data to the Model

Table 3 displays an overview of the primitive data. The models were constructed using a forward/backward stepwise regression method and then regression of all subsets with the minimized Bayesian information criterion (BIC) to determine the model with the best prediction capability. The non-statistically significant effects were eliminated, and the quadratic polynomial equations were simplified during the regression analysis (Table 4). The regression models produced satisfactory results, with coefficients (R^2) greater than 0.9. The predicted R^2 values were reasonably close to the adjusted R^2 values, indicating that the experimental data were well matched by the proposed models. All lack-of-fit tests were not significant relative to the pure error (p-value of the lack-of-fit test greater than 0.05), which means the results of the non-significant lack-of-fit test are good. The equations in terms of coded factors are listed in Equations (3)–(6), corresponding to Y_1 to Y_4, which can be used to make predictions about the response for given levels of each factor. By default, the high levels of the factors are coded as +1 and the low levels are coded as -1. The coded equation is useful for identifying the relative impact of the factors by comparing the factor coefficients.

Table 3. The responses of I-optimal design.

NO.	Y_1: Moisture Content	Y_2: D_{50}	Y_3: Span	Y_4: Yield
	%	μm		%
1	30.49	498.5	2.16	81.19
2	31.74	463.1	1.24	92.12
3	28.14	450.9	1.58	88.18
4	35.86	609.1	1.14	92.20
5	33.57	441.9	1.60	88.64
6	29.50	498.6	1.37	91.31
7	32.53	592.0	1.24	91.58
8	29.65	438.3	1.74	84.82
9	32.20	481.8	1.48	90.12
10	30.78	464.4	1.88	85.00
11	31.22	476.8	1.42	90.87
12	27.90	446.9	1.71	86.61
13	29.63	485.6	1.47	89.45
14	28.39	389.6	2.25	78.37
15	33.47	555.4	1.38	89.27
16	32.19	508.4	1.30	92.73
17	29.12	454.9	1.87	83.81
18	31.99	521.9	1.22	90.81
19	32.02	515.4	1.11	95.05
20	32.28	528.5	1.34	91.86
21	33.57	586.5	1.32	90.08
22	32.80	517.2	1.36	91.13
23	30.67	461.8	1.54	88.40
24	30.52	438.9	1.73	86.66
25	29.50	438.0	1.68	85.38
26	29.01	440.0	1.53	89.37
27	32.80	536.2	1.24	93.65

Table 4. The regression equation of the response surface quadratic model and their ANOVA results.

Response Variable	Regression Equation		R^2	$Adj\ R^2$ *	$Pre\ R^2$ *	p (Lack-of-Fit Test)	PRESS *
Y_1	$=32.15 - 0.25X_1 + 2.07X_2 - 0.86X_3 + 0.33X_4 - 0.23X_5 - 0.46X_1{}^2 - 0.67X_5{}^2$	(3)	0.9402	0.9182	0.8740	0.0615	12.47
Y_2	$=512.16 + 22.95X_1 + 47.68X_2 - 18.24X_3 + 14.75X_1X_2 + 21.91X_1X_3 - 9.96X_2X_3 - 13.30X_3X_4 - 22.42X_1{}^2$	(4)	0.9072	0.8659	0.7844	0.2183	16,088.62
Y_3	$=1.31 - 0.12X_1 - 0.23X_2 + 0.068X_3 + 0.088X_4 + 0.14X_5 - 0.062X_2X_4 - 0.11X_2X_5 - 0.054X_3X_5 + 0.069X_4X_5 + 0.12X_1{}^2 + 0.093X_2{}^2 + 0.11X_3{}^2 - 0.058X_4{}^2 + 0.051X_5{}^2$	(5)	0.9706	0.9364	0.7750	0.4604	0.4937
Y_4	$=91.90 + 1.85X_1 + 2.99X_2 - 1.09X_3 - 1.17X_4 - 2.06X_5 - 0.62X_1X_2 - 0.38X_1X_3 + 0.51X_2X_4 + 0.92X_2X_5 + 0.77X_3X_5 - 0.66X_4X_5 - 1.54X_1{}^2 - 1.54X_2{}^2 - 1.55X_3{}^2$	(6)	0.9791	0.9548	0.8633	0.7291	52.71

* $Adj\ R^2$: Adjusted R^2; $Pre\ R^2$: Predicted R^2; and PRESS: Predicted residual sum of squares.

3.2. The Effect of Factors on the Moisture Content

Response contour plots were utilized for illustrating the correlation between the independent and dependent variables. These plots allow for the simultaneous examination of the impact of two factors on the response while maintaining the other factors constantly in a two-dimensional setting. By varying the axis variables while maintaining other factors at stationary levels, we observed the change rules intuitively, as shown in Figure 2. The regression equation and ANOVA results are shown in Tables 4 and 5.

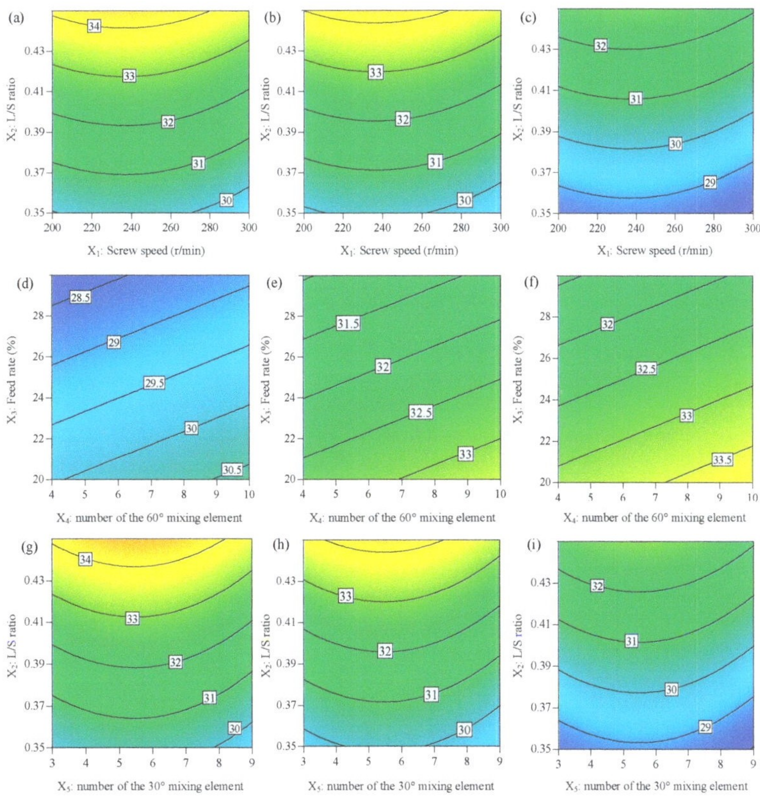

Figure 2. The response contour plot displays the effects of independent variables on the moisture content: (**a–c**) effects of X_1 and X_2, (**d–f**) X_3 and X_4, and (**g–i**) X_2 and X_5 on the moisture content while other factors were maintained at low, middle, and high levels, respectively.

Table 5. Regression coefficients and associated probability values (*p*-values) for the models of granule CQAs.

Model Terms	Y_1: Moisture Content		Y_2: D$_{50}$		Y_3: Span		Y_4: Yield	
	Coef.	*p*-Values	Coef.	*p*-Values	Coef.	*p*-Values	Coef.	*p*-Values
Intercept	32.15		512.16		1.31		91.90	
X_1	−0.25	0.0968	22.95	0.0003	−0.12	<0.0001	1.85	< 0.0001
X_2	2.07	<0.0001	47.68	<0.0001	−0.23	<0.0001	2.99	<0.0001
X_3	−0.86	<0.0001	−18.24	0.0017	0.07	0.0043	−1.09	0.0003
X_4	0.33	0.0215			0.09	0.0003	−1.17	<0.0001
X_5	−0.23	0.1070			0.14	<0.0001	−2.06	<0.0001
$X_1 X_2$			14.75	0.0364			−0.62	0.0477
$X_1 X_3$			21.91	0.0026			−0.38	0.2213
$X_1 X_4$								
$X_1 X_5$								
$X_2 X_3$			−9.96	0.1270				
$X_2 X_4$					−0.06	0.0131	0.51	0.069
$X_2 X_5$					−0.11	0.0008	0.92	0.0066
$X_3 X_4$			−13.30	0.0463				
$X_3 X_5$					−0.05	0.0413	0.77	0.015
$X_4 X_5$					0.07	0.0070	−0.66	0.0171
$X_1{}^2$	−0.46	0.0569	−22.42	0.0133	0.12	0.0033	−1.54	0.001
$X_2{}^2$					0.09	0.0140	−1.54	0.0012
$X_3{}^2$					0.11	0.0044	−1.55	0.001
$X_4{}^2$					−0.06	0.1333		
$X_5{}^2$	−0.67	0.0085			0.05	0.1302		

In this study, it can be observed from Table 5 that the factor screw speed (X_1) and the quadratic term $X_1{}^2$ had negative effects on the moisture content (Y_1). Similarly, the number of 60° kneading elements (X_5) and the quadratic term $X_5{}^2$ had negative effects on the moisture content (Y_1). The L/S ratio (X_2) and number of 30° kneading elements (X_4) had positive effects on the moisture content (Y_1). The powder feed rate (X_3) had negative effects on the moisture content (Y_1).

The L/S ratio (X_2) was found to be the most influential factor for moisture content (Y_1). This makes sense since a higher L/S ratio (X_2) provides a greater liquid amount, leading to greater liquid distribution and providing more surface wetting of granules [19]. However, under the assumption of uniform material mixing at a fixed L/S ratio (X_2), increasing the screw speed and powder feed rate will not change the moisture content of particles theoretically. A possible reason is that the increased screw speed and powder feed rate result in insufficient mechanical dispersion between liquid and particles to form a homogenous liquid distribution [11].

3.3. The Effect of Factors on the Mean Particle Size

Particle size and distribution were regarded as some of the most important attributes of the granules because granules with appropriate particle size and distribution can significantly increase the blend uniformity, flowability, and compactibility of the product [8]. Response contour plots for D$_{50}$ (Y_2) are shown in Figure 3. Analysis of variance (ANOVA) was performed and results are shown in Tables 4 and 5, which provide information about the model of D$_{50}$ (Y_2).

In this study, it can be observed from Table 4 that the factor screw speed (X_1), L/S ratio (X_2), and powder feed rate (X_3) had significant effects on the mean granule size D$_{50}$ (Y_2). The L/S ratio (X_2) was found to be the most influential factor in achieving granules with the desired mean granule size D$_{50}$ (Y_2), and the screw speed (X_1) and L/S ratio (X_2) had a positive effect on granule size. The powder feed rate (X_3) had a negative effect on granule size. However, the number of 30° (X_4) and the number of 60° kneading elements (X_5) were not found as having a significant effect on granule size in this study. Interaction effects between X_1 and X_2, and X_1 and X_3 influenced the granule size positively, while interaction effects between X_2 and X_3, and X_3 and X_4 influenced the granule size negatively. The quadratic term X_1^2 had negative effects on D_{50} (Y_2).

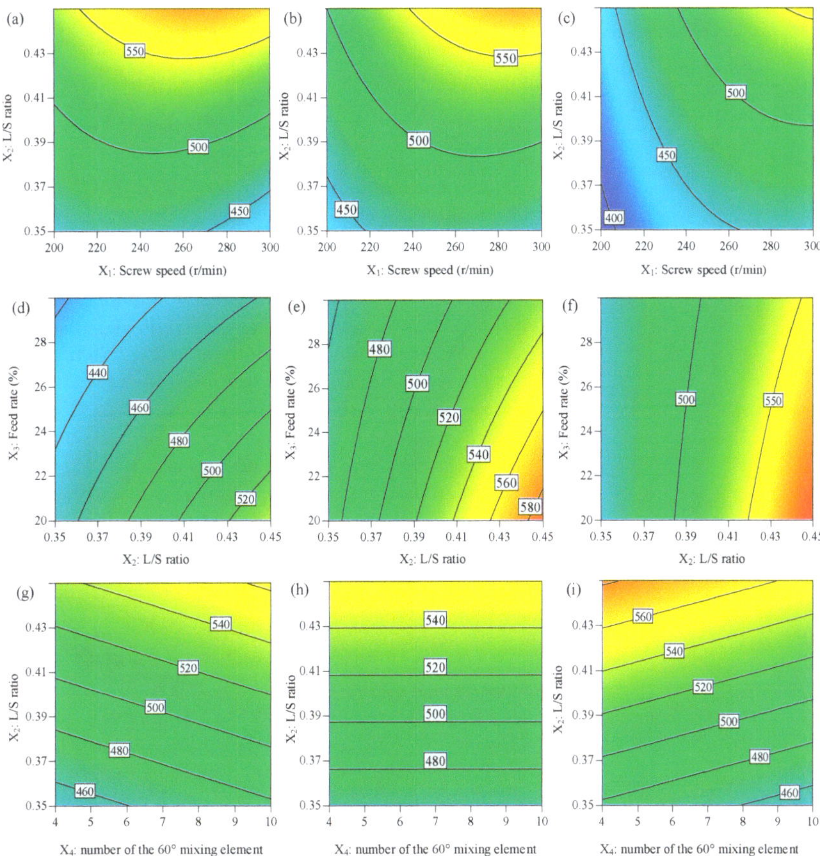

Figure 3. The response contour plot displays the effects of independent variables on the granule size D_{50}: (**a–c**) effects of X_1 and X_2, (**d–f**) X_2 and X_3, and (**g–i**) X_2 and X_4 on the granule size D_{50} while other factors were maintained at low, middle, and high levels, respectively.

In this study, it can be seen that the higher screw speed can produce a larger granule diameter. It is reported that high screw speeds can increase the conveying capacity of the screws, resulting in a lower barrel fill level and lower residence times within the granulator [6,10,20]. However, this appears to contradict some studies, suggesting that the granule size slightly increased at low screw speeds [7]. They found that the lower barrel fill level results in low compaction and particle interaction and is not conducive to the growth of granules. However, at low screw speeds, the longer residence times of particles in the barrel allow for greater growth of the granules. A possible reason is that increasing the screw speed can increase the throughput and provide a greater compaction force, which enhances the consolidation and compaction between granules in the barrel; this could promote granulation and produce granules with larger sizes. However, at the same time, the intense mechanical forces generated by the high-speed rotation of the screws cause more aggressive particle breakage and attrition, resulting in smaller granules. These two effects are opposite for particle growth, and under different conditions, one of the factors may play a dominant role. It may be that, as reported within typical operation limits (e.g., screw speed and throughput ranges, kneading element configuration), screw speed has a positive effect on granule growth, but outside the upper and lower ranges, the effect on granule size becomes the opposite [10]. A possible reason is the high fill levels at low screw speeds, which lead to material compaction where blockages can form at high mass loads.

In the present study, the effect of the L/S ratio (X_2) on particle size was in agreement with the research report that increasing the L/S ratio (X_2) can reduce the number of fines and produce particles with superior flow properties [5,7]. A possible reason is that an increased L/S ratio (X_2) can provide more liquid to form a higher liquid distribution and more surface wetting of granules [19].

It is found that increasing material feed rate (X_3) at a constant L/S ratio (X_2) mainly had a negative effect on the particle size of granules. That is, an increasing throughput decreased the average granule diameter in the present study. It is explained that the higher barrel fill level at high throughput leads to restricted liquid distribution along with an increase in friction between the barrel wall and granules, resulting in higher attrition of granules [21]. However, some research showed that the average granule diameter was higher at increased material throughput due to a higher filling degree of the barrel [7,9], and this may be attributed to the presence of a kneading screw element in the screw configuration, which could promote granulation.

It was observed that the number of 30° and 60° kneading elements, as well as their interaction, were found to have no significant effect on the granule size. A possible reason was assumed that the increased number of kneading elements provided stronger mixing ability and compaction forces, which enhanced the consolidation and compaction between granules and also led to a more homogeneous distribution of the granulation liquid within the setting range [11].

3.4. The Effect of Factors on the Span

The results of the ANOVA analysis of span (Y_3) are shown in Tables 4 and 5. Response contour plots are shown in Figure 4. The results showed that screw speed (X_1), L/S ratio (X_2), and powder feed rate (X_3) had significant impacts on span (Y_3), and the L/S ratio (X_2) was found to be the most significant term among them. The three main terms were reported to be critical factors that determine the granule properties [3]. These parameters have a great influence on the fill level within the granulator. The variation in fill level can make considerable differences in binder distribution and granule properties [5,22]. The effect of the L/S ratio (X_2) on the span is similar to its effect on D_{50} (Y_2). An appropriate amount of liquid is a necessary condition for particles to grow to the target size and distribution. Within the scope of the experiment, a higher L/S ratio (X_2) can produce granules with a more uniform distribution. The number of 30° (X_4) and 60° kneading elements (X_5), as well as their interaction, were found to have positive significant effects on the span (Y_3), which indicates that increasing X_5 can broaden the span of the final granules. Kneading elements can make the liquid distribution become more uniform in screw configurations and provide more densification [23]. We found a notably higher improvement in liquid distribution homogeneity when increasing the kneading block length [24]. When the number of kneading elements increased, it can provide stronger compaction forces, which enhanced the consolidation and compaction between the particles. The uniform distribution during nucleation makes the granules' growth less reliant on mechanical dispersion provided by the kneading elements [24].

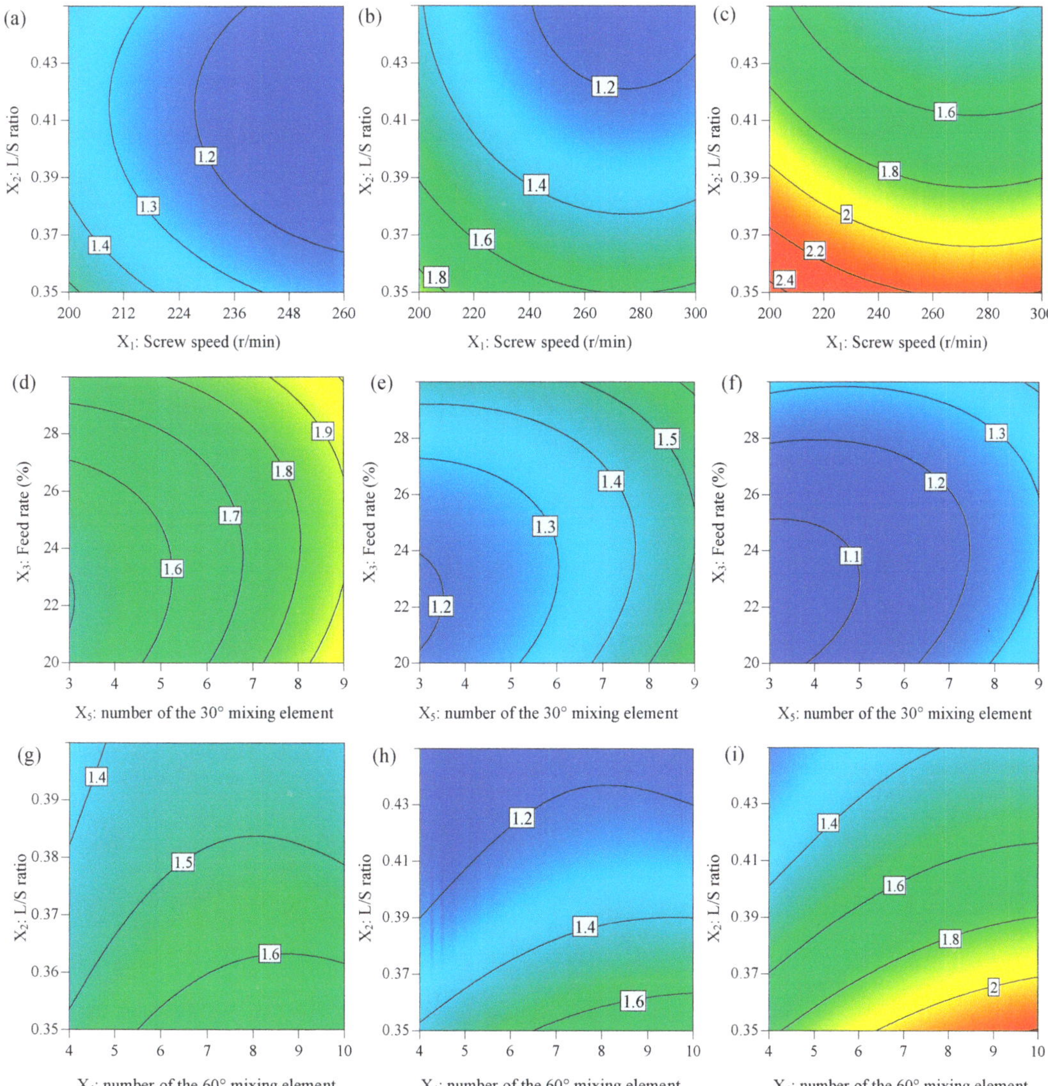

Figure 4. The response contour plot displays the effects of independent variables on the span: (**a–c**) effects of X_1 and X_2, (**d–f**) X_3 and X_5, and (**g–i**) X_2 and X_4 on the span while other factors were maintained at low, middle, and high levels, respectively.

3.5. The Effect of Factors on the Production Yield

Tables 4 and 5 display the results from the ANOVA analysis on yield (Y_4). Response contour plots are shown in Figure 5. It can be observed from Table 5 that the screw speed (X_1) and L/S ratio (X_2) had positive effects on the production yield. The powder feed rate (X_3), the number of 30° (X_4), and the number of 60° kneading elements (X_5) had negative effects on the production yield. There was an interaction between factors X_1X_2, X_1X_3, and X_4X_5, which had a negative effect on yield (Y_4), while X_2X_4, X_2X_5, and X_3X_5 had a positive effect on yield (Y_4). Quadratic terms $X_1{}^2$, $X_2{}^2$, and $X_3{}^2$ had a negative effect on yield (Y_4).

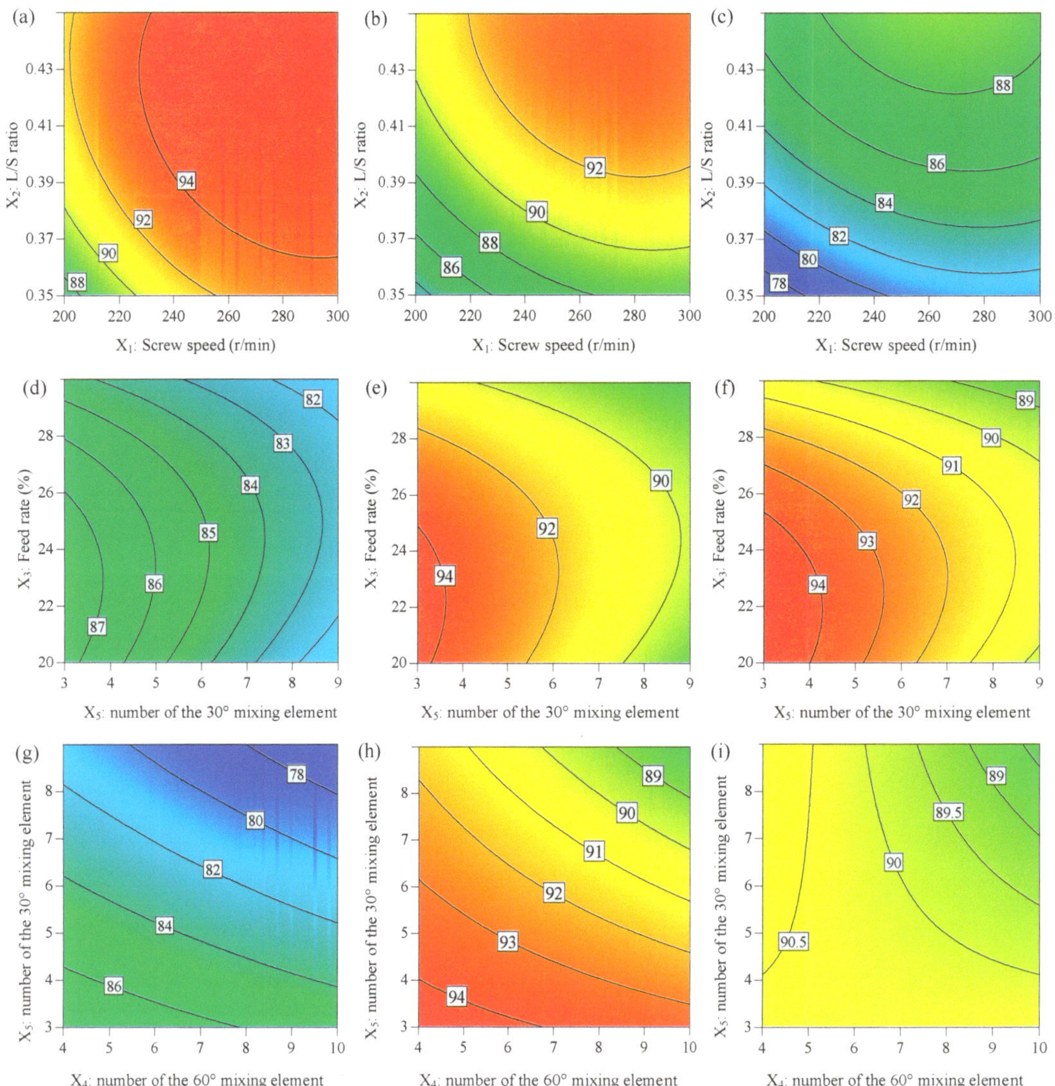

Figure 5. The response contour plot displays the effects of independent variables on the production yield: (**a–c**) effects of X_1 and X_2, (**d–f**) X_3 and X_5, and (**g–i**) X_4 and X_5 on the production yield while other factors were maintained at low, middle, and high levels, respectively.

The L/S ratio (X_2) was the most significant positive factor that contributed to the production yield model. It was seen that the yield increased as the L/S ratio (X_2) increased, and granules produced at a low L/S ratio (X_2) had a high span and broad size distribution, including the fine and large agglomerates. However, granules produced at a high L/S ratio (X_2) had low span and narrow size distribution [20]. It is inferred that the low L/S ratio (X_2) results in concentrated wetted areas by direct injection through liquid inlet ports. The particles suffer from insufficient mechanical dispersion to form homogenous liquid distribution, which results in small dry fine and large wetted agglomerates [25].

3.6. Defining a Design Space and Validation of the TWSG Process

In order to produce granule products that meet the quality requirements, the TWSG process parameters should be operated within a defined design space. The design space represents a combination and interaction of input variables and process parameters that are designed to meet the quality attributes [26]. Therefore, the process parameters and the screw element configuration can be optimized to produce a granular product with different physical properties.

The granules obtained by TSWG need to be dried downstream; in the drying process, the amount of water needed to be reduced during the drying phase. To maximize production efficiency and minimize energy consumption, it was encouraging to produce qualified products with minimum water [8]. Therefore, the minimal L/S ratio required for granulation needs to be determined. The granules with a *span* value less than 1.50 were considered to have a narrow size distribution that can enable efficient downstream tableting processing. Meanwhile, the median particle size D_{50} needs to be close to the desired granule size, and the particle size distribution should be as narrow as possible to obtain the maximum yield. Therefore, the granule diameter (400 µm < D_{50} < 600 µm), span (span < 1.5), and production yield (150 µm < granule size < 1200 µm, target: maximize) were used as representative CQAs of granules in the design space determination based on the previous experiments and data reported in the literature [8,27]. Table 6 displays the constraints regarding the CQAs. We navigated the design space via the regression models of D_{50}, span, and production yield. Figure 6 illustrates the design space from various perspectives.

Table 6. Constraints on dependent responses.

Dependent Responses	Constraints	Optimum
X_1: L/S ratio	$0.35 \leq X_1 \leq 0.45$	minimum
Y_2: Mean particle size D_{50}	$400 \text{ µm} \leq Y_2 \leq 600 \text{ µm}$	-
Y_3: Span of granule	$Y_3 \leq 1.5$	minimum
Y_4: Yield	$90 \leq Y_4 \leq 100$	maximize

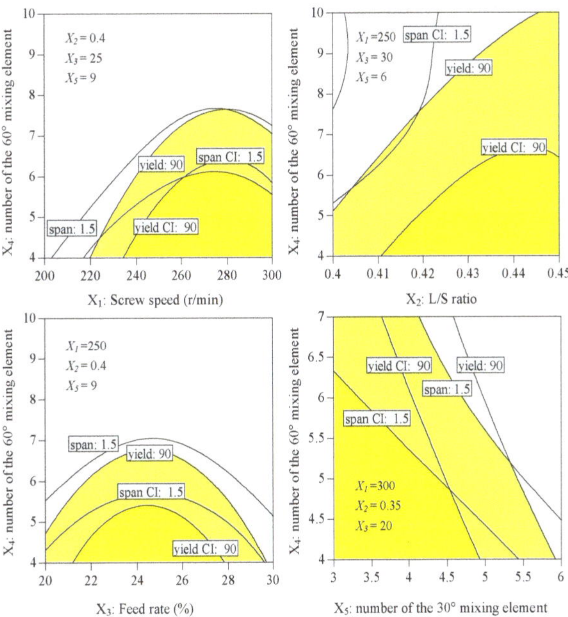

Figure 6. Design space for TSWG from different viewpoints.

The overlay plots depict the range of two factors while the remaining factors are held at a specified level. The bright yellow section represents the design space for TSWG with a 90% confidence interval. The dark yellow section indicates a portion of the design space where there is a 10% probability of failing to meet process objectives. The dark gray area is outside of the design space and does not meet technical requirements. The successful application of the I-optimal design helped define the design space, ensuring the desired quality of the final granule by considering the operating parameter ranges of the five factors being investigated.

Additional validation experiments were carried out to evaluate the accuracy of the established regression models. The operating parameters of one experiment were within the design space, and the other experimental parameters were outside the design space. The operating conditions and experimental and predicted values are shown in Table 7. The results showed that the prediction error percentage, comparing the actual values obtained from validation experiments with the anticipated values from the responses, was below 10%, which is considered acceptable. This suggests that the models are reliable and possess an effective predictive capability.

Table 7. Process parameters and experimental results of validation test points.

CPP (X_1; X_2; X_3; X_4; X_5)	CQA	Experimental Value (y)	Predicted Value ($\hat{y}$)	ARD (%)
(240; 0.45; 30; 4; 7)	Y_1	33.1	30.5	7.7
	Y_2	532.1	494.2	7.1
	Y_3	1.3	1.3	5.2
	Y_4	91.2	97.4	6.8
(200; 0.4; 25; 7; 7)	Y_1	33.2	31.8	4.3
	Y_2	466.8	499.1	6.9
	Y_3	1.6	1.5	4.9
	Y_4	87.8	94.8	7.9

4. Conclusions

The results of this study demonstrate the effective application of the I-optimal design method in analyzing the factors influencing TWSG related to target granule attributes. Five investigated process parameters—screw speed, L/S ratio, powder feed rate, and numbers of the 60° and 30° mixing elements—were studied via modeling using the I-optimal design. A design space that defined the ranges of operating parameters for TWSG was determined based on constraints for the target granule quality: mean particle size, span, and yield. Validation experiments confirmed the accuracy and reliability of the mathematical models. The application of the I-optimal design provides a precise estimation of the parameters and prediction of the response to determine optimum operating conditions. This study enhances comprehension of TWSG cost-effectively and efficiently, confirming that an I-optimal design is a viable approach for exploring the complicated TWSG process.

Author Contributions: Methodology, G.T.; Validation, G.T.; Investigation, J.Z.; Writing–original draft, J.Z.; Supervision, H.Q. All authors have read and agreed to the published version of the manuscript.

Funding: This work was supported by the Innovation Center in Zhejiang University, State Key Laboratory of Component-Based Chinese Medicine (ZYYCXTD-D-2020002), and Postdoctoral Science Foundation of Zhejiang Province (520000-X82101).

Data Availability Statement: All data have been reflected in the paper, and no additional data are required.

Conflicts of Interest: The authors declare no conflict of interest.

References

1. Muddu, S.V.; Kotamarthy, L.; Ramachandran, R. A Semi-Mechanistic Prediction of Residence Time Metrics in Twin Screw Granulation. *Pharmaceutics* **2021**, *13*, 393. [CrossRef] [PubMed]
2. Lee, S.L.; O'Connor, T.F.; Yang, X.; Cruz, C.N.; Chatterjee, S.; Madurawe, R.D.; Moore, C.M.V.; Yu, L.X.; Woodcock, J. Modernizing Pharmaceutical Manufacturing: From Batch to Continuous Production. *J. Pharm. Innov.* **2015**, *10*, 191–199. [CrossRef]
3. Gorringe, L.J.; Kee, G.S.; Saleh, M.F.; Fa, N.H.; Elkes, R.G. Use of the channel fill level in defining a design space for twin screw wet granulation. *Int. J. Pharm.* **2017**, *519*, 165–177. [CrossRef]
4. Vercruysse, J.; Peeters, E.; Fonteyne, M.; Cappuyns, P.; Delaet, U.; Van Assche, I.; De Beer, T.; Remon, J.P.; Vervaet, C. Use of a continuous twin screw granulation and drying system during formulation development and process optimization. *Eur. J. Pharm. Biopharm.* **2015**, *89*, 239–247. [CrossRef] [PubMed]
5. Portier, C.; Vervaet, C. Continuous Twin Screw Granulation: A Review of Recent Progress and Opportunities in Formulation and Equipment Design. *Pharmaceutics* **2021**, *13*, 668. [CrossRef]
6. Morrissey, J.P.; Hanley, K.J.; Ooi, J.Y. Conceptualisation of an Efficient Particle-Based Simulation of a Twin-Screw Granulator. *Pharmaceutics* **2021**, *13*, 2136. [CrossRef]
7. Liu, H.; Ricart, B.; Stanton, C.; Smith-Goettler, B.; Verdi, L.; O'Connor, T.; Lee, S.; Yoon, S. Design space determination and process optimization in at-scale continuous twin screw wet granulation. *Comput. Chem. Eng.* **2019**, *125*, 271–286. [CrossRef]
8. Fülp, G.; Domokos, A.; Galata, D.; Szabó, E.; Nagy, Z.K. Integrated Twin-Screw Wet Granulation, Continuous Vibrational Fluid Drying and Milling: A fully continuous powder to granule line. *J. Int. J. Pharm.* **2020**, *594*, 120126. [CrossRef]
9. Fonteyne, M.; Vercruysse, J.; Díaz, D.C.; Gildemyn, D.; Vervaet, C.; Remon, J.P.; Beer, T.D. Real-time assessment of critical quality attributes of a continuous granulation process. *Pharm. Dev. Technol.* **2013**, *18*, 85–97. [CrossRef]
10. Seem, T.C.; Rowson, N.A.; Ingram, A.; Huang, Z.; Yu, S.; Matas, M.D.; Gabbott, I.; Reynolds, G.K. Twin Screw Granulation—A Literature Review. *Powder Technol.* **2015**, *276*, 89–102. [CrossRef]
11. Kumar, A.; Alakarjula, M.; Vanhoorne, V.; Toiviainen, M.; De Leersnyder, F.; Vercruysse, J.; Juuti, M.; Ketolainen, J.; Vervaet, C.; Remon, J.P.; et al. Linking granulation performance with residence time and granulation liquid distributions in twin-screw granulation: An experimental investigation. *Eur. J. Pharm. Sci.* **2016**, *90*, 25–37. [CrossRef] [PubMed]
12. Meng, W.; Kotamarthy, L.; Panikar, S.; Sen, M.; Pradhan, S.; Marc, M.; Litster, J.D.; Muzzio, F.J.; Ramachandran, R. Statistical analysis and comparison of a continuous high shear granulator with a twin screw granulator: Effect of process parameters on critical granule attributes and granulation mechanisms. *Int. J. Pharm.* **2016**, *513*, 357–375. [CrossRef]
13. Ozdemir, A. An I -optimal experimental design-embedded nonlinear lexicographic goal programming model for optimization of controllable design factors. *Eng. Optim.* **2020**, *53*, 392–407. [CrossRef]
14. Jones, B.; Goos, P. I-Optimal Versus D-Optimal Split-Plot Response Surface Designs. *J. Qual. Technol.* **2012**, *44*, 85–101. [CrossRef]
15. Willecke, N.; Szepes, A.; Wunderlich, M.; Remon, J.P.; Vervaet, C.; De Beer, T. A novel approach to support formulation design on twin screw wet granulation technology: Understanding the impact of overarching excipient properties on drug product quality attributes. *Int. J. Pharm.* **2018**, *545*, 128–143. [CrossRef]
16. Stauffer, F.; Vanhoorne, V.; Pilcer, G.; Chavez, P.-F.; Vervaet, C.; De Beer, T. Managing API raw material variability during continuous twin-screw wet granulation. *Int. J. Pharm.* **2019**, *561*, 265–273. [CrossRef]
17. Meng, W.; Rao, K.S.; Snee, R.D.; Ramachandran, R.; Muzzio, F.J. A comprehensive analysis and optimization of continuous twin-screw granulation processes via sequential experimentation strategy. *Int. J. Pharm.* **2019**, *556*, 349–362. [CrossRef]
18. Zhao, J.; Li, W.; Qu, H.; Tian, G.; Wei, Y. Application of definitive screening design to quantify the effects of process parameters on key granule characteristics and optimize operating parameters in pulsed-spray fluid-bed granulation. *Particuology* **2019**, *43*, 56–65. [CrossRef]
19. Dhenge, R.M.; Cartwright, J.J.; Hounslow, M.J.; Salman, A.D. Twin screw granulation: Steps in granule growth. *Int. J. Pharm.* **2012**, *438*, 20–32. [CrossRef]
20. Dhenge, R.M.; Fyles, R.S.; Cartwright, J.J.; Doughty, D.G.; Hounslow, M.J.; Salman, A.D. Twin screw wet granulation: Granule properties. *Chem. Eng. J.* **2010**, *164*, 322–329. [CrossRef]
21. Dhenge, R.M.; Washino, K.; Cartwright, J.J.; Hounslow, M.J.; Salman, A.D. Twin screw granulation using conveying screws: Effects of viscosity of granulation liquids and flow of powders. *Powder Technol.* **2013**, *238*, 77–90. [CrossRef]
22. Djuric, D.; Kleinebudde, P. Technology, Continuous granulation with a twin-screw extruder: Impact of material throughput. *Pharm. Dev.* **2010**, *15*, 518–525. [CrossRef]
23. El Hagrasy, A.; Litster, J. Granulation rate processes in the kneading elements of a twin screw granulator. *AIChE J.* **2013**, *59*, 4100–4115. [CrossRef]
24. Yu, S.; Reynolds, G.K.; Huang, Z.; de Matas, M.; Salman, A.D. Granulation of increasingly hydrophobic formulations using a twin screw granulator. *Int. J. Pharm.* **2014**, *475*, 82–96. [CrossRef] [PubMed]
25. El Hagrasy, A.; Hennenkamp, J.; Burke, M.; Cartwright, J.; Litster, J. Twin screw wet granulation: Influence of formulation parameters on granule properties and growth behavior. *Powder Technol.* **2013**, *238*, 108–115. [CrossRef]

26. Guideline, I. Pharmaceutical development Q8. *Curr. Step* **2005**, *4*, 11.
27. Shekunov, B.Y.; Chattopadhyay, P.; Tong, H.; Chow, A. Particle Size Analysis in Pharmaceutics: Principles, Methods and Applications. *J. Pharm. Res.* **2007**, *24*, 203–227.

Review

Cell-Penetrating Peptide-Based Delivery of Macromolecular Drugs: Development, Strategies, and Progress

Zhe Sun [1,†], Jinhai Huang [1,†], Zvi Fishelson [2], Chenhui Wang [3] and Sihe Zhang [3,*]

1   School of Life Sciences, Tianjin University, Tianjin 300072, China; sz2019226008@tju.edu.cn (Z.S.); jinhaih@tju.edu.cn (J.H.)
2   Department of Cell and Developmental Biology, Faculty of Medicine, Tel Aviv University, Tel Aviv 69978, Israel; lifish@tauex.tau.ac.il
3   Department of Cell Biology, School of Medicine, Nankai University, Tianjin 300071, China; 15737371232@163.com
*   Correspondence: sihezhang@nankai.edu.cn
†   These authors contributed equally to this work.

Abstract: Cell-penetrating peptides (CPPs), developed for more than 30 years, are still being extensively studied due to their excellent delivery performance. Compared with other delivery vehicles, CPPs hold promise for delivering different types of drugs. Here, we review the development process of CPPs and summarize the composition and classification of the CPP-based delivery systems, cellular uptake mechanisms, influencing factors, and biological barriers. We also summarize the optimization routes of CPP-based macromolecular drug delivery from stability and targeting perspectives. Strategies for enhanced endosomal escape, which prolong its half-life in blood, improved tar-geting efficiency and stimuli-responsive design are comprehensively summarized for CPP-based macromolecule delivery. Finally, after concluding the clinical trials of CPP-based drug delivery systems, we extracted the necessary conditions for a successful CPP-based delivery system. This review provides the latest framework for the CPP-based delivery of macromolecular drugs and summarizes the optimized strategies to improve delivery efficiency.

Keywords: cell-penetrating peptide; macromolecular drug delivery; cellular uptake mechanism; biological barrier; optimized strategy

Citation: Sun, Z.; Huang, J.; Fishelson, Z.; Wang, C.; Zhang, S. Cell-Penetrating Peptide-Based Delivery of Macromolecular Drugs: Development, Strategies, and Progress. *Biomedicines* **2023**, *11*, 1971. https://doi.org/10.3390/biomedicines11071971

Academic Editors: Ali Nokhodchi and Bernard Lebleu

Received: 30 May 2023
Revised: 11 July 2023
Accepted: 11 July 2023
Published: 12 July 2023

1. Introduction

How macromolecular drugs can be efficiently delivered to target cells remains a big challenge. There are two ways to place macromolecular drugs into target cells: one is to directly introduce the macromolecular drugs into cells based on membrane disruption; the other is to deliver them by using delivery vehicles [1], which are mainly divided into either viral vehicles or non-viral vehicles. Various macromolecular drug delivery systems are shown in Figure 1.

Delivery, based on membrane disruption, can quickly and directly deliver almost all macromolecular drugs that can be dispersed in a solution; however, it must also avoid cell disturbance or death caused by membrane destruction [1]. The delivery methods based on membrane disruption are mainly classified into three approaches: (1) electro-magnetic/thermal; (2) mechanical; and (3) chemical. These approaches usually include electroporation [2], microinjection [3], osmotic pressure [4], nanoneedles [1], pore-forming agents [5], and more (Figure 1). The principle underlying these approaches is to briefly destroy the cell membrane so that macromolecular drugs can enter the cell, after which the cell membrane is repaired and can restore cell homeostasis. However, such an approach has strict requirements regarding the degree and time of membrane disruption, and it requires chemically modifying the macromolecular drugs to prevent their degradation [6]. Currently, the selection of a suitable membrane disruption method and its precise implementation in a high-throughput way remains challenging [1].

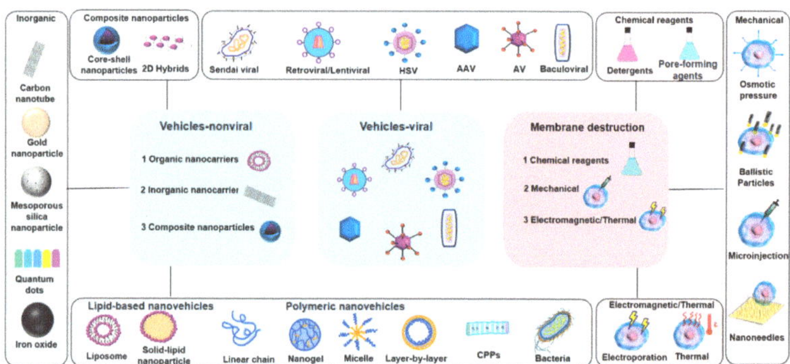

Figure 1. Delivery systems for macromolecular drugs.

Viral delivery vehicles can overcome cellular barriers and can effectively deliver macromolecular drugs into cells. The most commonly used types of viral delivery vehicles are the retrovirus (RV), lentivirus (LV) [7], adenovirus (AV), and adeno-associated virus (AAV) [8]. LVs are a subtype of RVs and can be stably integrated into the host genome of mammalian cells [9]. Compared with the RV and AV, the LV and AAV are the most used vehicles in clinical trials because of their lower immunogenicity. However, the LV is prone to bring about higher off-target effects [10], and the drug-loading capacity of the AAV is also limited [11]. Aside from these vehicles, there are also herpes simplex virus (HSV) vehicles, with a larger packaging capacity [9], and the Epstein-Barr virus (EBV); EBV is subordinate to HSV, but it is also used as a delivery vehicle for CRISPR/Cas [12,13]. In addition, there are Sendai virus vehicles, which are used as a negative-strand RNA virus vehicle system [14], as well as the baculovirus, which is used as a gene expression vehicle [15] (Figure 1). Although viral vehicles have high delivery efficiency for macromolecular drugs, their limited drug-loading capacity, high production costs, uncontrollable viral release, limited viral tropism, and safety issues regarding viral gene integration into the host genome have still not been completely resolved [16].

Non-viral delivery vehicles have been widely explored in recent years because of their lower immunogenicity, unrestricted drug-loading, high flexibility, simple synthesis, low cost, and easy storage [17–20]. Non-viral delivery vehicles can be divided into three categories: (1) organic nanovehicles; (2) inorganic nanovehicles; and (3) composite nanoparticles [21]. Of these vehicles, organic nanovehicles include cell-penetrating peptides (CPPs), bacteria-derived or lipid-based nanovehicles, and polymeric nanovehicles (Figure 1). Since low delivery efficiency is a problem for most non-viral delivery vehicles [22], CPPs that can traverse the biological barrier (cell membranes) with high efficiency have been extensively explored. CPPs are not only simple to synthesize but are diverse in their types and are widely used [23]. They can also be combined with other delivery vehicles to further improve the delivery efficiency of macromolecular drugs [24–26]. Compared with the usual way of membrane disruption, the use of CPPs causes less damage to the cell membrane and is more effective and safe [27]. Therefore, CPPs are very useful tools for macromolecular drug delivery.

2. The History of CPP Development

CPPs usually comprise 5–30 amino acids with good biological safety and can efficiently carry macromolecular drugs through the cell membrane. The history of CPPs can be traced back to 1988. The first CPP was discovered when researching potential targets for the treatment of AIDS, namely, the Trans-Activator of Transcription (Tat) protein of human immunodeficiency virus type 1 (HIV-1) [28,29]. Then, in 1991, the homologous protein

Penetratin/Antp from the antennae of Drosophila melanogaster was discovered [30]. The amphipathic penetrating peptide MPG [31], truncated HIV-Tat [32], and VP22, derived from HSV structural protein, were discovered in 1997 [33]. Then, the first chimeric peptide Transportan (TP) was reported in 1998 [34]. Subsequently, Oligo R [35], Pep-1 [36], and SS-31 [37] were discovered one after the other.

Transportan is a combination of the peptide mastoparan, isolated from wasp venom, and the N-terminal fragment of the human neuropeptide galanin. Transportan 10 (TP10), a chimeric peptide derived from mastoparan, comprises a 14-residue peptide from wasp (*Vespula lewisii*) venom linked to a 6-residue sequence from the neuropeptide galanin through an extra lysine residue, was developed in 2007 [38]. The tumor-targeting penetrating peptide iRGD was reported in 2009 [39]. In 2011, PepFect14 was reported to have improved the stability in serum [40]; PepFect15 was reported a year later [41]. Further research in 2013 found that the cancer-specific penetration peptide, BR2, can effectively deliver a single-chain antibody (scFv) [42]. Thereafter, PepM and PepR from the dengue virus capsid protein were discovered [43]. In recent years, the mitochondrial-penetrating peptide MtCPP-1 [44]; skin-permeable IMT-P8 [45]; bioreducible B-mR9 [46]; Tat-NTS, which can inhibit the transcriptional activity of *p53* [47]; VP1, derived from the VP1 protein of chicken anemia virus [48]; Ara-27, from Arabidopsis thaliana [49]; and peptide SOX2-iPep have been reported as successful [50]. Notably, various types of cyclic CPPs and intracellular organelle targeting CPPs have been newly developed [51–53]. Milestones in the development process of CPPs are depicted in Figure 2.

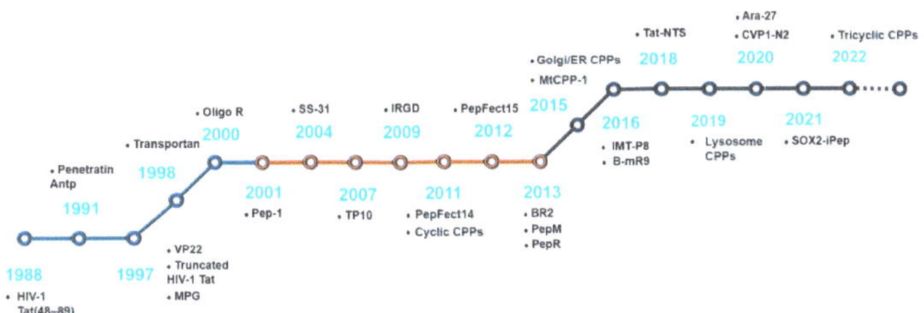

Figure 2. Milestones for the discovery and development of CPPs.

3. Classification of the CPP-Based Macromolecular Drug Delivery System

The components of the CPP-based delivery system include CPPs and delivered drug cargo. There are many types of CPPs and drug cargoes. CPPsite 2.0, a CPP database, currently holds around 1850 peptide entries and maintains more detailed information about different types of CPPs and delivered cargoes. The CPPsite 2.0 website can also be used to predict the tertiary structures of CPPs [54,55]. Below, we summarize the type, classification, and connection between the CPPs and the delivered drug cargoes (Figure 3).

3.1. Classification of CPPs

Generally, CPPs can be divided into three categories by their source, physical and chemical properties, and composition. According to the literature, CPPs can be divided into synthetic CPPs, chimeric CPPs, and protein-derived CPPs [56]. According to their physical and chemical properties, CPPs can be divided into cationic CPPs, amphipathic CPPs, and hydrophobic CPPs [57]. In addition, according to their composition, CPPs can be divided into cyclic CPPs and linear CPPs. Although they only occupy a small proportion (about 5%) compared with linear CPPs, cyclic CPPs reduce the degree of conformation freedom due to the peptide cyclization, which greatly improves their binding specificity with targeted molecules [58]. This stable structure of cyclic CPPs has greater resistance to proteolytic degradation [59]. Cyclic CPPs also perform better in terms of oral bioavailability [60]. In addition, cyclic CPPs can more efficiently interfere with protein–protein interactions (PPIs) than can traditional small-molecule drugs (MW $\leq$ 500) [61].

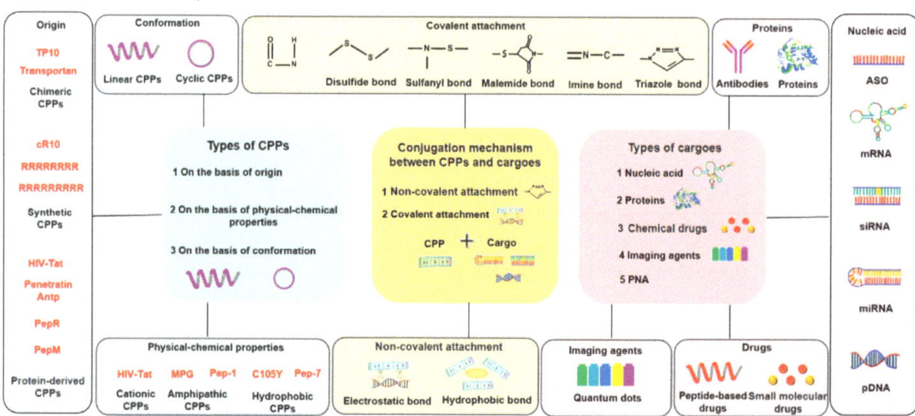

Figure 3. The components and classification of the CPP-based macromolecular drug delivery.

3.2. Classification of Cargos

Generally, macromolecular drug cargoes can be divided into nucleic acids, proteins, polymer chemicals or imaging agents, PNAs (peptide nucleic acids), and more (Figure 3). Small-molecule drugs include peptides and different types of chemicals. Macromolecular nucleic acid drugs mainly include antisense oligonucleotides (ASOs), mRNA, siRNA, miRNA, and pDNA. Macromolecular protein drugs mainly include different types of proteins and antibodies with functional activity. Protein-based therapy has a faster effect, better functional strength, and durable controllability; its lack of genetic toxicity has been extensively highlighted [62]. Moreover, the use of CPPs can promote the delivery of protein cargoes into cells non-invasively [63]. Because the delivery only by single CPPs has shortcomings of non-specificity, easy hydrolysis, and a short half-life [35,64], combining CPPs with other vehicles increases the delivery efficiency for macromolecular drugs [63]. In addition, accumulating research indicates that CPP-modified vehicles can be used to deliver other types of macromolecular-complexed drugs, including nanoparticles, liposomes, micelles, and cellulose-based superabsorbent hydrogels [65–67].

3.3. CPP and Cargo Connection Types

The CPP and cargo connection types can be divided into those with covalent bonds and those with non-covalent connections. The non-covalent connection methods mainly include hydrophobic interactions and electrostatic interactions [68]. The electrostatic interaction connection method has poor stability, which is highly dependent on the ionic strength and pH [69]. There are many types of covalent bond connection methods; these bonds include peptide bonds [70], disulfide bonds [71,72], sulfanyl bonds [35], maleimide bonds [73], imine bonds [73], and triazole bonds [74]. Covalent linkages of CPPs to cargoes provide high stability and precise control over site selectivity [21]; however, sometimes, the covalent connection method is not conducive to the release of cargo. Notably, ConjuPepDB, a database of peptide-drug conjugates, has been recently released. It covers more than 1600 peptide-conjugated drug molecules, along with basic information about their biomedical application and the type of chemical conjugation employed [75].

4. Cellular Uptake Mechanisms, Influencing Factors, and Biological Barriers

Cellular uptake plays a key role in the efficient delivery of macromolecular drugs. CPPs can interact with the plasma membrane phospholipids, proteoglycans, and protein receptors on the cell surface that are taken up into cells [61]. The cellular uptake mechanisms of CPP-based macromolecular drug delivery mainly include two categories: direct penetration and endocytic routes. The penetrating peptide and the smaller penetrating peptide–cargo complex penetrate directly into the cell. The methods of direct penetration mainly include the carpet model [1], barrel-stave model [76], toroidal pore model [77], inverted micelle model [78,79], and membrane thinning [80]. Different macromolecular drugs have different endocytic pathways. The endocytic pathways of CPP-based macromolecular drug delivery mainly include five types: clathrin-mediated endocytosis (CME), caveolin-mediated endocytosis, autophagy pathway, macropinocytosis pathway, and caveolin- and clathrin-independent mode of endocytosis (Figure 4). These energy-dependent endocytic pathways transport most of the uptaken macromolecular drug cargoes to the early endosome, the late endosome, and finally to the lysosome. For the CPP cargoes to function after their uptake into cells, the delivered proteins and RNAs inside the endosomes must escape into the cytoplasm, whereas the delivered pDNA needs to enter the nucleus in order to function.

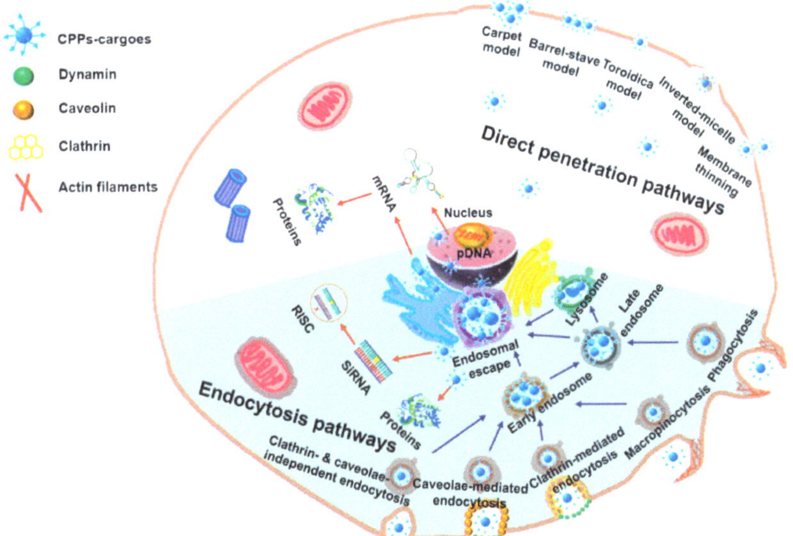

Figure 4. Cellular uptake mechanisms of CPP-based macromolecular drug delivery. The mechanisms mainly include two categories: direct penetration pathways and endocytotic pathways. Direct penetration pathways are usually energy-independent, whereas endocytotic pathways are energy-dependent. In order for the CPP-cargos to function after entering the target cells, proteins and RNA endosomes must escape into the cytoplasm, whereas pDNA must enter the nucleus.

Different pathways of cellular uptake have distinct characteristics. It was reported that some viruses use caveolin-mediated endocytic pathways to avoid lysosome degradation [81–83]. CME is very important because it is the main way for cells to obtain nutrients; for example, by promoting the absorption of cholesterol and iron. However, the average diameter of clathrin-coated vesicles is about 100 nm, which limits the size of the cargo transported via this route [84]. Some studies also indicate that the macropinocytosis pathway can enhance the cellular uptake of macromolecular cargoes; this is conducive to the delivery of therapeutic genes [85]. The spread of the SARS-CoV-2 virus highlights the importance of research on the cellular uptake mechanism because a better understanding of the cellular uptake mechanism will help develop antiviral therapies [86]. However, research on the cellular uptake routes is still in the preliminary stage, and the reported uptake pathways of many macromolecular particles remain controversial.

Such complicated cellular uptake routes are also affected by many factors: the nature of the delivery system (size, shape, charge, and surface rigidity), the cell types, and the experimental factors (conditions such as the drug concentration, action time, and treatment temperature) [87]. Generally, small nanoparticles are easier for cellular uptake, and nanoparticles above 200 nm tend to accumulate in the spleen and liver [88,89]. However, if the size of the nanoparticles is less than 10 nm, they will be rapidly cleared by the kidneys and will not easily accumulate at the tumor site through the enhanced permeability and retention (EPR) effect [90–92]. In addition, positively charged nanomaterials exhibit higher cellular uptake than neutral or negatively charged nanomaterials. The harder the nanoparticle, the easier it is for it to be taken up by the cells [83]. Moreover, when the hydrodynamic size is the same, nanorods can penetrate better than nanospheres [93].

Aside from the complex cellular uptake mechanisms, the transport of macromolecule-contained vesicles is also hindered by a variety of biological barriers in the delivery process, such as tissue pressure [93], opsonization, rapid kidney filtration, serum endonucleases, the macrophage system, hemorheological limitations [94], encapsulation of endosomal vesicles, lysosomal degradation, and intracellular transmission difficulties (Figure 5). The delivery system, loaded with pDNA, must enter the nucleus to function. As the gatekeeper of the cell nucleus, large protein assemblies, called nuclear pore complexes (NPCs), control the entry of genetic material into the nucleus [95]. NPCs are the last barrier to the pDNA delivery system. Therefore, the resolution of these obstacles needs to be further studied using optimization strategies.

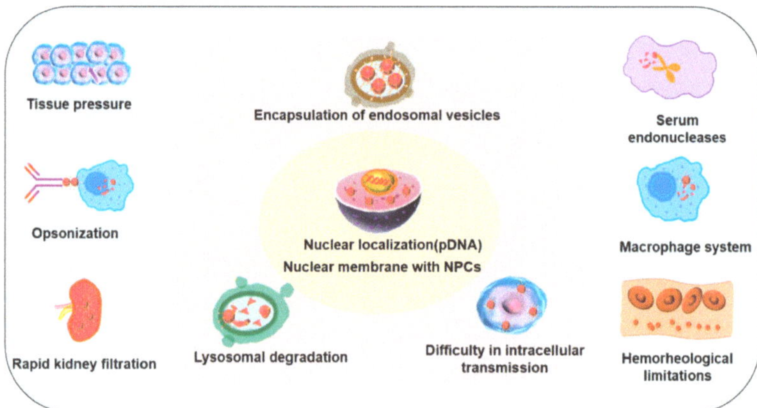

Figure 5. Biological barriers to CPP-based delivery systems. Biological barriers include tissue pressure, opsonization, rapid kidney filtration, serum endonucleases, a macrophage system, and hemorheological limitations. After the delivered macromolecular drugs enter the target cells, the biological barriers mainly include the encapsulation of endosomal vesicles, lysosome degradation, and the difficulty in intracellular transmission. The delivery system loaded with pDNA must enter the nucleus to function; thus, the NPC is another barrier to the pDNA delivery system.

5. Optimization Strategies of CPP-Based Delivery Systems

5.1. Enhancing the Endosomal Escape

Owing to the low endosomal escape efficiency of most non-viral delivery vehicles [96], encapsulation has been recognized as a key challenge for macromolecular drug delivery systems [97]. Even though one study reported that non-viral vehicles modified with CPP may inhibit drug degradation in lysosomes [98], the endosomal escape ability of most non-viral vehicles remains limited. Generally, the pH buffering effect of the proton group causes instability of the endosomal membrane or fusion with the lipid bilayer of the endosome. This increases the endosomal escape of non-viral vehicles such as membrane cleavage peptides and polymers, lysosome agents, and fusion lipids [99]. Among them, pH-sensitive peptides and polymers are widely used for endosomal escape [100–103]. In addition, Shiroh Futaki's team developed a lipid-sensitive endosomal lytic peptide that is totally different from the pH-sensitive cleavage peptide, and it was successfully used for antibody delivery [85].

Using the CPP delivery strategy, combined with endosomal escape, Xia Ningshao's team proposed a multifunctional chimeric peptide eTAT bio-macromolecular delivery system that greatly enhances the lysosome escape efficiency and delivery efficiency; it was successfully used in a mouse model with acute liver failure [104]. The protein delivery system consists of four modules in series: a CPP (TAT), a pH-dependent membrane-active peptide (INF7), endosome-specific protease sites, and a leucine zipper. The acidification allosteric of INF7 can destroy the endocytic vesicle membrane, and then with endosome-specific protease hydrolysis and cleavage, macromolecular protein escapes from the destroyed endosomal membrane and dimerizes through the leucine zipper, enhancing the serum tolerance, and further increasing the number of macromolecular cargos that escaped from the endosome [104]. In addition, polyethyleneimine (PEI) plays a unique proton sponge role that promotes endosome escape to a certain extent; it is usually used in combination with other components in the delivery system. In the CPP-based delivery system, PEI combined with other effectors can deliver siRNA successfully and can inhibit various kinds of tumors [105]. Since PEI still has cytotoxic and non-biodegradable properties [106,107], further optimization strategies have been proposed to overcome them [108,109].

5.2. Extending Half-Life in Blood

The shape and size of CPP-based vehicles can be designed to extend their half-life in the blood. It is believed that complex nanoparticles or nanovehicles at around 100 nm are an ideal choice for prolonging their half-life in the blood. Exosomes derived from living cells have excellent biocompatibility, low immunogenicity, and good delivery capabilities [110,111]. Exosomes hold great promise to serve as vehicles for targeted drug delivery. Loading superparamagnetic iron oxide nanoparticles (SPIONs) and curcumin (Cur) into exosomes and then conjugating the exosome membrane with neuropilin-1-targeted peptide (RGERPPR, RGE) by click chemistry can be used to obtain glioma-targeting exosomes with imaging and therapeutic functions [110,111]. Mechanical extrusion of approximately 10^7 cells grafted with lipidated ligands can generate cancer cell-targeting extracellular nanovesicles (ENV). And aptamer-conjugated nanovesicles have been used for targeted drug delivery [110,111]. By using arginine-rich CPPs to induce active macropinocytosis, the cellular uptake of CPP-modified EV can be enhanced. The induction of macropinocytosis via a simple modification to the exosomal membrane using stearylated octaarginine, which is a representative CPP, significantly enhances cellular EV uptake efficacy. Consequently, effective EV-based intracellular delivery of an artificially encapsulated ribosome-inactivating protein, saporin, in EVs was attained [112]. By designing a novel peptide-equipped exosome platform that can be assembled under convenient and mild reaction conditions, the Bangce Ye team was able to bind the HepG2 cell-derived exosome surface to CPP (R9) [113]. This not only enhances the penetrating capacity of exosomes but also assists exosomes in loading ASOs. Interestingly, such a CPP-equipped delivery system remarkably increases the delivery of ASO G3139 to silence anti-apoptotic protein Bcl-2 [113].

5.3. Targeting CPPs

The targeting CPPs contain peptide sequences that have natural targeting abilities and can penetrate the cell membrane by interacting with the cell surface receptors when they reach a specific site [114]. Adding a targeting moiety could reduce the side effects by reducing the accumulation of drugs in non-targeted tissues; however, it may increase the complexity of peptides, the difficulty of synthesis, and the price [115]. Therefore, the target sequence should be added with caution. Several targeting CPPs are introduced below, including R6LRVG, tLyP-1, and iRGD.

5.3.1. R6LRVG Targeting CPP

Oral medication is simple and convenient for patients, and the slow-release absorption of the drug reduces the side effects [116]. R6LRVG targeting CPP is an intestinal targeting penetrating peptide that consists of CPP (R6) and LRVG peptide [117]. Usually, LRVG peptide is used to target the intestinal epithelial cells in order to improve the oral bioavailability of drugs. Tyroserleutide (YSL), a tripeptide extracted from pig spleen, can suppress the invasion of mouse cancer cells [118]. Based on this evidence, Liefeng Zhang's team designed YSL-PLGA nanoparticles modified by R6LRVG (YSLPLGA/R6LRVG nanoparticles) and enhanced the absorption of YSL for oral anti-tumor therapy [119].

5.3.2. tLyP-1 Targeting CPP

Tumor-specific CendR peptides contain a tumor-homing motif and a cryptic CendR (C-terminal C-end Rule) motif that is proteolytically unmasked in tumor tissue. The cyclic tumor-homing peptide, LyP-1 (CGNKRTRGC), contains a CendR element and can penetrate tissue. In the truncated form of LyP-1, in which the CendR motif is exposed (CGNKRTR; tLyP-1), both LyP-1 and tLyP-1 internalize into cells through the neuropilin-1-dependent CendR internalization pathway. The targeted penetrating peptide tLyP-1 can promote tissue penetration through NRP-1-dependent Cend rule (CendR) internalization, thereby increasing the delivery of macromolecular drugs to the target tissue [120,121]. Guo Yuan and Qiu Zeng's team designed a peptide-functionalized dual-targeted delivery system that encapsulates paclitaxel and GANT61 (an inhibitor of SHH signaling pathway) in a tLyP-1 peptide-modified reconstituted high-density lipoprotein nanoparticle (tLyP-1-rHDL-PTX/GANT61 NP) for metastatic triple-negative breast cancer (TNBC) treatment. The apolipoprotein A-1 and tLyP-1 peptide, modified on the surface of nanoparticles, enable the delivery system to target tumor cells by binding to the overexpressed scavenger receptor B type I and neuropilin-1 receptor. Moreover, the tLyP-1 peptide also enables the deep tumor penetration of nanoparticles, thus further facilitating paclitaxel and GANT61 delivery. In metastatic TNBC model, such a nanoparticle delivery system successfully inhibited the growth and metastasis of a primary tumor, and it has good biological safety [122].

5.3.3. iRGD Targeting CPP

iRGD (CRGDKGPDC) is a cyclic tumor-targeting penetrating peptide, composed of a tumor-specific RGD motif (recognizing highly expressed integrins in cancer vasculature) and a CendR motif (which can enhance vascular permeability) [39]. Because iRGD has tumor-targeting properties and can efficiently penetrate the blood-brain barrier (BBB) [123,124], it has been extensively studied. Yunfeng Lin's team used iRGD to enhance targeted delivery to tumor sites. Briefly, Doxorubicin (DOX), iRGD-NH2, and tetrahedral framework nucleic acid (tFNA) are mixed in sequence to prepare tFNA/DOX@iRGD, which can target αv integrin-positive breast cancer cells and transfer them to lysosomes to release cytotoxicity [125]. The iRGD modification enhances the penetration of tFNA, which is used to target the delivered chemotherapeutics to triple-negative breast cancer. iRGD can also be used to improve the radiotherapy of glioblastoma [124]. The iRGD-modified ultrasmall single-crystal Fe nanoparticles, combined with immune checkpoint blocking immunotherapy, promoted ferroptosis and immunotherapy [126]. In addition, the Ppa-iRGDC-BK01 (iRGD derivative) can exert a multiple-quenching effect on the photosensitivity. This effectively prevented non-specific phototoxicity in the light-responsive nano-delivery system [127].

5.4. Stimuli-Responsive Strategies

Stimuli-responsive strategies are designed for the characteristic differences between target organizations and non-target organizations. Different factors, including pH value, enzyme activity, light radiation, reactive oxygen species (ROS), redox potential, adenosine triphosphate, temperature gradient, ultrasonic energy, and magnetic field, are utilized to design activatable CPPs (ACPPs) [128,129]. Combining stimulus-responsive strategies can greatly improve the targeting delivery system and can enhance the therapeutic effect of diseases. Several stimuli-responsive strategies have been integrated with CPP-based macromolecular drug delivery systems (Figure 6 and Table 1).

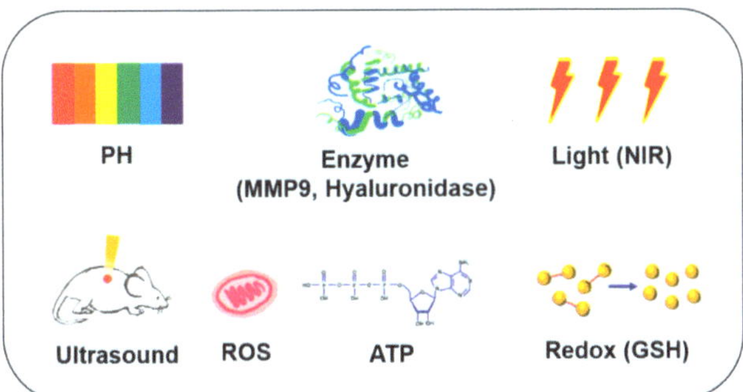

Figure 6. The main stimuli-responsive strategies of CPP-based macromolecular drug delivery. This figure lists the main stimulus factors used in stimuli-responsive strategies of CPP-based delivery systems in recent years, including pH, enzyme, light, ultrasound, ROS, ATP, and redox.

5.4.1. pH-Responsive Strategy

Generally, the acidity of tissues in pathological conditions (e.g., malignancy) is higher than that of normal tissues. The pH values of different intracellular substructures also differ. For example, endosomes and lysosomes have a higher acidity than the remaining subcellular sites [130]. Therefore, pH-sensitive nanoparticles can utilize this low-pH microenvironment to deliver therapeutic macromolecular drugs to target tissues [131]. This pH-responsive strategy has been extensively studied in the field of therapeutic delivery. Generally, pH-responsive nanoparticles can be synthesized by introducing ionizable pH-sensitive functional groups (such as amines or carboxylic acids) or acid-labile chemical bonds (such as hydrazone or ester bonds) with a specific acidity [132,133]. In a CPP-based macromolecular drug delivery system, Chunmeng Sun's team combined CPPs (R6) with pH-responsive nanoliposomes to deliver Artemisinin to inhibit tumor growth in mice with breast cancer [134]. Four CPP-activation strategies are frequently used to design pH-responsive macromolecule delivery: (1) pH-sensitive linker; (2) pH-sensitive conformational change; (3) pH-sensitive charge conversion; and (4) pH-sensitive side-chain modification (Table 1). These strategies mainly aim to remove the inhibitory domains or change the residue modifications that sterically mask the CPP.

5.4.2. Enzyme-Responsive Strategy

Enzymes play an important role in biological metabolism, with a high degree of selectivity and specificity [135]. Owing to the high expression of matrix metalloproteinase (MMP), hyaluronidase (HAase), proteolytic enzymes, and other extracellular enzymes in tumor tissues, developing macromolecular drug delivery systems that respond to different enzyme concentrations have been promoted [136].

Through integrating an MMP-responsive strategy into a CPP-based delivery system, Gao Jing's team designed an integrated hybrid nanovesicle composed of cancer cell membranes (Cm) and matrix metallopeptidase 9 (MMP-9)-switchable peptide-based charge-reversal liposome membranes (Lipm) that coat lipoic acid-modified polypeptides (LC) co-loaded with phosphoglycerate mutase 1 (*PGAM1*) siRNA (siPGAM1) and DTX. This hybrid nanovesicle comprises a cancer cell membrane (enhancing the ability of homologous targeting) and a CPP-bound MMP-9 sensitive peptide (R9-PVGLIG-EGGEGGEGG) that carries DTX, lipoic acid-modified Polypeptide (LC) and siRNA of *PGAM1* (a key aerobic glycolytic enzyme in cancer metabolism) to treat lung cancer [137]. This delivery system exhibits a favorable negative surface charge that maintains the stability of systemic circulation. It accumulates in tumor sites overexpressing MMP-9, exposes positively charged CPPs after MMP lysis, and leads to enhanced internalization in the target cells [138]. Moreover, this delivery system has no obvious toxicity and successfully prolongs the life span of tumor-xenografted mice. Hyaluronic acid (HA) is the main endogenous ligand of CD44. It is usually used to construct hyaluronidase-responsive vehicles and to deliver siRNA [139], pDNA [140], and other genetic materials. Nanoparticles modified with HA can effectively target cancer cells that overexpress CD44 [81]. To integrate the hyaluronidase-responsive strategy into a CPP delivery system, Jianping Liu's team developed hyaluronic acid (HA)-coated LOX-1-specific siRNA-condensed CPP nanocomplexes (NCs) and showed that HA-coated CPP NCs were promising as nanovehicles for efficient macrophage-targeted gene delivery and antiatherogenic therapy [141]. They also revealed that macrophages internalized these types of NCs via the caveolae-mediated endocytosis pathway [141]. Tingjie Yin's team presented a biomineralization-inspired dasatinib (DAS) nano drug (CIPHD/DAS) that sequentially overcomes all the abovementioned hindrances for the efficient treatment of solid tumors. CIPHD/DAS exhibited a robust hybrid structure constructed from an iRGD-modified hyaluronic acid-deoxycholic acid organic core and a calcium phosphate mineral shell. This optimized strategy, with sequential permeabilization of the capital "leakage obstacles", validates a promising paradigm that overcomes the "impaired delivery and penetration" bottleneck associated with nano drugs in the clinical treatment of solid tumors [142]. Aside from the MMPs and hyaluronidase, cathepsin, elastase, autophagy-specific enzymes, and aminopeptidase have also been utilized to trigger the removal of an inhibitory domain or to change the residue modification that masks the CPP activity (Table 1).

5.4.3. Light-Responsive Strategy

The tumor vascular system is characterized by a poor structure, uneven branching, and uneven distribution, which makes the spatial distribution of nanovehicle-based macromolecular drugs disorderly [143]. Usually, a light-responsive strategy has a good effect on tumor treatment by directly acting on the tumor site. The stimuli used for light response strategies usually include ultraviolet light (UV; 10–400 nm), visible light (400–750 nm), and near-infrared light (NIR; 750–900 nm), among which NIR is widely used to integrate with CPP-based drug delivery systems. Yang Ding's team proposed functionalized peptide–lipid hybrid particles, which were applied to encapsulate a PLGA polymeric core together with indocyanine green (ICG) and packaged by a lipoprotein-inspired structural shell. To initiate the precision tumor-penetrating performance, tLyP-1-fused apolipoprotein A-I-mimicking peptides (D4F) were utilized to impart tumor-homing and tumor-penetrating biological functions. The sub-100 nm drug vehicle possessed a long circulation time with uniform mono-disparity; however, it was stable enough to navigate freely, penetrate deeply into tumors, and deliver its cargo to the targeted sites. Moreover, ICG-encapsulated penetrable polymeric lipoprotein particles (PPL/ICG) could achieve real-time fluorescence/photoacoustic imaging and could monitor the in vivo dynamic distribution. Upon NIR laser irradiation, PPL/ICG exhibited a highly efficient phototherapeutic effect that eradicated the orthotopic xenografted tumors with good biosafety [144]. These external triggers, such as UV light, have also been explored to activate different types of CPPs by removing the inhibitory domain or changing the residue modification or conformational transition; however, they were validated relatively less when in vivo (Table 1).

5.4.4. ROS-Responsive Strategy

Generally, ROS levels would be significantly increased due to an abnormal metabolism in a variety of pathological conditions, including diabetes, cancer, premature senescence, and neurodegenerative diseases [145]. The ROS-responsive strategy for the site-specific delivery of macromolecular drugs was developed by exploiting the differences in ROS levels between healthy and pathological tissues. Tumor cells overexpress FGL1 and PD-L1, which, respectively, bind to LAG-3 and PD-1 on T cells, forming important signaling pathways (FGL1/LAG-3 and PD-1/PD-L1) that negatively regulate immune responses. In order to interfere with the inhibitory function of the FGL1 and PD-L1 proteins, Xuenong Zhang's team designed a new type of ROS-sensitive nanoparticle to load *FGL1* siRNA (siFGL1) and *PD-L1* siRNA (siPD-L1), which formed a stimuli-responsive polymer with a poly-l-lysine-thioketal and modified cis-aconitate to facilitate endosomal escape. Moreover, the tumor-penetrating peptide iRGD and the ROS-responsive nanoparticles were co-administered to further enhance the delivery efficiency of si*FGL1* and si*PD-L1*, thereby significantly reducing the protein levels of FGL1 and PD-L1 in tumor cells. This nanoparticle construction had good tumor microenvironment responsiveness, and the delivery efficiency was sharply enhanced [146]. Most ROS-responsive strategies are designed by introducing a 4-boronic mandelic acid moiety between a cationic CPP and an anionic inhibitory domain (Table 1).

5.4.5. Other Responsive Strategies

The integration of an ultrasound-responsive strategy into a CPP-based delivery system has also rapidly developed in recent years. Among the ultrasound-responsive strategies, low-intensity focused ultrasound (LIFU) is widely used. JianLi Ren's team successfully developed novel tumor-homing-penetrating peptide-functionalized drug-loaded phase-transformation nanoparticles (tLyP-1-10-HCPT-PFP NPs) for LIFU-assisted tumor ultrasound molecular imaging and precise therapy. Induced by the tLyP-1 peptide with targeting and penetrating efficiency, the tLyP-1-10-HCPT-PFP NPs could increase tumor accumulation and penetrate deeply into the extravascular tumor tissue, penetrating through the extracellular matrix and the cellular membrane into the cytoplasm. With LIFU assistance, the tLyP-1-10-HCPT-PFP NPs could phase-transform into microbubbles and enhance the tumor ultrasound molecular imaging for tumor diagnosis [147]. The ATP-responsive strategy has also been integrated with a CPP-based macromolecular drug delivery system. For example, Kaiyong Cai's team reported an adenosine triphosphate (ATP)-responsive nanovehicle with zeolitic imidazolate framework-90 (ZIF-90) for breast cancer combination therapy. Briefly, Atovaquone (AVO) and hemin are loaded into ZIF-90; then, the iRGD peptide is modified on the ZIF-90 nanoplatform. Hemin can specifically degrade BTB and CNC homology1 (BACH1), consequently changing the mitochondrial metabolism; AVO acts as the inhibitor of the electron transport chain (ETC). The degradation of BACH1 using Hemin can effectively improve the anti-tumor efficiency of the mitochondrial metabolism inhibitor, AVO, by increasing its dependency on mitochondrial respiration. This nano platform displays both tumor-targeting and mitochondria-targeting capacities, along with a high level of ATP-responsive drug release behavior and limited side effects [148]. In addition, it was reported that the strategies regarding the hypoxia-responsive fusion/conjugate constructs and the GSH-responsive release of the inhibitory domain could be used to design an ACPP-based macromolecular delivery system (Table 1).

5.5. Multiple Stimuli-Responsive Strategies

Monotherapy, with a continuous low dose, usually cannot completely suppress tumor growth; consequently, it is easy to increase the risk of metastasis and drug resistance [149,150]. Therefore, the combined use of multiple stimuli-responsive therapies may have a better therapeutic effect. For example, Tingjie Yin's team combined hyaluronidase-responsive, light-responsive, and tumor-targeted peptide strategies to achieve a focus-specific penetrated delivery with photothermal therapy-mediated chemosensitization and photothermal therapy-strengthened Integrin targeting. By combining polyethylene glycol (PEG), hyaluronic acid (HA), and iRGD-modified graphene oxide (GO), they constructed iRGD-modified GO nanosheets (IPHG). The IPHG can be actively transported through the vasculature, significantly improving the infiltration of drugs in tumors. After the tumors infiltrated by IPHG-DOX are exposed to NIR stimulation, the induced photothermal effect makes the tumors susceptible to chemotherapy and inhibits cytoskeleton remodeling. Consequently, IPHG-DOX significantly inhibited the growth and metastasis of breast cancer in situ, and it could be used to prevent tumor resistance and metastasis caused by poor chemotherapy targeting [151].

The combination of redox-responsive and light-responsive nanoparticles with CPPs has also been developed to enhance chemo-photodynamic therapy. For example, by co-administering tumor-penetrating peptide iRGD and GSH-responsive SN38-dimer (d-SN38)-loaded nanoparticles, a gradual stimulation response strategy was developed [152]. These nanoparticles were transformed into nanofibers concomitantly when the tumor site was irradiated by laser to promote their retention in the tumor and the burst release of ROS for photodynamic therapy. D-SN38, loaded with disulfide bonds, responds to high levels of GSH at the tumor site, resulting in the release of SN38 and excellent chemo-photodynamic effects. This enhanced chemo-photodynamic therapy effect produced high anti-tumor effects in breast tumor models [152]. In addition, to achieve synergistic chemodynamic therapy and chemotherapy, the iRGD-modified theranostic nano drug (iRPPA@TMZ/MnO) contains pH-responsive and redox-responsive manganese oxide, which provides a new type of therapeutic, diagnostic nanomedicine for brain MRI diagnosis and the treatment of glioma [123]. There are also additional stimuli-responsive strategies based on the ultrasound response. For example, Jianping Liu's team developed a TAT-based hyaluronidase-responsive strategy combined with ultrasound to provide a new method for the precise treatment of liver cancer [141].

All in all, CPPs can be unidirectionally inactivated by introducing inhibitory domains or when bulky groups are undesired, side-chain modifications can be used to mask the CPP activity. Masking groups can be removed by local triggers, such as enzymes or a low pH, as well as external triggers, such as light. Aside from these temporary activations, reversible activation has also been achieved by controlling the conformation of CPPs. Notably, the triggers used to activate CPPs are not so binary when in vivo. Some enzymes that are overexpressed in diseased tissue are also expressed in lower amounts in healthy tissue, and gradient pH variations can also be observed between tissues. Light-triggered CPP activation is beneficial to create temporal and special control; however, it is challenged by the poor tissue penetration depth as well as the potential cellular toxicity induced by the harmful wavelengths. Considering the heterogeneity and complexity of a disease microenvironment, multiple stimuli-responsive strategies may be more promising for designing CPP-based macromolecular drug delivery systems (Table 1).

Table 1. Stimuli-responsive strategies integrated with CPP-based macromolecular drug delivery.

Responding Strategy	Activatable/Specific Moiety	Loaded Drugs	CPPs	Disease Model	Refs
pH-responsive	pH-sensitive linker	Irinotecan- and *miR-200*-loaded liposomes and lipid nanoparticles	RF CPP: LKARFH. NG2 targeting the H peptide. Mitochondria targeting the K peptide. (PEG-lipid derivative with an imine bond confers pH-responsive release, internalization, and intracellular distribution in acidic microenvironment)	Colon cancer	[153]
	pH-sensitive linker	*PLK-1* siRNA-loaded liposome	ehGehGehGehG-(hydrazone)-RRRRRRRR. (Low pH triggers hydrazones to hydrolyse, resulting in loss of the inhibitory domain)	N/A	[154]
	pH-sensitive conformational change	DOX-loaded micelle	PLA-PEG -polyHis-GCGGGYGRKKRRQRRR. (Imidazole confers histidines that act as a pH trigger. Low pH protonates histidine, causing it to lose its hydrophobic interactions and the exposed Tat)	Ovarian cancer, breast cancer, and lung cancer	[155]
	pH-sensitive conformational change	PTX-loaded liposome	AGYLLGHINLHHLAHL(Aib)HHILC. (the H side-chain charges. Endowed pH responsiveness after complete replacement of all lysine in the sequence with histidine)	Colon cancer	[156]
	pH-sensitive conformational change	PET- and SPECT-probes, gold nanoparticles, and magnetic nanoparticles	pHLIP-var3: ACDDQNPWRAYLDLLFPTDTLLLDLLW. pHLIP-var7: ACEEQNPWARYLEWLFPTETLLLEL. (low pH insertion peptide pHLIP reversibly folds and is inserted across membranes in response to pH changes)	Cervix cancer, lung cancer, pancreatic cancer	[157,158]
	pH-sensitive conformational change	PTX	(LHHLCHLLHHLCHLAG)$_2$. (Disulfide oxidation forms LH$_2$ dimeric peptide. Lysine is substituted for histidine for endosomal escape, and the dimeric form of amphipathic CPPs shows enhanced CPP activities)	Breast cancer	[159]
	pH-sensitive charge conversion	ART-loaded liposome	HEHEHEHEHEHEHEHEHEGGGGGRRRRRR. (the histidine-glutamic acid-based masking peptide is modified to R6 via a spacer of 5-mer glycine)	Breast cancer	[134]

Table 1. *Cont.*

Responding Strategy	Activatable/Specific Moiety	Loaded Drugs	CPPs	Disease Model	Refs
Enzyme-responsive	pH-sensitive charge conversion and structure shift	TRAIL- and PTX-co-delivered liposomes	c(RGDfK)-AGYLLGHINLHHLAHL(Aib)HHIL-Lys-C18. (a histidine-rich peptide for pH responsiveness, c(RGDfK) peptide for $\alpha v\beta 3$ binding, and stearyl chain C18 for membrane anchoring)	Melanoma	[160]
	pH-sensitive side-chain modification	DOX	CRRRRRRRRGGGPKKKKKK. (Conjugated DMA to lysine induces intramolecular electrostatic interactions with arginine, thereby inactivating ACPP. Low pH triggers labile amides that are hydrolyzed)	Liver cancer	[161]
	pH-sensitive side-chain modification	DOX-loaded PEG-PCL micelle	YGRᵃKᵃKRRQRRRC. (Amidized CPPs. Conjugated succinyl moieties to the glutamine and both lysine residues of Tat)	Ovarian cancer	[162]
	MMP-9-sensitive linker	DNase I- and PTX prodrug-loaded NET-regulated nanoparticle	GRKKRRQRRRPQPLGLAGGC. (MMP-9 substrate peptide linked to Tat)	Breast cancer, lung cancer	[163]
	MMP9-sensitive linker	CsA-loaded, MMP-9-sensitive CPP-decorated reconstituted lipoprotein nanoparticles	ACFAEKFKEAVKDYFAKFWDGSGRRRR-RRRRRPVGLIGEGGEGGEGG. (MMP-9 substrate peptide conjugating with APOA-I mimics α-helix peptide through a GSG linker)	Traumatic brain injury	[138]
	MMP9-sensitive linker	*PGAM1*-siRNA- and DTX-loaded nanovesicles	RRRRRRRRRPVGLIGEGGEGGEGG.	Lung cancer	[137]
	MMP-2 & -9-sensitive linker	Cy5, Gadolinium chelates	EEEEEEEE-PLGLAG-RRRRRRRRR. EEEEEE-PLGLAG-RRRRRRRRR. (polycationic CPP is coupled via a cleavable linker to a neutralizing peptide)	Image-guided surgery of different kinds of tumors	[164,165]
	HAase-sensitive linker	HA-coated, *LOX-1*-siRNA-loaded nanocomplexes	RQIKIWFQNRRMKWKK.	Atherosclerosis	[141]
	Cathepsin-sensitive linker	Dox-loaded SiO2 nanoparticles	EEEEEEPGFKYGRKKRRQRRR.	Lung cancer, ovarian cancer	[166]
	Elastase-sensitive linker	Cy5	EEEEEEEEE-RLQLK(Ac)L-RRRRRRRRR.	Breast cancer	[167]
	PSA-sensitive linker	*PLK-1* siRNA-loaded liposomes	DGGDGGDGGDGG-HSSKYQ-RRRRRRRR. (PSA is serine protease)	Prostate cancer	[168]

Table 1. *Cont.*

Responding Strategy	Activatable/Specific Moiety	Loaded Drugs	CPPs	Disease Model	Refs
	ATG4B-sensitive linker	DTX and CQ—loaded nanoparticles	GTFGFRRRRRRRRR. (Autophagy-specific enzyme ATG4B substrate linked to R9)	Melanoma	[169]
	APN-DPP4-sensitive side-chain modification	FITC	GRKKRRQRRRAhxC (Side chain modifications. Aminopeptidase N dipeptidyl peptidase IV)	N/A	[170]
Hypoxia-responsive	Oxygen-sensitive degradation of fusion protein	ODD-beta-Gal	YGRKKRRQRRR-ODD-Casp3(wt) (Tat-oxygen-dependent degradation domain-Caspase 3 fusion protein is selectively stabilized in hypoxic tumors)	Pancreatic cancer	[171]
	Azoreductase-sensitive modification	Peptide nucleic acid (PNA)	MVTVLFRRLRIRRACGPPRVRV-azo-PEG (Activatable CPP-PEG conjugates. Azoreductase-triggered CPP-inactivation through functionalization with a self-immolative azobenzene moiety)	Colon mucosa	[172]
ROS-responsive	ROS-sensitive polymer	*FGL1*-siRNA, PD-L1-siRNA	c(CRGDKGPDC) (Proteolysis of iRGD peptide exposes a new motif that can bind to NRP-1 and activate neuropilin, allowing drugs or nanoparticles to leak out from tumor blood vessels and penetrate the tumor tissue.)	Liver cancer	[146]
	ROS-sensitive linker	FITC, Cy5	EEEEEEEEE-cleavable linker-RRRRRRRRR. (H_2O_2-activated CPP. A boronic acid-containing cleavable linker between polycationic CPP and polyanionic fragments)	Lung inflammation	[173]
ATP-responsive	ATP-sensitive release of guest molecules	Photosensitizers	Ac-QYFMpTEpYVA (ATP-triggered release of phosphopeptides from the pegylated GC5A-12C nanocarrier (12C-NC) system. Host-guest ATP-responsive system)	N/A	[174]
	ATP-sensitive disintegration	Atovaquone (AVO), hemin	c(CRGDKGPDC)-ZIF-90 (iRGD peptide-modified ZI-90/protein nanoparticles disintegrate in the presence of ATP to release protein as a result of the competitive coordination between Zn2+ and ATP)	Breast cancer	[148]

Table 1. *Cont.*

Responding Strategy	Activatable/Specific Moiety	Loaded Drugs	CPPs	Disease Model	Refs
Ultrasound-responsive	Ultrasound-assisted phase-Transformation	Hydroxycamptothecin	CGNKRTR. (Tumor homing-penetrating peptide-functionalized drug-loaded phase-transformation nanoparticles tLyP-1-10-HCPT-PFP)	Breast cancer	[147]
	Ultrasound-activated cavitation effect	Pefluoropentane, 10-Hydroxycamptothecin-loaded liposome nanoparticle	CGNKRTR. (Truncated form of LyP-1 CPP (CGNKRTRGC))	Breast cancer	[175]
	Ultrasound-dependent endosomal escape	shRNA	Tat-U1A-rose bengal conjugate (Tat cell-penetrating peptide, U1A RNA-binding protein, and rose bengal as a sonosensitizer)	N/A	[176]
GSH-responsive	GSH-sensitive disulfide linker	Podophyllotoxin (PPT), Doxorubicin	PRASHANT. (anti-mitotic PRA octapeptide-linked PPT conjugate that can self-assemble into a vesicle via water and targeted synergistic drug delivery)	N/A	[177]
Light-responsive	UV light-sensitive self-immolative linker	Doxorubicin	ARTKQTARKSTGGKAPRKQLATKAARKSA-PATGGC^{35}KKPHRYRPGTVALREIRRYQKST-ELLIRKLPFQRLVREIAQDFKTDLRFQSSAV-MALQEASEAYLVALFEDTNLAAIHAKRVTI-MPKDIQLARRIRGERA. (H3-^{35}PC4AP. PC4AP (a photo-caged C4'-oxidized abasic site) as a light-responsive, self-immolative linker to conjugate drugs to a CPP)	N/A	[178]
	UV light-sensitive linker	Quantum dots, polystyrene particles, Au nanostars, and liposomes	RRRRRRR-o-nitrobenzyl-GGGEEEEEEE. (a photo-caged peptide that undergoes a structural transition from an antifouling ligand to CPP upon photo-irradiation)	N/A	[179]
	NIR-sensitive side-chain modification	VB-loaded liposome	CGRRMK^{PG}WK^{PG}K^{PG}. NGR peptide: CYGGRGNG; Synergistic effect (light-released photolabile-protective group PG (4,5-dimethoxy-2-nitrobenzene chloroformate))	Fibrosarcoma	[180]
	UV-sensitive side-chain modification	Proapoptotic peptide (KLAKLAK)$_2$	Ac-KRRMK^{Nvov}WK^{Nvoc}K^{nvoc}. (Nvoc=6-nitroveratrylcarbonyl; light-activated caged Pen CPP; photo-cleavable groups)	N/A	[181]
	UV-sensitive conformational change	Tamra	cis-Ab-LK. Trans-Ab-LK azobenzene (Ab) linker	N/A	[182]

Table 1. *Cont.*

Responding Strategy	Activatable/Specific Moiety	Loaded Drugs	CPPs	Disease Model	Refs
	UV/Vis-sensitive conformational change	RhoB	RRRRRRRRR-AB-EEEEEEEEE. (cis-to-trans isomerization of azobenzene (AB) moiety; photoswitchable)	N/A	[183]
	UV light-sensitive linker inhibitory domain	Atto655-loaded liposome	YGAKKARQRRAGC-PEG-loop. (modified on both termini of Tat with an alkyl chain; UV-cleavable linker)	N/A	[184]
Multiple-responsive	NIR- and pH-dual sensitive linker	*EGFR* siRNA	CGRRMKWKK-DMNB-EEEERRRR. (CPP is quenched by a pH-sensitive inhibitory peptide, which is linked via a photo-cleavable group DMNB)	Breast cancer	[185]

6. Clinical Challenges of CPP-Based Macromolecular Drug Delivery

Several types of clinical trials have involved CPP-based macromolecular drug delivery (Table 2). For example, as a c-Jun N-terminal Kinase (JNK) inhibitor, the AM-111 penetrating peptide (brimapitide) drug is being tested to treat acute unilateral sudden deafness. The XG-102 drug (brimapitide), a TAT-coupled dextrogyre peptide that selectively inhibits the c-Jun N-terminal kinase for treating postoperative ocular inflammation, has entered Phase III clinical trials [186,187]. In addition, TAT-based class A botulinum toxin drugs have also been extensively explored in clinical practice [188,189]. Although several trials of initial CPP-based macromolecular drugs have been discontinued, advanced CPP-integrated macromolecules are still actively tested in clinics (Table 2).

Table 2. CPP-based therapeutics in clinical trials.

CPPs	Cargos	Recruitment Status	Application	Gov ID	Year	Refs
TAT	Botulinum toxin A	Phase IIIb (completed)	Cervical dystonia	NCT01753310	2012	[190]
TAT	JNKI-1	Phase III (completed)	Postoperative ocular inflammation	NCT02508337, 02235272	2015 2017	[187]
TAT	PSD-95 protein inhibitor	Phase III (completed)	Ischemic stroke	NCT02930018	2016	[191]
TAT	D-JNKI-1 gel	Phase III (completed)	Hearing loss, idiopathic sudden sensorineural	NCT02809118, 02561091	2016 2015	[186]
TAT	Botulinum toxin A	Phase II (completed)	Cervical dystonia	NCT02706795	2016	[188]
TAT	δ-PKC inhibitor	Phase II (completed)	Myocardial infarction	NCT00785954	2008	N/A
TAT	ε-PKC inhibitor	Phase II (completed)	Pain: postherpetic neuralgia, spinal cord injury, postoperative	NCT01106716, 01135108, 01015235	2010 2011 2013	[192,193]
TAT	PKC inhibitor	Phase II (completed)	Acute myocardial infarction	NCT00093197	2004	[194]
TAT	Botulinum toxin A	Phase I/II	Glabellar lines	N/A	2015	[189]
TAT	Dextrogyre peptide	Phase I (completed)	Intraocular inflammation and pain	NCT01570205	2012	[195]
TAT	MAGE-A3,HPV-16	Phase I (completed)	Head and neck carcinoma	NCT00257738	2005	[196]
TAT	Cu, Zn-Superoxide dismutase	Phase I	Obesity	N/A	2011	[197]
ATX-101	N/A	Phase Ib/IIa (recruiting)	Several cancers	NCT04814875	2021	[198]
AM-111	D-JNKI-1 gel	Phase II (completed)	Acute sensorineural hearing loss	NCT00802425	2008	[199]
P28	P28GST	Phase II (completed)	Intestinal inflammation	NCT02281916	2014	[200]
P28	P28, Non-HDM2-mediated peptide inhibitor of p53	Phase I	Central nervous system tumors	NCT01975116	2016	[201]
P28	P28, Non-HDM2-mediated peptide inhibitor of p53	Phase I (completed)	P53 ubiquitination in patients with advanced solid tumors	NCT00914914	2013	[202]
ALRN-6924	Palbociclib	Phase IIa (completed)	Solid tumor, Lymphoma, Peripheral T-cell lymphoma	NCT02264613	2014	[203]
ALRN-6924	Cytarabine	Phase I (completed)	Acute myeloid leukemia, Myelodysplastic syndromes	NCT02909972	2016	

Table 2. *Cont.*

CPPs	Cargos	Recruitment Status	Application	Gov ID	Year	Refs
ALRN-6924	Paclitaxel	Phase 1 (active)	Advanced, metastatic or unresectable solid tumors	NCT03725436	2018	
ALRN-6924	Cytarabine	Phase 1 (active)	leukemia, brain tumor, pediatric lymphoma	NCT03654716	2018	
ALRN-6924	Topotecan	Phase 1a (terminated)	Small cell lung cancer	NCT04022876	2019	
R7	Cyclosporin A	Phase II	Psoriasis	N/A	2003	[204]
(R-Ahx-R)$_4$	PMO	Phase III (terminated)	Cardiovascular disease, coronary artery bypass	NCT00451256	2007	[205]
(R-Ahx-R)$_4$	PMO targeted to human c-Myc	Phase II	Obstruction of vein graft after cardiovascular bypass surgery	N/A	2009	[206]
TransMTS	Botulinumtoxin A	Phase III (completed)	Cervical dystonia	NCT03608397	2018	[188]
MTS	Botulinumtoxin A	Phase III, Phase II, Phase II (completed)	Skin aging, hyperhidrosis, lateral canthal lines, crow's feet, and facial wrinkles	NCT02580370, 02565732	2015	[207,208]
AVB-620 (ACPP)	Cy5, Cy7	Phase II (completed)	Breast cancer	NCT03113825	2017	[209]
Pepducin	PZ-128	Phase II	Coronary artery disease	N/A	2015	[210]
AVB-620 (ACPP)	Cy5, Cy7	Phase I (completed)	Interpretative tumor detection using a ratiometric activatable fluorescent peptide	NCT02391194	2015	[211]
BT1718	Toxic DM1	Phase I/Iia (active)	Targeting MT1-MMP for treatment of solid tumors	NCT03486730	2018	[212]
PEP-010	Paclitaxel	Phase 1 (recruiting)	Metastatic solid tumor	NCT04733027	2021	
ATP128	BI 754091	Phase 1b (recruiting)	Stage IV colorectal cancer	NCT04046445	2019 2022	
PTD4	HSP20 phosphopeptide	Phase II (recruiting)	Scar prevention, reduction	NCT00825916	2009	[213]
Charged Oligo peptide	SN38	Phase I	Tumor	N/A	2016	[214]

Refs: References. Gov ID: Gov Identifier from Clinical trials.gov.

However, to date, the macromolecular drugs formulated with a CPP-based delivery system have not been formally approved by the FDA for marketing. Therefore, we proposed that a successful CPP-based macromolecular drug delivery system must be safe (low cytotoxicity, no residues after its action, and biodegradable), effective (a strong specificity and good effects), manufacturable (high drug-loaded, with a clear structure for scale-up, and uniform and stable materials), and cost suitable (Figure 7).

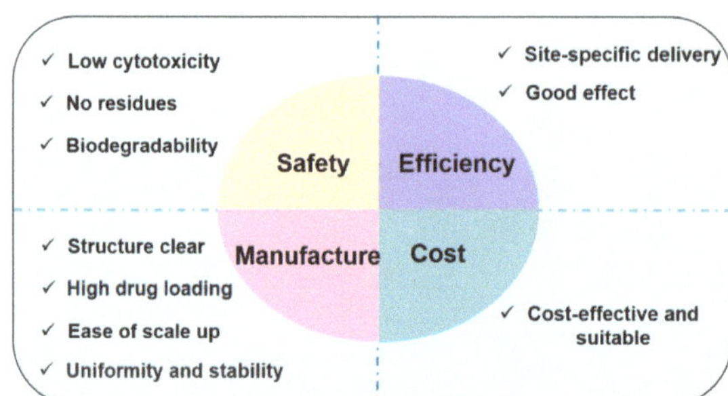

Figure 7. Conditions for successful CPP-based macromolecular drug delivery. In the experimental phase of the delivery system, safety and efficiency are the main concerns. In the final application to the clinic, manufacturing and cost will be essential factors to be considered. Successful CPP-based delivery of macromolecular drugs must concomitantly satisfy the conditions of safety, efficiency, manufacturing, and cost.

7. Challenges and Future Directions

By comparing different CPP-based macromolecular drug delivery platforms, this review deconstructs the literature into a comprehensive and understandable framework. (1) The source and classification of CPPs are introduced in detail; then, the cellular uptake mechanisms and influencing factors are analyzed based on the CPP-based delivery system. (2) The optimization strategies of CPP-based delivery systems, including improving endosomal escape, prolonging the half-life in blood, targeting CPPs, as well as single and multiple stimuli-responsive strategies, are examined. (3) The CPP-based delivery systems have achieved good curative effects in refractory diseases, including glioma, solid tumors, and triple-negative breast cancer. After reviewing the clinical attempts, the conditions for a successful CPP-based macromolecular drug delivery system are summarized to provide reference significance.

Discoveries from a CPP-based delivery system, combined with therapeutic macromolecular drugs, have prompted renewed attention in this field. However, a successful CPP-based delivery system has not yet been developed. In recent years, research on macromolecular drug therapy has mainly focused on breakthroughs in solving the stability of and targeting CPP-based delivery systems. Specifically, targeting without off-target toxicity is the most important key to ensuring the safety and efficacy of CPP-based macromolecular drug delivery, whereas good stability is an unavoidable key that should be scaled up. Although the current review proposes the latest optimization strategies, the complex structure of CPP-based macromolecular drug delivery systems, when combined with some targeting moiety, may bring about new safety and production issues. Thus, the CPP-based drug delivery platforms remain immature and need to be further improved in the future.

Author Contributions: S.Z.: conception and design, interpretation, and manuscript revision; Z.S., J.H., and C.W.: literature searching and reviewing, figure illustrations, and manuscript draft; Z.F.: manuscript discussion and revision. All authors have read and agreed to the published version of the manuscript.

Funding: This work was supported by the National Translational Science Center for Molecular Medicine Fund (F1034361), the Chinese Pharmaceutical Association-Yiling Pharmaceutical Innovation Fund (CPAYLJ202003), the Key Research Fund of the Tianjin Project and Team (XB202010), the Key Research and Development Program of Tianjin (20YFZCSY00450), and the Key Program of the Natural Science Fund of Tianjin (21JCZDJC00230).

Institutional Review Board Statement: Not applicable.

Informed Consent Statement: Not applicable.

Data Availability Statement: Not applicable.

Conflicts of Interest: The authors declare no conflict of interest.

References

1. Stewart, M.P.; Langer, R.; Jensen, K.F. Intracellular Delivery by Membrane Disruption: Mechanisms, Strategies, and Concepts. *Chem. Rev.* **2018**, *118*, 7409–7531. [CrossRef] [PubMed]
2. Hapala, I. Breaking the barrier: Methods for reversible permeabilization of cellular membranes. *Crit. Rev. Biotechnol.* **1997**, *17*, 105–122. [CrossRef] [PubMed]
3. Stephens, D.J.; Pepperkok, R. The many ways to cross the plasma membrane. *Proc. Natl. Acad. Sci. USA* **2001**, *98*, 4295–4298. [CrossRef] [PubMed]
4. Sinha, B.; Köster, D.; Ruez, R.; Gonnord, P.; Bastiani, M.; Abankwa, D.; Stan, R.V.; Butler-Browne, G.; Vedie, B.; Johannes, L.; et al. Cells respond to mechanical stress by rapid disassembly of caveolae. *Cell* **2011**, *144*, 402–413. [CrossRef]
5. Peraro, M.D.; van der Goot, G. Pore-forming toxins: Ancient, but never really out of fashion. *Nat. Rev. Genet.* **2015**, *14*, 77–92. [CrossRef]
6. Shi, H.; Xue, T.; Yang, Y.; Jiang, C.; Huang, S.; Yang, Q.; Lei, D.; You, Z.; Jin, T.; Wu, F.; et al. Microneedle-mediated gene delivery for the treatment of ischemic myocardial disease. *Sci. Adv.* **2020**, *6*, eaaz3621. [CrossRef]
7. Chen, S.; Haam, J.; Walker, M.; Scappini, E.; Naughton, J.; Martin, N.P. Recombinant Viral Vectors as Neuroscience Tools. *Curr. Protoc. Neurosci.* **2019**, *87*, e67. [CrossRef]
8. Naso, M.F.; Tomkowicz, B.; Perry, W.L., 3rd; Strohl, W.R. Adeno-Associated Virus (AAV) as a Vector for Gene Therapy. *BioDrugs* **2017**, *31*, 317–334. [CrossRef]
9. Davidson, B.L.; Breakefield, X.O. Viral vectors for gene delivery to the nervous system. *Nat. Rev. Neurosci.* **2003**, *4*, 353–364. [CrossRef]
10. Chen, X.; Gonçalves, M.A.F.V. Engineered Viruses as Genome Editing Devices. *Mol. Ther.* **2016**, *24*, 447–457. [CrossRef]
11. Wu, Z.; Yang, H.; Colosi, P. Effect of genome size on AAV vector packaging. *Mol. Ther.* **2010**, *18*, 80–86. [CrossRef] [PubMed]
12. Farrell, P.J. Epstein-Barr Virus and Cancer. *Annu. Rev. Pathol. Mech. Dis.* **2019**, *14*, 29–53. [CrossRef] [PubMed]
13. Sacks, D.; Baxter, B.; Campbell, B.C.V.; Carpenter, J.S.; Cognard, C.; Dippel, D.; Eesa, M.; Fischer, U.; Hausegger, K.; Hirsch, J.A.; et al. Multisociety Consensus Quality Improvement Revised Consensus Statement for Endovascular Therapy of Acute Ischemic Stroke. *Int. J. Stroke* **2018**, *13*, 612–632. [CrossRef] [PubMed]
14. Bitzer, M.; Armeanu, S.; Lauer, U.M.; Neubert, W.J. Sendai virus vectors as an emerging negative-strand RNA viral vector system. *J. Gene Med.* **2003**, *5*, 543–553. [CrossRef]
15. Ono, C.; Okamoto, T.; Abe, T.; Matsuura, Y. Baculovirus as a Tool for Gene Delivery and Gene Therapy. *Viruses* **2018**, *10*, 510. [CrossRef]
16. Zhou, Z.; Liu, X.; Zhu, D.; Wang, Y.; Zhang, Z.; Zhou, X.; Qiu, N.; Chen, X.; Shen, Y. Nonviral cancer gene therapy: Delivery cascade and vector nanoproperty integration. *Adv. Drug Deliv. Rev.* **2017**, *115*, 115–154. [CrossRef]
17. Chalbatani, G.M.; Dana, H.; Gharagouzloo, E.; Grijalvo, S.; Eritja, R.; Logsdon, C.D.; Memari, F.; Miri, S.R.; Rad, M.R.; Marmari, V. Small interfering RNAs (siRNAs) in cancer therapy: A nano-based approach. *Int. J. Nanomed.* **2019**, *14*, 3111–3128. [CrossRef]
18. Picanço-Castro, V.; Pereira, C.G.; Covas, D.T.; Porto, G.S.; Athanassiadou, A.; Figueiredo, M.L. Emerging patent landscape for non-viral vectors used for gene therapy. *Nat. Biotechnol.* **2020**, *38*, 151–157. [CrossRef]
19. Mintzer, M.A.; Simanek, E.E. Nonviral vectors for gene delivery. *Chem. Rev.* **2008**, *109*, 259–302. [CrossRef]
20. Wang, H.-X.; Li, M.; Lee, C.M.; Chakraborty, S.; Kim, H.-W.; Bao, G.; Leong, K.W. CRISPR/Cas9-Based Genome Editing for Disease Modeling and Therapy: Challenges and Opportunities for Nonviral Delivery. *Chem. Rev.* **2017**, *117*, 9874–9906. [CrossRef]
21. Gessner, I.; Neundorf, I. Nanoparticles Modified with Cell-Penetrating Peptides: Conjugation Mechanisms, Physicochemical Properties, and Application in Cancer Diagnosis and Therapy. *Int. J. Mol. Sci.* **2020**, *21*, 2536. [CrossRef] [PubMed]
22. De Laporte, L.; Rea, J.C.; Shea, L.D. Design of modular non-viral gene therapy vectors. *Biomaterials* **2006**, *27*, 947–954. [CrossRef] [PubMed]
23. Oliveira, G., Jr.; Zigon, E.; Rogers, G.; Davodian, D.; Lu, S.; Jovanovic-Talisman, T.; Jones, J.; Tigges, J.; Tyagi, S.; Ghiran, I.C. Detection of Extracellular Vesicle RNA Using Molecular Beacons. *iScience* **2019**, *23*, 100782. [CrossRef]
24. Habault, J.; Poyet, J.-L. Recent Advances in Cell Penetrating Peptide-Based Anticancer Therapies. *Molecules* **2019**, *24*, 927. [CrossRef] [PubMed]
25. Silva, S.; Almeida, A.J.; Vale, N. Combination of Cell-Penetrating Peptides with Nanoparticles for Therapeutic Application: A Review. *Biomolecules* **2019**, *9*, 22. [CrossRef] [PubMed]
26. Evans, B.C.; Fletcher, B.; Kilchrist, K.V.; Dailing, E.A.; Mukalel, A.J.; Colazo, J.M.; Oliver, M.; Cheung-Flynn, J.; Brophy, C.M.; Tierney, J.W.; et al. An anionic, endosome-escaping polymer to potentiate intracellular delivery of cationic peptides, biomacromolecules, and nanoparticles. *Nat. Commun.* **2019**, *10*, 5012–5019. [CrossRef]
27. Trabulo, S.; Cardoso, A.; Morais, C.; Jurado, A.S.; de Lima, M.P. Cell-penetrating peptides as nucleic acid delivery systems: From biophysics to biological applications. *Curr. Pharm. Des.* **2013**, *19*, 2895–2923. [CrossRef]
28. Frankel, A.D.; Pabo, C.O. Cellular uptake of the tat protein from human immunodeficiency virus. *Cell* **1988**, *55*, 1189–1193. [CrossRef]
29. Green, M.; Loewenstein, P.M. Autonomous functional domains of chemically synthesized human immunodeficiency virus tat trans-activator protein. *Cell* **1988**, *55*, 1179–1188. [CrossRef]

30. Joliot, A.; Pernelle, C.; Deagostini-Bazin, H.; Prochiantz, A. Antennapedia homeobox peptide regulates neural morphogenesis. *Proc. Natl. Acad. Sci. USA* **1991**, *88*, 1864–1868. [CrossRef]
31. Morris, M.C.; Vidal, P.; Chaloin, L.; Heitz, F.; Divita, G. A new peptide vector for efficient delivery of oligonucleotides into mammalian cells. *Nucleic Acids Res.* **1997**, *25*, 2730–2736. [CrossRef] [PubMed]
32. Vivès, E.; Brodin, P.; Lebleu, B. A Truncated HIV-1 Tat protein basic domain rapidly translocates through the plasma membrane and accumulates in the cell nucleus. *J. Biol. Chem.* **1997**, *272*, 16010–16017. [CrossRef] [PubMed]
33. Elliott, G.; O'Hare, P. Intercellular Trafficking and Protein Delivery by a Herpesvirus Structural Protein. *Cell* **1997**, *88*, 223–233. [CrossRef] [PubMed]
34. Pooga, M.; Hällbrink, M.; Zorko, M.; Langel, U. Cell penetration by transportan. *FASEB J.* **1998**, *12*, 67–77. [CrossRef] [PubMed]
35. Kardani, K.; Milani, A.; Shabani, S.H.; Bolhassani, A. Cell penetrating peptides: The potent multi-cargo intracellular carriers. *Expert Opin. Drug Deliv.* **2019**, *16*, 1227–1258. [CrossRef]
36. Morris, M.C.; Depollier, J.; Mery, J.; Heitz, F.; Divita, G. A peptide carrier for the delivery of biologically active proteins into mammalian cells. *Nat. Biotechnol.* **2001**, *19*, 1173–1176. [CrossRef]
37. Zhao, K.; Zhao, G.-M.; Wu, D.; Soong, Y.; Birk, A.V.; Schiller, P.W.; Szeto, H.H. Cell-permeable peptide antioxidants targeted to inner mitochondrial membrane inhibit mitochondrial swelling, oxidative cell death, and reperfusion injury. *J. Biol. Chem.* **2004**, *279*, 34682–34690. [CrossRef]
38. Yandek, L.E.; Pokorny, A.; Florén, A.; Knoelke, K.; Langel, U.; Almeida, P.F. Mechanism of the cell-penetrating peptide transportan 10 permeation of lipid bilayers. *Biophys. J.* **2007**, *92*, 2434–2444. [CrossRef]
39. Sugahara, K.N.; Teesalu, T.; Karmali, P.P.; Kotamraju, V.R.; Agemy, L.; Girard, O.M.; Hanahan, D.; Mattrey, R.F.; Ruoslahti, E. Tissue-penetrating delivery of compounds and nanoparticles into tumors. *Cancer Cell* **2009**, *16*, 510–520. [CrossRef]
40. Ezzat, K.; EL Andaloussi, S.; Zaghloul, E.M.; Lehto, T.; Lindberg, S.; Moreno, P.M.D.; Viola, J.R.; Magdy, T.; Abdo, R.; Guterstam, P.; et al. PepFect 14, a novel cell-penetrating peptide for oligonucleotide delivery in solution and as solid formulation. *Nucleic Acids Res.* **2011**, *39*, 5284–5298. [CrossRef]
41. Lindberg, S.; Muñoz-Alarcón, A.; Helmfors, H.; Mosqueira, D.; Gyllborg, D.; Tudoran, O.; Langel, U. PepFect15, a novel endosomolytic cell-penetrating peptide for oligonucleotide delivery via scavenger receptors. *Int. J. Pharm.* **2013**, *441*, 242–247. [CrossRef] [PubMed]
42. Lim, K.J.; Sung, B.H.; Shin, J.R.; Lee, Y.W.; Kim, D.J.; Yang, K.S.; Kim, S.C. A cancer specific cell-penetrating peptide, BR2, for the efficient delivery of an scFv into cancer cells. *PLoS ONE* **2013**, *8*, e66084. [CrossRef]
43. Freire, J.M.; Veiga, A.S.; de Figueiredo, I.R.; de la Torre, B.G.; Santos, N.; Andreu, D.; Da Poian, A.; Castanho, M.A.R.B. Nucleic acid delivery by cell penetrating peptides derived from dengue virus capsid protein: Design and mechanism of action. *FEBS J.* **2013**, *281*, 191–215. [CrossRef]
44. Cerrato, C.P.; Pirisinu, M.; Vlachos, E.N.; Langel, U. Novel cell-penetrating peptide targeting mitochondria. *FASEB J.* **2015**, *29*, 4589–4599. [CrossRef] [PubMed]
45. Gautam, A.; Nanda, J.S.; Samuel, J.S.; Kumari, M.; Priyanka, P.; Bedi, G.; Nath, S.K.; Mittal, G.; Khatri, N.; Raghava, G.P.S. Topical Delivery of Protein and Peptide Using Novel Cell Penetrating Peptide IMT-P8. *Sci. Rep.* **2016**, *6*, 26278. [CrossRef] [PubMed]
46. Yoo, J.; Lee, D.; Gujrati, V.; Rejinold, N.S.; Lekshmi, K.M.; Uthaman, S.; Jeong, C.; Park, I.-K.; Jon, S.; Kim, Y.-C. Bioreducible branched poly(modified nona-arginine) cell-penetrating peptide as a novel gene delivery platform. *J. Control. Release* **2017**, *246*, 142–154. [CrossRef]
47. Li, X.; Zheng, L.; Xia, Q.; Liu, L.; Mao, M.; Zhou, H.; Zhao, Y.; Shi, J. A novel cell-penetrating peptide protects against neuron apoptosis after cerebral ischemia by inhibiting the nuclear translocation of annexin A1. *Cell Death Differ.* **2018**, *26*, 260–275. [CrossRef]
48. Hu, G.; Miao, Y.; Luo, X.; Chu, W.; Fu, Y. Identification of a novel cell-penetrating peptide derived from the capsid protein of chicken anemia virus and its application in gene delivery. *Appl. Microbiol. Biotechnol.* **2020**, *104*, 10503–10513. [CrossRef]
49. Min, S.; Kim, K.; Ku, S.; Park, J.; Seo, J.; Roh, S. Newly synthesized peptide, Ara-27, exhibits significant improvement in cell-penetrating ability compared to conventional peptides. *Biotechnol. Prog.* **2020**, *36*, e3014. [CrossRef]
50. Gandhi, N.S.; Wang, E.; Sorolla, A.; Kan, Y.J.; Malik, A.; Batra, J.; Young, K.A.; Tie, W.J.; Blancafort, P.; Mancera, R.L. Design and Characterization of a Cell-Penetrating Peptide Derived from the SOX2 Transcription Factor. *Int. J. Mol. Sci.* **2021**, *22*, 9354. [CrossRef]
51. Cerrato, C.P.; Langel, U. An update on cell-penetrating peptides with intracellular organelle targeting. *Expert Opin. Drug Deliv.* **2022**, *19*, 133–146. [CrossRef]
52. Mandal, D.; Shirazi, A.N.; Parang, K. Cell-penetrating homochiral cyclic peptides as nuclear-targeting molecular transporters. *Angew. Chem. Int. Ed.* **2011**, *50*, 9633–9637. [CrossRef] [PubMed]
53. Tietz, O.; Cortezon-Tamarit, F.; Chalk, R.; Able, S.; Vallis, K.A. Tricyclic cell-penetrating peptides for efficient delivery of functional antibodies into cancer cells. *Nat. Chem.* **2022**, *14*, 284–293. [CrossRef] [PubMed]
54. Agrawal, P.; Bhalla, S.; Usmani, S.S.; Singh, S.; Chaudhary, K.; Raghava, G.P.S.; Gautam, A. CPPsite 2.0: A repository of experimentally validated cell-penetrating peptides. *Nucleic Acids Res.* **2015**, *44*, D1098–D1103. [CrossRef] [PubMed]
55. Kardani, K.; Bolhassani, A. Cppsite 2.0: An Available Database of Experimentally Validated Cell-Penetrating Peptides Predicting their Secondary and Tertiary Structures. *J. Mol. Biol.* **2020**, *433*, 166703. [CrossRef] [PubMed]

56. Mandal, S.; Mann, G.; Satish, G.; Brik, A. Enhanced Live-Cell Delivery of Synthetic Proteins Assisted by Cell-Penetrating Peptides Fused to DABCYL. *Angew. Chem. Int. Ed.* **2021**, *60*, 7333–7343. [CrossRef]
57. Sajid, M.I.; Mandal, D.; El-Sayed, N.S.; Lohan, S.; Moreno, J.; Tiwari, R.K. Oleyl Conjugated Histidine-Arginine Cell-Penetrating Peptides as Promising Agents for siRNA Delivery. *Pharmaceutics* **2022**, *14*, 881. [CrossRef]
58. Gentilucci, L.; De Marco, R.; Cerisoli, L. Chemical modifications designed to improve peptide stability: Incorporation of non-natural amino acids, pseudo-peptide bonds, and cyclization. *Curr. Pharm. Des.* **2010**, *16*, 3185–3203. [CrossRef]
59. Qian, Z.; LaRochelle, J.R.; Jiang, B.; Lian, W.; Hard, R.L.; Selner, N.G.; Luechapanichkul, R.; Barrios, A.M.; Pei, D. Early endosomal escape of a cyclic cell-penetrating peptide allows effective cytosolic cargo delivery. *Biochemistry* **2014**, *53*, 4034–4046. [CrossRef]
60. Qian, Z.; Martyna, A.; Hard, R.L.; Wang, J.; Appiah-Kubi, G.; Coss, C.; Phelps, M.A.; Rossman, J.S.; Pei, D. Discovery and Mechanism of Highly Efficient Cyclic Cell-Penetrating Peptides. *Biochemistry* **2016**, *55*, 2601–2612. [CrossRef]
61. Dougherty, P.G.; Sahni, A.; Pei, D. Understanding Cell Penetration of Cyclic Peptides. *Chem. Rev.* **2019**, *119*, 10241–10287. [CrossRef]
62. Leader, B.; Baca, Q.J.; Golan, D.E. Protein therapeutics: A summary and pharmacological classification. *Nat. Rev. Drug Discov.* **2008**, *7*, 21–39. [CrossRef]
63. Bolhassani, A.; Jafarzade, B.S.; Mardani, G. In vitro and in vivo delivery of therapeutic proteins using cell penetrating peptides. *Peptides* **2017**, *87*, 50–63. [CrossRef] [PubMed]
64. Zhang, D.; Wang, J.; Xu, D. Cell-penetrating peptides as noninvasive transmembrane vectors for the development of novel multifunctional drug-delivery systems. *J. Control. Release* **2016**, *229*, 130–139. [CrossRef] [PubMed]
65. Torchilin, V.P. Tat peptide-mediated intracellular delivery of pharmaceutical nanocarriers. *Adv. Drug Deliv. Rev.* **2008**, *60*, 548–558. [CrossRef] [PubMed]
66. Sabbagh, F.; Muhamad, I.I.; Pa'e, N.; Hashim, Z. Strategies in Improving Properties of Cellulose-Based Hydrogels for Smart Applications. In *Cellulose-Based Superabsorbent Hydrogels*; Mondal, M.I.H., Ed.; Springer International Publishing: Cham, Switzerland, 2018; pp. 1–22. [CrossRef]
67. Nam, S.H.; Park, J.; Koo, H. Recent advances in selective and targeted drug/gene delivery systems using cell-penetrating peptides. *Arch. Pharmacal Res.* **2023**, *46*, 18–34. [CrossRef] [PubMed]
68. Kim, G.C.; Cheon, D.H.; Lee, Y. Challenge to overcome current limitations of cell-penetrating peptides. *Biochim. Biophys. Acta (BBA)—Proteins Proteom.* **2021**, *1869*, 140604. [CrossRef] [PubMed]
69. Kumar, S.; Aaron, J.; Sokolov, K. Directional conjugation of antibodies to nanoparticles for synthesis of multiplexed optical contrast agents with both delivery and targeting moieties. *Nat. Protoc.* **2008**, *3*, 314–320. [CrossRef]
70. Bode, S.A.; Timmermans, S.B.P.E.; Eising, S.; van Gemert, S.P.W.; Bonger, K.M.; Löwik, D.W.P.M. Click to enter: Activation of oligo-arginine cell-penetrating peptides by bioorthogonal tetrazine ligations. *Chem. Sci.* **2018**, *10*, 701–705. [CrossRef]
71. Chen, X.; Zaro, J.L.; Shen, W.-C. Fusion protein linkers: Property, design and functionality. *Adv. Drug Deliv. Rev.* **2012**, *65*, 1357–1369. [CrossRef]
72. Turner, J.J.; Arzumanov, A.A.; Gait, M.J. Synthesis, cellular uptake and HIV-1 Tat-dependent trans-activation inhibition activity of oligonucleotide analogues disulphide-conjugated to cell-penetrating peptides. *Nucleic Acids Res.* **2005**, *33*, 27–42. [CrossRef]
73. Gayraud, F.; Klußmann, M.; Neundorf, I. Recent Advances and Trends in Chemical CPP–Drug Conjugation Techniques. *Molecules* **2021**, *26*, 1591. [CrossRef]
74. Shabanpoor, F.; McClorey, G.; Saleh, A.F.; Järver, P.; Wood, M.J.; Gait, M.J. Bi-specific splice-switching PMO oligonucleotides conjugated via a single peptide active in a mouse model of Duchenne muscular dystrophy. *Nucleic Acids Res.* **2014**, *43*, 29–39. [CrossRef] [PubMed]
75. Balogh, B.; Ivánczi, M.; Nizami, B.; Beke-Somfai, T.; Mándity, I.M. ConjuPepDB: A database of peptide-drug conjugates. *Nucleic Acids Res.* **2020**, *49*, D1102–D1112. [CrossRef] [PubMed]
76. Shai, Y. Mode of action of membrane active antimicrobial peptides. *Biopolymers* **2002**, *66*, 236–248. [CrossRef] [PubMed]
77. Mishra, A.; Gordon, V.D.; Yang, L.; Coridan, R.; Wong, G.C.L. HIV TAT forms pores in membranes by inducing saddle-splay curvature: Potential role of bidentate hydrogen bonding. *Angew. Chem. Int. Ed.* **2008**, *47*, 2986–2989. [CrossRef] [PubMed]
78. Derossi, D.; Calvet, S.; Trembleau, A.; Brunissen, A.; Chassaing, G.; Prochiantz, A. Cell internalization of the third helix of the antennapedia homeodomain is receptor-independent. *J. Biol. Chem.* **1996**, *271*, 18188–18193. [CrossRef] [PubMed]
79. Derossi, D.; Joliot, A.H.; Chassaing, G.; Prochiantz, A. The third helix of the Antennapedia homeodomain translocates through biological membranes. *J. Biol. Chem.* **1994**, *269*, 10444–10450. [CrossRef]
80. Sun, T.-L.; Sun, Y.; Lee, C.-C.; Huang, H.W. Membrane permeability of hydrocarbon-cross-linked peptides. *Biophys. J.* **2013**, *104*, 1923–1932. [CrossRef]
81. Zhang, F.; Guo, H.; Zhang, J.; Chen, Q.; Fang, Q. Identification of the caveolae/raft-mediated endocytosis as the primary entry pathway for aquareovirus. *Virology* **2018**, *513*, 195–207. [CrossRef] [PubMed]
82. Staring, J.; Raaben, M.; Brummelkamp, T.R. Viral escape from endosomes and host detection at a glance. *J. Cell Sci.* **2018**, *131*, jcs216259. [CrossRef]
83. Makvandi, P.; Chen, M.; Sartorius, R.; Zarrabi, A.; Ashrafizadeh, M.; Moghaddam, F.D.; Ma, J.; Mattoli, V.; Tay, F.R. Endocytosis of abiotic nanomaterials and nanobiovectors: Inhibition of membrane trafficking. *Nano Today* **2021**, *40*, 101279. [CrossRef] [PubMed]
84. Ehrlich, M.; Boll, W.; van Oijen, A.; Hariharan, R.; Chandran, K.; Nibert, M.L.; Kirchhausen, T. Endocytosis by random initiation and stabilization of clathrin-coated pits. *Cell* **2004**, *118*, 591–605. [CrossRef] [PubMed]

85. Akishiba, M.; Takeuchi, T.; Kawaguchi, Y.; Sakamoto, K.; Yu, H.-H.; Nakase, I.; Takatani-Nakase, T.; Madani, F.; Gräslund, A.; Futaki, S. Cytosolic antibody delivery by lipid-sensitive endosomolytic peptide. *Nat. Chem.* **2017**, *9*, 751–761. [CrossRef] [PubMed]
86. Rennick, J.J.; Johnston, A.P.R.; Parton, R.G. Key principles and methods for studying the endocytosis of biological and nanoparticle therapeutics. *Nat. Nanotechnol.* **2021**, *16*, 266–276. [CrossRef] [PubMed]
87. Desale, K.; Kuche, K.; Jain, S. Cell-penetrating peptides (CPPs): An overview of applications for improving the potential of nanotherapeutics. *Biomater. Sci.* **2020**, *9*, 1153–1188. [CrossRef]
88. Mandl, H.K.; Quijano, E.; Suh, H.W.; Sparago, E.; Oeck, S.; Grun, M.; Glazer, P.M.; Saltzman, W.M. Optimizing biodegradable nanoparticle size for tissue-specific delivery. *J. Control. Release* **2019**, *314*, 92–101. [CrossRef]
89. Hickey, J.W.; Santos, J.L.; Williford, J.-M.; Mao, H.-Q. Control of polymeric nanoparticle size to improve therapeutic delivery. *J. Control. Release* **2015**, *219*, 536–547. [CrossRef]
90. Zhang, Q.; Jiang, Q.; Li, N.; Dai, L.; Liu, Q.; Song, L.; Wang, J.; Li, Y.; Tian, J.; Ding, B.; et al. DNA origami as an in vivo drug delivery vehicle for cancer therapy. *ACS Nano* **2014**, *8*, 6633–6643. [CrossRef]
91. Li, Y.; Zhai, Y.; Liu, W.; Zhang, K.; Liu, J.; Shi, J.; Zhang, Z. Ultrasmall nanostructured drug based pH-sensitive liposome for effective treatment of drug-resistant tumor. *J. Nanobiotechnol.* **2019**, *17*, 117. [CrossRef]
92. Peer, D.; Karp, J.M.; Hong, S.; Farokhzad, O.C.; Margalit, R.; Langer, R. Nanocarriers as an emerging platform for cancer therapy. *Nat. Nanotechnol.* **2007**, *2*, 751–760. [CrossRef] [PubMed]
93. Chauhan, V.P.; Popović, Z.; Chen, O.; Cui, J.; Fukumura, D.; Bawendi, M.G.; Jain, R.K. Fluorescent nanorods and nanospheres for real-time in vivo probing of nanoparticle shape-dependent tumor penetration. *Angew. Chem. Int. Ed.* **2011**, *50*, 11417–11420. [CrossRef] [PubMed]
94. Blanco, E.; Shen, H.; Ferrari, M. Principles of nanoparticle design for overcoming biological barriers to drug delivery. *Nat. Biotechnol.* **2015**, *33*, 941–951. [CrossRef] [PubMed]
95. Boehmer, T.; Jeudy, S.; Berke, I.C.; Schwartz, T.U. Structural and functional studies of Nup107/Nup133 interaction and its implications for the architecture of the nuclear pore complex. *Mol. Cell* **2008**, *30*, 721–731. [CrossRef]
96. Erazo-Oliveras, A.; Muthukrishnan, N.; Baker, R.; Wang, T.-Y.; Pellois, J.-P. Improving the endosomal escape of cell-penetrating peptides and their cargos: Strategies and challenges. *Pharmaceuticals* **2012**, *5*, 1177–1209. [CrossRef]
97. Allen, J.K.; Brock, D.J.; Kondow-McConaghy, H.M.; Pellois, J.-P. Efficient Delivery of Macromolecules into Human Cells by Improving the Endosomal Escape Activity of Cell-Penetrating Peptides: Lessons Learned from dfTAT and its Analogs. *Biomolecules* **2018**, *8*, 50. [CrossRef]
98. Yamano, S.; Dai, J.; Hanatani, S.; Haku, K.; Yamanaka, T.; Ishioka, M.; Takayama, T.; Yuvienco, C.; Khapli, S.; Moursi, A.M.; et al. Long-term efficient gene delivery using polyethylenimine with modified Tat peptide. *Biomaterials* **2014**, *35*, 1705–1715. [CrossRef]
99. Varkouhi, A.K.; Scholte, M.; Storm, G.; Haisma, H.J. Endosomal escape pathways for delivery of biologicals. *J. Control. Release* **2011**, *151*, 220–228. [CrossRef]
100. Li, W.; Nicol, F.; Szoka, F.C., Jr. GALA: A designed synthetic pH-responsive amphipathic peptide with applications in drug and gene delivery. *Adv. Drug Deliv. Rev.* **2004**, *56*, 967–985. [CrossRef]
101. Kyriakides, T.R.; Cheung, C.Y.; Murthy, N.; Bornstein, P.; Stayton, P.S.; Hoffman, A.S. pH-Sensitive polymers that enhance intracellular drug delivery in vivo. *J. Control. Release* **2001**, *78*, 295–303. [CrossRef]
102. Kobayashi, S.; Nakase, I.; Kawabata, N.; Yu, H.-H.; Pujals, S.; Imanishi, M.; Giralt, E.; Futaki, S. Cytosolic targeting of macromolecules using a pH-dependent fusogenic peptide in combination with cationic liposomes. *Bioconjugate Chem.* **2009**, *20*, 953–959. [CrossRef] [PubMed]
103. Shi, G.; Guo, W.; Stephenson, S.M.; Lee, R.J. Efficient intracellular drug and gene delivery using folate receptor-targeted pH-sensitive liposomes composed of cationic/anionic lipid combinations. *J. Control. Release* **2002**, *80*, 309–319. [CrossRef] [PubMed]
104. Yu, S.; Yang, H.; Li, T.; Pan, H.; Ren, S.; Luo, G.; Jiang, J.; Yu, L.; Chen, B.; Zhang, Y.; et al. Efficient intracellular delivery of proteins by a multifunctional chimaeric peptide in vitro and in vivo. *Nat. Commun.* **2021**, *12*, 5131. [CrossRef] [PubMed]
105. Wang, J.; Meng, F.; Kim, B.-K.; Ke, X.; Yeo, Y. In-vitro and in-vivo difference in gene delivery by lithocholic acid-polyethylenimine conjugate. *Biomaterials* **2019**, *217*, 119296. [CrossRef]
106. He, Y.; Guo, S.; Wu, L.; Chen, P.; Wang, L.; Liu, Y.; Ju, H. Near-infrared boosted ROS responsive siRNA delivery and cancer therapy with sequentially peeled upconversion nano-onions. *Biomaterials* **2019**, *225*, 119501. [CrossRef]
107. Alipour, M.; Hosseinkhani, S. Design, Preparation, and Characterization of Peptide-Based Nanocarrier for Gene Delivery. *Methods Mol. Biol.* **2019**, *2000*, 59–69. [CrossRef]
108. Xie, L.; Tan, Y.; Wang, Z.; Liu, H.; Zhang, N.; Zou, C.; Liu, X.; Liu, G.; Lu, J.; Zheng, H. ε-Caprolactone-Modified Polyethylenimine as Efficient Nanocarriers for siRNA Delivery in Vivo. *ACS Appl. Mater. Interfaces* **2016**, *8*, 29261–29269. [CrossRef]
109. Degors, I.M.S.; Wang, C.; Rehman, Z.U.; Zuhorn, I.S. Carriers Break Barriers in Drug Delivery: Endocytosis and Endosomal Escape of Gene Delivery Vectors. *Accounts Chem. Res.* **2019**, *52*, 1750–1760. [CrossRef]
110. Jia, G.; Han, Y.; An, Y.; Ding, Y.; He, C.; Wang, X.; Tang, Q. NRP-1 targeted and cargo-loaded exosomes facilitate simultaneous imaging and therapy of glioma in vitro and in vivo. *Biomaterials* **2018**, *178*, 302–316. [CrossRef]
111. Wan, Y.; Wang, L.; Zhu, C.; Zheng, Q.; Wang, G.; Tong, J.; Fang, Y.; Xia, Y.; Cheng, G.; He, X.; et al. Aptamer-Conjugated Extracellular Nanovesicles for Targeted Drug Delivery. *Cancer Res.* **2018**, *78*, 798–808. [CrossRef]

112. Nakase, I.; Noguchi, K.; Fujii, I.; Futaki, S. Vectorization of biomacromolecules into cells using extracellular vesicles with enhanced internalization induced by macropinocytosis. *Sci. Rep.* **2016**, *6*, 34937. [CrossRef]
113. Xu, H.; Liao, C.; Liang, S.; Ye, B.-C. A Novel Peptide-Equipped Exosomes Platform for Delivery of Antisense Oligonucleotides. *ACS Appl. Mater. Interfaces* **2021**, *13*, 10760–10767. [CrossRef]
114. Liu, D.; Angelova, A.; Liu, J.; Garamus, V.M.; Angelov, B.; Zhang, X.; Li, Y.; Feger, G.; Li, N.; Zou, A. Self-assembly of mitochondria-specific peptide amphiphiles amplifying lung cancer cell death through targeting the VDAC1–hexokinase-II complex. *J. Mater. Chem. B* **2019**, *7*, 4706–4716. [CrossRef] [PubMed]
115. Cerrato, C.P.; Künnapuu, K.; Langel, U. Cell-penetrating peptides with intracellular organelle targeting. *Expert Opin. Drug Deliv.* **2016**, *14*, 245–255. [CrossRef] [PubMed]
116. Date, A.A.; Hanes, J.; Ensign, L.M. Nanoparticles for oral delivery: Design, evaluation and state-of-the-art. *J. Control. Release* **2016**, *240*, 504–526. [CrossRef] [PubMed]
117. Fievez, V.; Plapied, L.; Plaideau, C.; Legendre, D.; Rieux, A.D.; Pourcelle, V.; Freichels, H.; Jérôme, C.; Marchand, J.; Préat, V.; et al. In vitro identification of targeting ligands of human M cells by phage display. *Int. J. Pharm.* **2010**, *394*, 35–42. [CrossRef]
118. Yao, Z.; Che, X.-C.; Lu, R.; Zheng, M.-N.; Zhu, Z.-F.; Li, J.-P.; Jian, X.; Shi, L.-X.; Liu, J.-Y.; Gao, W.-Y. Inhibition by tyroserleutide (YSL) on the invasion and adhesion of the mouse melanoma cell. *Mol. Med.* **2007**, *13*, 14–21. [CrossRef] [PubMed]
119. Ma, C.; Wei, T.; Hua, Y.; Wang, Z.; Zhang, L. Effective Antitumor of Orally Intestinal Targeting Penetrating Peptide-Loaded Tyroserleutide/PLGA Nanoparticles in Hepatocellular Carcinoma. *Int. J. Nanomed.* **2021**, *16*, 4495–4513. [CrossRef]
120. Ruoslahti, E. Peptides as targeting elements and tissue penetration devices for nanoparticles. *Adv. Mater.* **2012**, *24*, 3747–3756. [CrossRef]
121. Roth, L.; Agemy, L.; Kotamraju, V.R.; Braun, G.; Teesalu, T.; Sugahara, K.N.; Hamzah, J.; Ruoslahti, E. Transtumoral targeting enabled by a novel neuropilin-binding peptide. *Oncogene* **2011**, *31*, 3754–3763. [CrossRef]
122. Jiang, C.; Wang, X.; Teng, B.; Wang, Z.; Li, F.; Zhao, Y.; Guo, Y.; Zeng, Q. Peptide-Targeted High-Density Lipoprotein Nanoparticles for Combinatorial Treatment against Metastatic Breast Cancer. *ACS Appl. Mater. Interfaces* **2021**, *13*, 35248–35265. [CrossRef] [PubMed]
123. Tan, J.; Duan, X.; Zhang, F.; Ban, X.; Mao, J.; Cao, M.; Han, S.; Shuai, X.; Shen, J. Theranostic Nanomedicine for Synergistic Chemodynamic Therapy and Chemotherapy of Orthotopic Glioma. *Adv. Sci.* **2020**, *7*, 2003036. [CrossRef] [PubMed]
124. Dong, C.; Hong, S.; Zheng, D.; Huang, Q.; Liu, F.; Zhong, Z.; Zhang, X. Multifunctionalized Gold Sub-Nanometer Particles for Sensitizing Radiotherapy against Glioblastoma. *Small* **2020**, *17*, e2006582. [CrossRef]
125. Liu, M.; Ma, W.; Zhao, D.; Li, J.; Li, Q.; Liu, Y.; Hao, L.; Lin, Y. Enhanced Penetrability of a Tetrahedral Framework Nucleic Acid by Modification with iRGD for DOX-Targeted Delivery to Triple-Negative Breast Cancer. *ACS Appl. Mater. Interfaces* **2021**, *13*, 25825–25835. [CrossRef]
126. Liang, H.; Wu, X.; Zhao, G.; Feng, K.; Ni, K.; Sun, X. Renal Clearable Ultrasmall Single-Crystal Fe Nanoparticles for Highly Selective and Effective Ferroptosis Therapy and Immunotherapy. *J. Am. Chem. Soc.* **2021**, *143*, 15812–15823. [CrossRef]
127. Cho, H.-J.; Park, S.-J.; Jung, W.H.; Cho, Y.; Ahn, D.J.; Lee, Y.-S.; Kim, S. Injectable Single-Component Peptide Depot: Autonomously Rechargeable Tumor Photosensitization for Repeated Photodynamic Therapy. *ACS Nano* **2020**, *14*, 15793–15805. [CrossRef] [PubMed]
128. Yu, C.; Li, L.; Hu, P.; Yang, Y.; Wei, W.; Deng, X.; Wang, L.; Tay, F.R.; Ma, J. Recent Advances in Stimulus-Responsive Nanocarriers for Gene Therapy. *Adv. Sci.* **2021**, *8*, 2100540. [CrossRef]
129. De Jong, H.; Bonger, K.M.; Löwik, D.W.P.M. Activatable cell-penetrating peptides: 15 years of research. *RSC Chem. Biol.* **2020**, *1*, 192–203. [CrossRef]
130. Roointan, A.; Farzanfar, J.; Mohammadi-Samani, S.; Behzad-Behbahani, A.; Farjadian, F. Smart pH responsive drug delivery system based on poly(HEMA-co-DMAEMA) nanohydrogel. *Int. J. Pharm.* **2018**, *552*, 301–311. [CrossRef]
131. Yang, Y.; Wang, Z.; Peng, Y.; Ding, J.; Zhou, W. A Smart pH-Sensitive Delivery System for Enhanced Anticancer Efficacy via Paclitaxel Endosomal Escape. *Front. Pharmacol.* **2019**, *10*, 10. [CrossRef]
132. Karimi, M.; Eslami, M.; Sahandi-Zangabad, P.; Mirab, F.; Farajisafiloo, N.; Shafaei, Z.; Ghosh, D.; Bozorgomid, M.; Dashkhaneh, F.; Hamblin, M.R. pH -Sensitive stimulus-responsive nanocarriers for targeted delivery of therapeutic agents. *WIREs Nanomed. Nanobiotechnol.* **2016**, *8*, 696–716. [CrossRef] [PubMed]
133. Li, J.; Mo, L.; Lu, C.-H.; Fu, T.; Yang, H.-H.; Tan, W. Functional nucleic acid-based hydrogels for bioanalytical and biomedical applications. *Chem. Soc. Rev.* **2016**, *45*, 1410–1431. [CrossRef] [PubMed]
134. Yu, Y.; Zu, C.; He, D.; Li, Y.; Chen, Q.; Chen, Q.; Wang, H.; Wang, R.; Chaurasiya, B.; Zaro, J.L.; et al. pH-dependent reversibly activatable cell-penetrating peptides improve the antitumor effect of artemisinin-loaded liposomes. *J. Colloid Interface Sci.* **2020**, *586*, 391–403. [CrossRef] [PubMed]
135. Lee, N.; Spears, M.E.; Carlisle, A.E.; Kim, D. Endogenous toxic metabolites and implications in cancer therapy. *Oncogene* **2020**, *39*, 5709–5720. [CrossRef]
136. Gordon, M.R.; Zhao, B.; Anson, F.; Fernandez, A.; Singh, K.; Homyak, C.; Canakci, M.; Vachet, R.W.; Thayumanavan, S. Matrix Metalloproteinase-9-Responsive Nanogels for Proximal Surface Conversion and Activated Cellular Uptake. *Biomacromolecules* **2018**, *19*, 860–871. [CrossRef]

137. Zhang, W.; Gong, C.; Chen, Z.; Li, M.; Li, Y.; Gao, J. Tumor microenvironment-activated cancer cell membrane-liposome hybrid nanoparticle-mediated synergistic metabolic therapy and chemotherapy for non-small cell lung cancer. *J. Nanobiotechnol.* **2021**, *19*, 339. [CrossRef]

138. Chen, L.; Song, Q.; Chen, Y.; Meng, S.; Zheng, M.; Huang, J.; Zhang, Q.; Jiang, J.; Feng, J.; Chen, H.-Z.; et al. Tailored Reconstituted Lipoprotein for Site-Specific and Mitochondria-Targeted Cyclosporine A Delivery to Treat Traumatic Brain Injury. *ACS Nano* **2020**, *14*, 6636–6648. [CrossRef]

139. Liu, L.; Cao, F.; Liu, X.; Wang, H.; Zhang, C.; Sun, H.; Wang, C.; Leng, X.; Song, C.; Kong, D.; et al. Hyaluronic Acid-Modified Cationic Lipid–PLGA Hybrid Nanoparticles as a Nanovaccine Induce Robust Humoral and Cellular Immune Responses. *ACS Appl. Mater. Interfaces* **2016**, *8*, 11969–11979. [CrossRef]

140. Liang, K.; Bae, K.H.; Lee, F.; Xu, K.; Chung, J.E.; Gao, S.J.; Kurisawa, M. Self-assembled ternary complexes stabilized with hyaluronic acid-green tea catechin conjugates for targeted gene delivery. *J. Control. Release* **2016**, *226*, 205–216. [CrossRef]

141. Zhao, Y.; He, Z.; Gao, H.; Tang, H.; He, J.; Guo, Q.; Zhang, W.; Liu, J. Fine Tuning of Core–Shell Structure of Hyaluronic Acid/Cell-Penetrating Peptides/siRNA Nanoparticles for Enhanced Gene Delivery to Macrophages in Antiatherosclerotic Therapy. *Biomacromolecules* **2018**, *19*, 2944–2956. [CrossRef]

142. Liu, Y.; Li, L.; Liu, J.; Yang, M.; Wang, H.; Chu, X.; Zhou, J.; Huo, M.; Yin, T. Biomineralization-inspired dasatinib nanodrug with sequential infiltration for effective solid tumor treatment. *Biomaterials* **2020**, *267*, 120481. [CrossRef]

143. Jain, R.K. Normalization of tumor vasculature: An emerging concept in antiangiogenic therapy. *Science* **2005**, *307*, 58–62. [CrossRef] [PubMed]

144. Zhou, S.; Ding, C.; Wang, C.; Fu, J. UV-light cross-linked and pH de-cross-linked coumarin-decorated cationic copolymer grafted mesoporous silica nanoparticles for drug and gene co-delivery in vitro. *Mater. Sci. Eng. C* **2019**, *108*, 110469. [CrossRef] [PubMed]

145. Ye, H.; Zhou, Y.; Liu, X.; Chen, Y.; Duan, S.; Zhu, R.; Liu, Y.; Yin, L. Recent Advances on Reactive Oxygen Species-Responsive Delivery and Diagnosis System. *Biomacromolecules* **2019**, *20*, 2441–2463. [CrossRef] [PubMed]

146. Wan, W.-J.; Huang, G.; Wang, Y.; Tang, Y.; Li, H.; Jia, C.-H.; Liu, Y.; You, B.-G.; Zhang, X.-N. Coadministration of iRGD peptide with ROS-sensitive nanoparticles co-delivering si*FGL1* and si*PD-L1* enhanced tumor immunotherapy. *Acta Biomater.* **2021**, *136*, 473–484. [CrossRef] [PubMed]

147. Zhu, L.; Zhao, H.; Zhou, Z.; Xia, Y.; Wang, Z.; Ran, H.; Li, P.; Ren, J. Peptide-Functionalized Phase-Transformation Nanoparticles for Low Intensity Focused Ultrasound-Assisted Tumor Imaging and Therapy. *Nano Lett.* **2018**, *18*, 1831–1841. [CrossRef]

148. Lu, L.; Liu, G.; Lin, C.; Li, K.; He, T.; Zhang, J.; Luo, Z.; Cai, K. Mitochondrial Metabolism Targeted Nanoplatform for Efficient Triple-Negative Breast Cancer Combination Therapy. *Adv. Health Mater.* **2021**, *10*, e2100978. [CrossRef]

149. Fan, W.; Yung, B.; Huang, P.; Chen, X. Nanotechnology for Multimodal Synergistic Cancer Therapy. *Chem. Rev.* **2017**, *117*, 13566–13638. [CrossRef]

150. Nastiuk, K.L.; Krolewski, J.J. Opportunities and challenges in combination gene cancer therapy. *Adv. Drug Deliv. Rev.* **2016**, *98*, 35–40. [CrossRef]

151. Wang, H.; Zhou, J.; Fu, Y.; Zheng, Y.; Shen, W.; Zhou, J.; Yin, T. Deeply Infiltrating iRGD-Graphene Oxide for the Intensive Treatment of Metastatic Tumors through PTT-Mediated Chemosensitization and Strengthened Integrin Targeting-Based Antimigration. *Adv. Health Mater.* **2021**, *10*, e2100536. [CrossRef]

152. Lin, C.; Tong, F.; Liu, R.; Xie, R.; Lei, T.; Chen, Y.; Yang, Z.; Gao, H.; Yu, X. GSH-responsive SN38 dimer-loaded shape-transformable nanoparticles with iRGD for enhancing chemo-photodynamic therapy. *Acta Pharm. Sin. B* **2020**, *10*, 2348–2361. [CrossRef] [PubMed]

153. Juang, V.; Chang, C.; Wang, C.; Wang, H.; Lo, Y. pH-Responsive PEG-Shedding and Targeting Peptide-Modified Nanoparticles for Dual-Delivery of Irinotecan and microRNA to Enhance Tumor-Specific Therapy. *Small* **2019**, *15*, e1903296. [CrossRef]

154. Xiang, B.; Jia, X.-L.; Qi, J.-L.; Yang, L.-P.; Sun, W.-H.; Yan, X.; Yang, S.-K.; Cao, D.-Y.; Du, Q.; Qi, X.-R. Enhancing siRNA-based cancer therapy using a new pH-responsive activatable cell-penetrating peptide-modified liposomal system. *Int. J. Nanomed.* **2017**, *12*, 2385–2405. [CrossRef] [PubMed]

155. Lee, E.S.; Gao, Z.; Kim, D.; Park, K.; Kwon, I.C.; Bae, Y.H. Super pH-sensitive multifunctional polymeric micelle for tumor pHe specific TAT exposure and multidrug resistance. *J. Control. Release* **2008**, *129*, 228–236. [CrossRef] [PubMed]

156. Zhang, Q.; Tang, J.; Fu, L.; Ran, R.; Liu, Y.; Yuan, M.; He, Q. A pH-responsive α-helical cell penetrating peptide-mediated liposomal delivery system. *Biomaterials* **2013**, *34*, 7980–7993. [CrossRef]

157. Weerakkody, D.; Moshnikova, A.; Thakur, M.S.; Moshnikova, V.; Daniels, J.; Engelman, D.M.; Andreev, O.A.; Reshetnyak, Y.K. Family of pH (low) insertion peptides for tumor targeting. *Proc. Natl. Acad. Sci. USA* **2013**, *110*, 5834–5839. [CrossRef]

158. Han, H.; Hou, Y.; Chen, X.; Zhang, P.; Kang, M.; Jin, Q.; Ji, J.; Gao, M. Metformin-Induced Stromal Depletion to Enhance the Penetration of Gemcitabine-Loaded Magnetic Nanoparticles for Pancreatic Cancer Targeted Therapy. *J. Am. Chem. Soc.* **2020**, *142*, 4944–4954. [CrossRef]

159. Nam, S.H.; Jang, J.; Cheon, D.H.; Chong, S.-E.; Ahn, J.H.; Hyun, S.; Yu, J.; Lee, Y. pH-Activatable cell penetrating peptide dimers for potent delivery of anticancer drug to triple-negative breast cancer. *J. Control. Release* **2020**, *330*, 898–906. [CrossRef]

160. Huang, S.; Zhang, Y.; Wang, L.; Liu, W.; Xiao, L.; Lin, Q.; Gong, T.; Sun, X.; He, Q.; Zhang, Z.; et al. Improved melanoma suppression with target-delivered TRAIL and Paclitaxel by a multifunctional nanocarrier. *J. Control. Release* **2020**, *325*, 10–24. [CrossRef]

161. Cheng, H.; Zhu, J.-Y.; Xu, X.-D.; Qiu, W.-X.; Lei, Q.; Han, K.; Cheng, Y.-J.; Zhang, X.-Z. Activable Cell-Penetrating Peptide Conjugated Prodrug for Tumor Targeted Drug Delivery. *ACS Appl. Mater. Interfaces* **2015**, *7*, 16061–16069. [CrossRef]

162. Jin, E.; Zhang, B.; Sun, X.; Zhou, Z.; Ma, X.; Sun, Q.; Tang, J.; Shen, Y.; Van Kirk, E.; Murdoch, W.J.; et al. Acid-active cell-penetrating peptides for in vivo tumor-targeted drug delivery. *J. Am. Chem. Soc.* **2012**, *135*, 933–940. [CrossRef] [PubMed]

163. Yin, H.; Lu, H.; Xiong, Y.; Ye, L.; Teng, C.; Cao, X.; Li, S.; Sun, S.; Liu, W.; Lv, W.; et al. Tumor-Associated Neutrophil Extracellular Traps Regulating Nanocarrier-Enhanced Inhibition of Malignant Tumor Growth and Distant Metastasis. *ACS Appl. Mater. Interfaces* **2021**, *13*, 59683–59694. [CrossRef] [PubMed]

164. Jiang, T.; Olson, E.S.; Nguyen, Q.T.; Roy, M.; Jennings, P.A.; Tsien, R.Y. Tumor imaging by means of proteolytic activation of cell-penetrating peptides. *Proc. Natl. Acad. Sci. USA* **2004**, *101*, 17867–17872. [CrossRef] [PubMed]

165. Nguyen, Q.T.; Olson, E.S.; Aguilera, T.A.; Jiang, T.; Scadeng, M.; Ellies, L.G.; Tsien, R.Y. Surgery with molecular fluorescence imaging using activatable cell-penetrating peptides decreases residual cancer and improves survival. *Proc. Natl. Acad. Sci. USA* **2010**, *107*, 4317–4322. [CrossRef]

166. Li, J.; Liu, F.; Shao, Q.; Min, Y.; Costa, M.; Yeow, E.K.L.; Xing, B. Enzyme-responsive cell-penetrating peptide conjugated mesoporous silica quantum dot nanocarriers for controlled release of nucleus-targeted drug molecules and real-time intracellular fluorescence imaging of tumor cells. *Adv. Health Mater.* **2014**, *3*, 1230–1239. [CrossRef]

167. Whitney, M.; Crisp, J.L.; Olson, E.S.; Aguilera, T.A.; Gross, L.A.; Ellies, L.G.; Tsien, R.Y. Parallel in vivo and in vitro selection using phage display identifies protease-dependent tumor-targeting peptides. *J. Biol. Chem.* **2010**, *285*, 22532–22541. [CrossRef]

168. Xiang, B.; Dong, D.-W.; Shi, N.-Q.; Gao, W.; Yang, Z.-Z.; Cui, Y.; Cao, D.-Y.; Qi, X.-R. PSA-responsive and PSMA-mediated multifunctional liposomes for targeted therapy of prostate cancer. *Biomaterials* **2013**, *34*, 6976–6991. [CrossRef]

169. Wang, F.; Xie, D.; Lai, W.; Zhou, M.; Wang, J.; Xu, R.; Huang, J.; Zhang, R.; Li, G. Autophagy responsive intra-intercellular delivery nanoparticles for effective deep solid tumor penetration. *J. Nanobiotechnol.* **2022**, *20*, 300. [CrossRef]

170. Bode, S.A.; Hansen, M.B.; Oerlemans, R.A.J.F.; van Hest, J.C.M.; Löwik, D.W.P.M. Enzyme-Activatable Cell-Penetrating Peptides through a Minimal Side Chain Modification. *Bioconjug. Chem.* **2015**, *26*, 850–856. [CrossRef]

171. Harada, H.; Hiraoka, M.; Kizaka-Kondoh, S. Antitumor effect of TAT-oxygen-dependent degradation-caspase-3 fusion protein specifically stabilized and activated in hypoxic tumor cells. *Cancer Res.* **2002**, *62*, 2013–2018.

172. Lee, S.H.; Moroz, E.; Castagner, B.; Leroux, J.-C. Activatable cell penetrating peptide-peptide nucleic acid conjugate via reduction of azobenzene PEG chains. *J. Am. Chem. Soc.* **2014**, *136*, 12868–12871. [CrossRef] [PubMed]

173. Weinstain, R.; Savariar, E.N.; Felsen, C.N.; Tsien, R.Y. In vivo targeting of hydrogen peroxide by activatable cell-penetrating peptides. *J. Am. Chem. Soc.* **2013**, *136*, 874–877. [CrossRef] [PubMed]

174. Han, B.-B.; Pan, Y.-C.; Li, Y.-M.; Guo, D.-S.; Chen, Y.-X. A host-guest ATP responsive strategy for intracellular delivery of phosphopeptides. *Chem. Commun.* **2020**, *56*, 5512–5515. [CrossRef] [PubMed]

175. Xie, Z.; Wang, J.; Luo, Y.; Qiao, B.; Jiang, W.; Zhu, L.; Ran, H.; Wang, Z.; Zhu, W.; Ren, J.; et al. Tumor-penetrating nanoplatform with ultrasound "unlocking" for cascade synergistic therapy and visual feedback under hypoxia. *J. Nanobiotechnol.* **2023**, *21*, 30. [CrossRef]

176. Sumi, N.; Nagahiro, S.; Nakata, E.; Watanabe, K.; Ohtsuki, T. Ultrasound-dependent RNAi using TatU1A-rose bengal conjugate. *Bioorg. Med. Chem. Lett.* **2022**, *68*, 128767. [CrossRef] [PubMed]

177. Wen, J.; Liu, F.; Tao, B.; Sun, S. GSH-responsive anti-mitotic cell penetrating peptide-linked podophyllotoxin conjugate for improving water solubility and targeted synergistic drug delivery. *Bioorganic Med. Chem. Lett.* **2019**, *29*, 1019–1022. [CrossRef]

178. Zang, C.; Wang, H.; Li, T.; Zhang, Y.; Li, J.; Shang, M.; Du, J.; Xi, Z.; Zhou, C. A light-responsive, self-immolative linker for controlled drug delivery via peptide- and protein-drug conjugates. *Chem. Sci.* **2019**, *10*, 8973–8980. [CrossRef]

179. Lin, Y.; Mazo, M.M.; Skaalure, S.C.; Thomas, M.R.; Schultz, S.R.; Stevens, M.M. Activatable cell-biomaterial interfacing with photo-caged peptides. *Chem. Sci.* **2018**, *10*, 1158–1167. [CrossRef]

180. Xie, X.; Yang, Y.; Yang, Y.; Zhang, H.; Li, Y.; Mei, X. A photo-responsive peptide- and asparagine-glycine-arginine (NGR) peptide-mediated liposomal delivery system. *Drug Deliv.* **2015**, *23*, 2445–2456. [CrossRef]

181. Shamay, Y.; Adar, L.; Ashkenasy, G.; David, A. Light induced drug delivery into cancer cells. *Biomaterials* **2011**, *32*, 1377–1386. [CrossRef]

182. Kim, G.C.; Ahn, J.H.; Oh, J.H.; Nam, S.; Hyun, S.; Yu, J.; Lee, Y. Photoswitching of Cell Penetration of Amphipathic Peptides by Control of α-Helical Conformation. *Biomacromolecules* **2018**, *19*, 2863–2869. [CrossRef] [PubMed]

183. Prestel, A.; Möller, H.M. Spatio-temporal control of cellular uptake achieved by photoswitchable cell-penetrating peptides. *Chem. Commun.* **2015**, *52*, 701–704. [CrossRef] [PubMed]

184. Hansen, M.B.; van Gaal, E.; Minten, I.; Storm, G.; van Hest, J.C.; Löwik, D.W. Constrained and UV-activatable cell-penetrating peptides for intracellular delivery of liposomes. *J. Control. Release* **2012**, *164*, 87–94. [CrossRef] [PubMed]

185. Yang, Y.; Xie, X.; Yang, Y.; Li, Z.; Yu, F.; Gong, W.; Li, Y.; Zhang, H.; Wang, Z.; Mei, X. Polymer Nanoparticles Modified with Photo- and pH-Dual-Responsive Polypeptides for Enhanced and Targeted Cancer Therapy. *Mol. Pharm.* **2016**, *13*, 1508–1519. [CrossRef] [PubMed]

186. Staecker, H.; Jokovic, G.; Karpishchenko, S.; Kienle-Gogolok, A.; Krzyzaniak, A.; Lin, C.-D.; Navratil, P.; Tzvetkov, V.; Wright, N.; Meyer, T. Efficacy and Safety of AM-111 in the Treatment of Acute Unilateral Sudden Deafness—A Double-blind, Randomized, Placebo-controlled Phase 3 Study. *Otol. Neurotol.* **2019**, *40*, 584–594. [CrossRef]

187. Chiquet, C.; Aptel, F.; Creuzot-Garcher, C.; Berrod, J.-P.; Kodjikian, L.; Massin, P.; Deloche, C.; Perino, J.; Kirwan, B.-A.; de Brouwer, S.; et al. Postoperative Ocular Inflammation: A Single Subconjunctival Injection of XG-102 Compared to Dexamethasone Drops in a Randomized Trial. *Am. J. Ophthalmol.* **2016**, *174*, 76–84. [CrossRef]

188. Jankovic, J.; Truong, D.; Patel, A.T.; Brashear, A.; Evatt, M.; Rubio, R.G.; Oh, C.K.; Snyder, D.; Shears, G.; Comella, C. Injectable DaxibotulinumtoxinA in Cervical Dystonia: A Phase 2 Dose-Escalation Multicenter Study. *Mov. Disord. Clin. Pract.* **2018**, *5*, 273–282. [CrossRef]

189. Garcia-Murray, E.; Villasenor, M.L.V.; Acevedo, B.; Luna, S.; Lee, J.; Waugh, J.M.; Hornfeldt, C.S. Safety and efficacy of RT002, an injectable botulinum toxin type A, for treating glabellar lines. *Dermatol. Surg.* **2015**, *41* (Suppl. 1), S47–S55. [CrossRef]

190. Patel, A.T.; Lew, M.F.; Dashtipour, K.; Isaacson, S.; Hauser, R.A.; Ondo, W.; Maisonobe, P.; Wietek, S.; Rubin, B.; Brashear, A. Sustained functional benefits after a single set of injections with abobotulinumtoxinA using a 2-mL injection volume in adults with cervical dystonia: 12-week results from a randomized, double-blind, placebo-controlled phase 3b study. *PLoS ONE* **2021**, *16*, e0245827. [CrossRef]

191. Hill, M.D.; Goyal, M.; Menon, B.K.; Nogueira, R.G.; McTaggart, R.A.; Demchuk, A.M.; Poppe, A.Y.; Buck, B.H.; Field, T.S.; Dowlatshahi, D.; et al. Efficacy and safety of nerinetide for the treatment of acute ischaemic stroke (ESCAPE-NA1): A multicentre, double-blind, randomised controlled trial. *Lancet* **2020**, *395*, 878–887. [CrossRef]

192. Cousins, M.J.; Pickthorn, K.; Huang, S.; Critchley, L.; Bell, G. The safety and efficacy of KAI-1678—An inhibitor of epsilon protein kinase C (εPKC)—Versus lidocaine and placebo for the treatment of postherpetic neuralgia: A crossover study design. *Pain Med.* **2013**, *14*, 533–540. [CrossRef] [PubMed]

193. Lincoff, A.M.; Roe, M.; Aylward, P.; Galla, J.; Rynkiewicz, A.; Guetta, V.; Zelizko, M.; Kleiman, N.; White, H.; McErlean, E.; et al. Inhibition of delta-protein kinase C by delcasertib as an adjunct to primary percutaneous coronary intervention for acute anterior ST-segment elevation myocardial infarction: Results of the PROTECTION AMI Randomized Controlled Trial. *Eur. Heart J.* **2014**, *35*, 2516–2523. [CrossRef] [PubMed]

194. Direct Inhibition of δ-Protein Kinase C Enzyme to Limit Total Infarct Size in Acute Myocardial Infarction (DELTA MI) Investigators; Bates, E.; Bode, C.; Costa, M.; Gibson, C.M.; Granger, C.; Green, C.; Grimes, K.; Harrington, R.; Huber, K.; et al. Intracoronary KAI-9803 as an adjunct to primary percutaneous coronary intervention for acute ST-segment elevation myocardial infarction. *Circulation* **2008**, *117*, 886–896. [CrossRef]

195. Deloche, C.; Lopez-Lazaro, L.; Mouz, S.; Perino, J.; Abadie, C.; Combette, J. XG-102 administered to healthy male volunteers as a single intravenous infusion: A randomized, double-blind, placebo-controlled, dose-escalating study. *Pharmacol. Res. Perspect.* **2014**, *2*, e00020. [CrossRef] [PubMed]

196. Kim, J.Y.; Ahn, J.; Kim, J.; Choi, M.; Jeon, H.; Choe, K.; Lee, D.Y.; Kim, P.; Jon, S. Nanoparticle-Assisted Transcutaneous Delivery of a Signal Transducer and Activator of Transcription 3-Inhibiting Peptide Ameliorates Psoriasis-like Skin Inflammation. *ACS Nano* **2018**, *12*, 6904–6916. [CrossRef]

197. Guo, J.; Chen, Y.; Yuan, B.; Liu, S.; Rao, P. Effects of intracellular superoxide removal at acupoints with TAT-SOD on obesity. *Free. Radic. Biol. Med.* **2011**, *51*, 2185–2189. [CrossRef]

198. Lemech, C.R.; Kichenadasse, G.; Marschner, J.-P.; Alevizopoulos, K.; Otterlei, M.; Millward, M. ATX-101, a cell-penetrating protein targeting PCNA, can be safely administered as intravenous infusion in patients and shows clinical activity in a Phase 1 study. *Oncogene* **2022**, *42*, 541–544. [CrossRef]

199. Suckfuell, M.; Lisowska, G.; Domka, W.; Kabacinska, A.; Morawski, K.; Bodlaj, R.; Klimak, P.; Kostrica, R.; Meyer, T. Efficacy and safety of AM-111 in the treatment of acute sensorineural hearing loss. *Otol. Neurotol.* **2014**, *35*, 1317–1326. [CrossRef]

200. Foligné, B.; Plé, C.; Titécat, M.; Dendooven, A.; Pagny, A.; Daniel, C.; Singer, E.; Pottier, M.; Bertin, B.; Neut, C.; et al. Contribution of the Gut Microbiota in P28GST-Mediated Anti-Inflammatory Effects: Experimental and Clinical Insights. *Cells* **2019**, *8*, 577. [CrossRef]

201. Lulla, R.R.; Goldman, S.; Yamada, T.; Beattie, C.W.; Bressler, L.; Pacini, M.; Pollack, I.F.; Fisher, P.G.; Packer, R.J.; Dunkel, I.J.; et al. Phase I trial of p28 (NSC745104), a non-HDM2-mediated peptide inhibitor of p53 ubiquitination in pediatric patients with recurrent or progressive central nervous system tumors: A Pediatric Brain Tumor Consortium Study. *Neuro-Oncology* **2016**, *18*, 1319–1325. [CrossRef]

202. Warso, M.A.; Richards, J.M.; Mehta, D.; Christov, K.; Schaeffer, C.; Bressler, L.R.; Yamada, T.; Majumdar, D.; Kennedy, S.A.; Beattie, C.W.; et al. A first-in-class, first-in-human, phase I trial of p28, a non-HDM2-mediated peptide inhibitor of p53 ubiquitination in patients with advanced solid tumours. *Br. J. Cancer* **2013**, *108*, 1061–1070. [CrossRef] [PubMed]

203. Zhou, X.; Singh, M.; Santos, G.S.; Guerlavais, V.; Carvajal, L.A.; Aivado, M.; Zhan, Y.; Oliveira, M.M.; Westerberg, L.S.; Annis, D.A.; et al. Pharmacologic Activation of p53 Triggers Viral Mimicry Response Thereby Abolishing Tumor Immune Evasion and Promoting Antitumor Immunity. *Cancer Discov.* **2021**, *11*, 3090–3105. [CrossRef] [PubMed]

204. Rothbard, J.B.; Garlington, S.; Lin, Q.; Kirschberg, T.; Kreider, E.; McGrane, P.L.; Wender, P.A.; Khavari, P.A. Conjugation of arginine oligomers to cyclosporin A facilitates topical delivery and inhibition of inflammation. *Nat. Med.* **2000**, *6*, 1253–1257. [CrossRef] [PubMed]

205. Hosseini, A.; Lattanzio, F.A., Jr.; Samudre, S.S.; DiSandro, G.; Sheppard, J.D., Jr.; Williams, P.B.; Shi, N.-Q.; Qi, X.-R.; Xiang, B.; Zhang, Y.; et al. Efficacy of a phosphorodiamidate morpholino oligomer antisense compound in the inhibition of corneal transplant rejection in a rat cornea transplant model. *J. Ocul. Pharmacol. Ther.* **2012**, *28*, 194–201. [CrossRef]

206. Abes, R.; Arzumanov, A.; Moulton, H.; Abes, S.; Ivanova, G.; Iversen, P.; Gait, M.; Lebleu, B. Cell-penetrating-peptide-based delivery of oligonucleotides: An overview. *Biochem. Soc. Trans.* **2007**, *35*, 775–779. [CrossRef]
207. Brandt, F.; O'Connell, C.; Cazzaniga, A.; Waugh, J.M. Efficacy and safety evaluation of a novel botulinum toxin topical gel for the treatment of moderate to severe lateral canthal lines. *Dermatol. Surg.* **2010**, *36* (Suppl. 4), 2111–2118. [CrossRef]
208. Glogau, R.; Blitzer, A.; Brandt, F.; Kane, M.; Monheit, G.D.; Waugh, J.M. Results of a randomized, double-blind, placebo-controlled study to evaluate the efficacy and safety of a botulinum toxin type A topical gel for the treatment of moderate-to-severe lateral canthal lines. *J. Drugs Dermatol.* **2012**, *11*, 38–45.
209. Miampamba, M.; Liu, J.; Harootunian, A.; Gale, A.J.; Baird, S.; Chen, S.L.; Nguyen, Q.T.; Tsien, R.Y.; González, J.E. Sensitive in vivo Visualization of Breast Cancer Using Ratiometric Protease-activatable Fluorescent Imaging Agent, AVB-620. *Theranostics* **2017**, *7*, 3369–3386. [CrossRef]
210. Gurbel, P.A.; Bliden, K.P.; Turner, S.E.; Tantry, U.S.; Gesheff, M.G.; Barr, T.P.; Covic, L.; Kuliopulos, A. Cell-Penetrating Pepducin Therapy Targeting PAR1 in Subjects with Coronary Artery Disease. *Arter. Thromb. Vasc. Biol.* **2016**, *36*, 189–197. [CrossRef]
211. Unkart, J.T.; Chen, S.L.; Wapnir, I.L.; González, J.E.; Harootunian, A.; Wallace, A.M. Intraoperative Tumor Detection Using a Ratiometric Activatable Fluorescent Peptide: A First-in-Human Phase 1 Study. *Ann. Surg. Oncol.* **2017**, *24*, 3167–3173. [CrossRef]
212. Gowland, C.; Berry, P.; Errington, J.; Jeffrey, P.; Bennett, G.; Godfrey, L.; Pittman, M.; Niewiarowski, A.; Symeonides, S.N.; Veal, G.J. Development of a LC–MS/MS method for the quantification of toxic payload DM1 cleaved from BT1718 in a Phase I study. *Bioanalysis* **2021**, *13*, 101–113. [CrossRef] [PubMed]
213. Lopes, L.B.; Furnish, E.J.; Komalavilas, P.; Flynn, C.R.; Ashby, P.; Hansen, A.; Ly, D.P.; Yang, G.P.; Longaker, M.T.; Panitch, A.; et al. Cell permeant peptide analogues of the small heat shock protein, HSP20, reduce TGF-β1-induced CTGF expression in keloid fibroblasts. *J. Investig. Dermatol.* **2009**, *129*, 590–598. [CrossRef] [PubMed]
214. Coriat, R.; Faivre, S.J.; Mir, O.; Dreyer, C.; Ropert, S.; Bouattour, M.; Desjardins, R.; Goldwasser, F.; Raymond, E. Pharmacokinetics and safety of DTS-108, a human oligopeptide bound to SN-38 with an esterase-sensitive cross-linker in patients with advanced malignancies: A Phase I study. *Int. J. Nanomed.* **2016**, *11*, 6207–6216. [CrossRef] [PubMed]

Review

Pharmaceutical Application of Process Understanding and Optimization Techniques: A Review on the Continuous Twin-Screw Wet Granulation

Jie Zhao, Geng Tian and Haibin Qu *

Pharmaceutical Informatics Institute, College of Pharmaceutical Sciences, Zhejiang University, Hangzhou 310058, China; zhaojie_1021@zju.edu.cn (J.Z.); iamtiangeng@zju.edu.cn (G.T.)
* Correspondence: quhb@zju.edu.cn; Tel./Fax: +86-571-8820-8428

Abstract: Twin-screw wet granulation (TSWG) is a method of continuous pharmaceutical manufacturing and a potential alternative method to batch granulation processes. It has attracted more and more interest nowadays due to its high efficiency, robustness, and applications. To improve both the product quality and process efficiency, the process understanding is critical. This article reviews the recent work in process understanding and optimization for TSWG. Various aspects of the progress in TSWG like process model construction, process monitoring method development, and the strategy of process control for TSWG have been thoroughly analyzed and discussed. The process modeling technique including the empirical model, the mechanistic model, and the hybrid model in the TSWG process are presented to increase the knowledge of the granulation process, and the influence of process parameters involved in granulation process on granule properties by experimental study are highlighted. The study analyzed several process monitoring tools and the associated technologies used to monitor granule attributes. In addition, control strategies based on process analytical technology (PAT) are presented as a reference to enhance product quality and ensure the applicability and capability of continuous manufacturing (CM) processes. Furthermore, this article aims to review the current research progress in an effort to make recommendations for further research in process understanding and development of TSWG.

Keywords: twin-screw wet granulation (TSWG); process understanding; process model; continuous manufacturing (CM); process analytical technology (PAT); quality control

Citation: Zhao, J.; Tian, G.; Qu, H. Pharmaceutical Application of Process Understanding and Optimization Techniques: A Review on the Continuous Twin-Screw Wet Granulation. *Biomedicines* **2023**, *11*, 1923. https://doi.org/10.3390/biomedicines11071923

Academic Editors: M. R. Mozafari and Giovanni Tosi

Received: 31 May 2023
Revised: 15 June 2023
Accepted: 5 July 2023
Published: 6 July 2023

1. Introduction

In the past decade, continuous manufacturing (CM) of pharmaceutical solid dosage forms has gained significant attention [1]. As an emerging technology, it has high potential to improve their product quality and reliability, reduce the manufacturing costs and waste, and increase manufacturing flexibility and agility in response to fluctuations in drug product demand [2]. The release of the "Continuous Manufacturing of Drug Substances and Drug Products, Q13" by the ICH, offers further evidence of the ongoing efforts of regulatory bodies to support the modernization of the pharmaceutical industry throughout the development and implementation of innovative continuous approaches.

Recently, twin-screw wet granulation (TSWG), as a typical application in continuous pharmaceutical industry, has gained significant attention. Granulation is an important process of solid dosage drug manufacturing. It could improve the properties of granule, such as particle size, dissolution rate, bulk density, powder flowability, compressibility, active pharmaceutical ingredient (API) uniformity, granular strength and density by wet granulation or dry granulation method [3,4]. Compared with the batch granulation techniques, such as the fluid-bed granulation, roller compaction, and high-shear wet granulation methods, TSWG could be seen as a promising tool for transforming traditional batch mode to CM in the pharmaceutical industry [5]. It has the advantages of processing high volume, higher production efficiency, short residence time, and better mixing and controlling processes [6]. It can be readily integrated into the CM of pharmaceutical dosage forms. In addition, TSWG provides easier process scale-up, better quality assurance, low production costs, and materials waste [7].

A deep understanding of the process is essential for the CM implementation of pharmaceutical products. Unlike traditional batch manufacturing, CM involves a series of integrated and interconnected unit operations that work together to produce a consistent quality product. In batch processes, material quality is assumed to be uniform throughout the entire batch, while material quality is subjected to variations in continuous processes if the process is not within a state of control [8]. The Quality by Design (QbD) guidelines promote the thorough generation of product and process understanding, which is necessary for the continuous and systematic operation of processes and quality control of products. This includes determining essential critical quality attributes (CQA), critical process parameters (CPP), and control strategy [9]. In the conventional batch production mode, the product quality control is carried out by sampling and analyzing after each unit step process, which is not suitable for the continuous process. With the development of advanced process analytical technology (PAT), it offers a path towards the successful deployment to maintenance a state of control for CM during production and allow real-time evaluation of system performance [10]. Moreover, the methods for monitoring, controlling, and optimizing in CM necessitate distinctive strategies. A QbD approach in conjunction with PAT has thus been identified as a promising pathway towards the successful industrial implementation of CM for pharmaceutical uses [11].

However, the technology transfer from batch granulation to continuous granulation based on TSWG is considered a challenge with technical difficulty due to its flexible geometry structure, diverse formulations, and complex granulation mechanism. At the same time, the product produced by TSWG needs to be conducive to the downstream processing, and the granule produced by TSWG need to meet the high regulatory standards in the pharmaceutical industry to ensure the safety and quality of the products. Facing the considerable demand of TSWG in CM for drugs, enhancing the knowledge of manufacturing processes and improving the control of the TSWG process are the key to establishing adaptive processing procedure for TSWG. Therefore, this work focuses on the process understanding for TSWG, discusses the recent progress in process understanding and optimization for TSWG, including some progress in process model construction, the method for process monitoring, and the progress of process control for TSWG. This article aims to enhance comprehension and advancement of processing by reviewing and presenting existing research on twin-screw granulation, which is gaining interest, and the crucial need for viable continuous processing.

2. Process Understanding and Optimization

Understanding the process is the first step for the application of TSWG. There are many factors that affect the TSWG process, which can be divided into six categories, i.e., apparatus, materials, people, process variables, measurement, and environment, as shown in Figure 1. The variation of variables makes the TSWG system an exceptionally large operating space. Therefore, because of the complexity and variability of the TSWG system, it is required a systematic understanding of the dynamic change in the characteristics of the material while progressing in the TSWG process to control. The process model development approach can be adopted through the statistical, mechanistic, and hybrid models in the TSWG process to increase the knowledge of the granulation process. The process design, control, and variable optimization of the TSWG process can also be developed by modeling techniques.

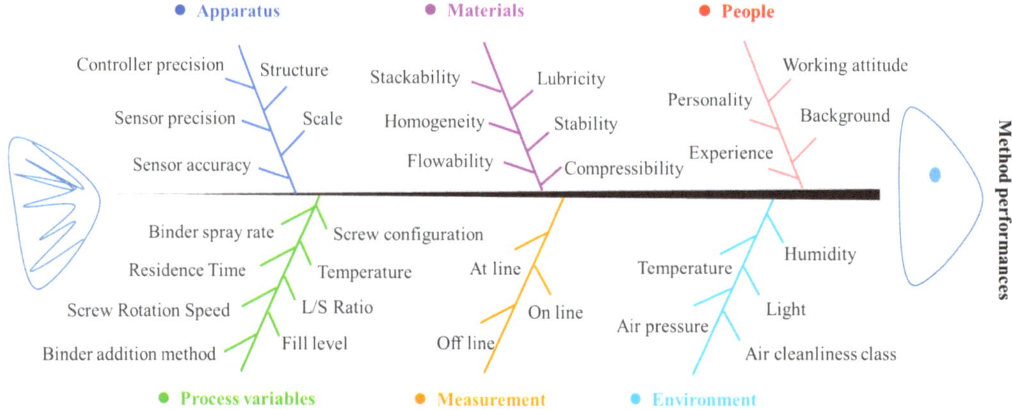

Figure 1. Ishikawa diagram for the twin-screw granulation process behavior.

2.1. Empirical Model

Empirical models are mathematical expressions that are created by analyzing observed data from experiments or observations. Empirical models are often used in scientific research to quantify relationships between variables in a specific context or system. Using the QbD principle, the empirical model developed based on the statistical data from design of the experiment (DoE) can evaluate the effect of CPP and critical material attributes (CMA) on the CQA simultaneously. The statistical data from the DoE could contribute to developing the model and defining the design space, thus ensuring desired quality of the product [12]. It is an effective method applied to TSWG for process understanding, designing, controlling, and optimizing continuous pharmaceutical manufacturing.

Studying the process settings and generalizing the conclusions contribute to better understanding of the TSWG. Many researchers have used the QbD principle to study the TSWG process. At present, the research on CPP in the TSWG process mainly focuses on the geometry and configuration of the screw elements, process variables, and product quality. Different experimental design methods have been performed to evaluate the influence of material properties (formulation, adhesive), process variables (type and number of screw elements, screw speed, liquid feed rate, and liquid to solid (L/S) ratio) on the characteristics of granules [1,13–17]. Seem et al. [18] summarized the comprehensive review of the experimental twin-screw granulator (TSG) literature, indicating the complex interplay between the role of screw element type, screw configuration, feed formulation and liquid flowrates on the granules. The specific parameters haven been investigated, as well as the CQA of granules defined in the corresponding paper, and the DoE method used to process development with corresponding modeling methods have been summarized in Table 1. The overview of research papers that used the QbD principle for process understanding and optimization on TSWG can make a reference for further research.

Table 1. Some examples that utilized the QbD principle to achieve process understanding and optimization on TSWG.

No.	Formula (*w/w*)	CPP	DoE Method	CQA for Granules	Modeling Method	Reference
1	Powder: API; povidone; hypromellose. Liquid: water	Screw speed; mill speed.	Central composite response surface statistical design	Granule chord length; particle size.	Bivariate fit	[19]
2	Powder: MCC; lactose monohydrate; starch. Liquid: HPC; HPMC; PVP.	Water-binding capacity; barrel temperature; powder feed rate; liquid addition; screw configuration.	A two-level full-factorial design.	PSD; Span; HR; bulk/tapped density; moisture content; flowability.	PLS	[20]
3	Powder: α-Lactose monohydrate. Liquid: water	Screw speed; throughput; L/S ratio; number and stagger angle of the kneading discs.	Full factorial experimental design.	RTD; mixing capacity.	PLS	[21]
4	Powder: acetaminophen;a-lactose monohydrate; MCC; PVP K 29-32. Liquid: water.	L/S ratio; throughput; rotation speed.	Face-centered cubic design.	Granule size, porosity, flowability, particle morphology.	ANOVA model	[22]
5	Powder: API; MCC; lactose monohydrate; HPMC; croscarmellose sodium; Calcium carbonate; Polysorbate 80. Liquid: water.	Screw speed; throughput; L/S ratio.	Central composite face-centered experimental design.	Torque; granule residence time; fines, yield; over-sized; volume average granule diameter; relative width; Carr Index.	Second-order polynomial models	[23]
6	Powder: Ibuprofen; dibasic calcium phosphate anhydrous. Liquid: ethanol	DCPA/Polymer ration; binder amount (%); L/S ratio.	Fractional factorial design of experiment.	Release (%), d_{50}; specific surface area.	Regression analysis	[24]
7	Powder: MCC; α-lactose monohydrate; mannitol. Liquid: HPMC; PVP.	3 PCs to include the overarching properties of 8 selected pharmaceutical fillers; binder type; binder concentration.	D-optimal interaction design	Bulk/tapped density; HR; $d_{63.2}$; fine fraction; yield fraction; coarse fraction; flowability; friability; specific surface area; torque.	Multiple linear regression	[25]
8	Powder: API; MCC; HPMC. Liquid: HPMC.	PC 1; PC 2; L/S ratio.	D-optimal screening design; D-optimal optimization design	Span; friability; fine fraction.	MLR model	[26]
9	Powder: API; Lactose monohydrate; croscarmellose Sodium; PVP K29/32. Liquid: water.	Throughput; screw speed; the screw element.	Box–Behnken experimental design	Fines; yield; over-sized; volume average granule diameter; relative width; Carr Index.	Non-linear quadratic mathematical model	[14]
10	Powder: caffeine anhydrous; α-lactose monohydrate; MCC; PVP K30. Liquid: water.	Barrel temperature; L/S ratio; throughput.	Sequential experimental strategy: D-optimal design and response surface design.	d_{10}, d_{50}, d_{90}, Span; Eccentricity; porosity; bulk density; tapped density; HR; torque.	Stepwise least squares regression	[27]

Table 1. *Cont.*

No.	Formula (*w/w*)	CPP	DoE Method	CQA for Granules	Modeling Method	Reference
11	Powder: APIs; MCC; HPMC. Liquid: water.	L/S ratio; screw speed.	Central composite circumscribed designs.	Fines; yield; HR; oversized; tapped density; friability; torque.	Quadratic polynomial models	[28]
12	Powder: metformin hydrochloride; mebendazole; MCC; α-lactose monohydrate; HPMC. Liquid: water.	Fraction of KE in the first; kneading zone; KE thickness; screw speed; throughput; L/S ratio.	D-optimal design	Yield; over-sized; fines; bulk density; HR; friability.	MLR models	[29]
13	Powder: Ibuprofen; MCC; lactose monohydrate; Croscarmellose sodium; Hydroxypropyl cellulose; Colloidal silicon dioxide. Liquid: water.	L/S; throughput; screw speed; screw configuration; barrel temperature.	A full factorial design	The angle of repose; bulk/tapped density; PSD; granule strength; granule morphology.	Forward stepwise regression	[30]
14	Powder: (1) lactose; PVP. (2) lactose; HPMC. (3) lactose; MCC; PVP. (4) lactose; MCC; HPMC. Liquid: water.	Nozzle diameter; nozzle orientation; throughput; screw speed; screw configuration; barrel temperature; L/S ratio; total binder content; the relative fraction of binder added dry.	Plackett–Burman design for screening.	Oversized; fines; HR; bulk density; tapped density; yield; the angle of repose, torque.	Linear models	[31]
15	Powder: metformin hydrochloride; α-Lactose monohydrate; MCC; HPMC. Liquid: water.	3 PCs to include the MCC properties; L/S ratio; screw speed; drying time.	Two-level full factorial design.	Fines; HR; torque; LOD.	MLR	[32]
16	Powder: lactose monohydrate; MCC; PVP K25. Liquid: water.	L/S; SFL; screw length; DFS	Two-level full factorial design.	RTD; particle Size; shape distributions.	Multiple linear regression	[33]
17	Powder: MCC; α-lactose monohydrate Liquid: PVP.	Liquid feed rate; powder feed rate; L/S ratio; screw speed; viscosity.	Full factorial DoE.	RTD; PSD; liquid content distribution.		[34]
18	Powder: lactose; copovidone, PVP, hyprolose, hypromellose Liquid: water.	L/S; SFL; screw speed; powder feed rate.	3^2 experimental design and single-factor experiment.	PSD; span.	-	[35]

Microcrystalline cellulose (MCC); polyvinylpyrollidone (PVP); hydroxypropyl methylcellulose (HPMC); water binding capacity (WBC); Hausner ratio (HR); kneading elements (KE); length to diameter (L/D); residence time distribution (RTD); high specific feed load (SFL); distributive feed screw (DFS); partial least squares (PLS); particle size distribution (PSD); hydroxypropyl cellulose (HPC); multiple linear regression (MLR); analysis of variance (ANOVA); loss on dry (LOD).

The screw configuration, screw speed, and throughput have a large impact on the granule properties. The degree of shear applied to the material during granulation process can be varied by adjusting of these parameters. The screw configuration includes the pitch and length of the conveying elements, the thickness and angle of the kneading elements. The process variables consist of the properties of the materials and binder, L/S ratio, screw speed, and material feed rates, etc. Due to the large diversity of the model formulations and equipment, it is difficult to draw general conclusions. However, based on the literature, the results of this research showed that the process parameters, such as L/S ratio, screw speed, and throughput, are the key process variables of the TSWG process. It is important for the process understanding to clarify the influence mechanism of the process conditions on the particle properties.

The L/S ratio has been extensively identified as the most predominant factor in preparing TSWG products with desired quality attributes [36,37]. It is reported that the low L/S ratio produces the granules with broad and bimodal size distribution, while the size distribution becomes narrow and monomodal at high L/S ratio that are too large to be directly used for tableting [18]. The increase of L/S ratio could promote the growth of granules, increase bulk and tapped density, and facilitate the strength and flow properties of the particles. A second-order polynomial model based on a central composite face-centered experimental design demonstrates that as the L/S ratio increases from 10% to 70%, the proportion of fines dramatically decreases due to the high aggregation rate resulting from the liquid [23]. Moreover, the initial multi-factor analysis of variance (ANOVA) based on a face-centered cubic design model indicated that bulk and tapped densities and the shape of granules were greatly affected by the L/S ratio [22].

Changes in throughput can impact the barrel filling, with a higher value leading to more material hold-up volume due to the compression and packing of primary particles. A PLS model that was built based on a full factorial experimental design showed that the increment of material throughput could quickly lower the yield due to insufficient mixing between powder and liquid [21]. A non-linear quadratic mathematical model based on a Box–Behnken experimental design showed that the increased throughput could produce granules with a narrower PSD by reducing the oversized granules and fines [14]. Finally, a stepwise least squares regression model built on a sequential experiment showed that throughput significantly influences particle porosity. The larger porosity could be largely due to the high upstream throughput force and reduced residence time, the materials were conveyed faster, and particles suffered from less interaction and compaction [27].

2.2. Mechanistic Model

Mechanistic models can be constructed to reveal the mechanism of material transformation in the TSWG process, and the model could then be coupled with process control systems to allow control of product specification in TSWG systems. Among the various modeling and simulation methods applicable for the wet granulation process, the population balance model (PBM), discrete element method (DEM), computational fluid dynamics (CFD), and the combination of these models have drawn widespread attention (Figure 2).

DEM, as a scientifically meaningful model method, is playing an increasingly important role in simulating a variety of phenomena including the mechanistic aggregation, consolidation, and breakage rate expressions in the TSWG process. DEM adopts Newton's second law and can track the spatial coordinate of each particle. It can obtain compartmental residence time data, particle velocities, collision rates, and provide macroscopic information and microscopic insights into the complex TSWG process [38]. With the improvement of high-performance computer capabilities such as the graphics processor unit (GPU) acceleration and parallel computing, the DEM can speed up the simulation and overcome the problem of computations further, indicating that the real-time simulations become possible for the TSWG implications in the near future.

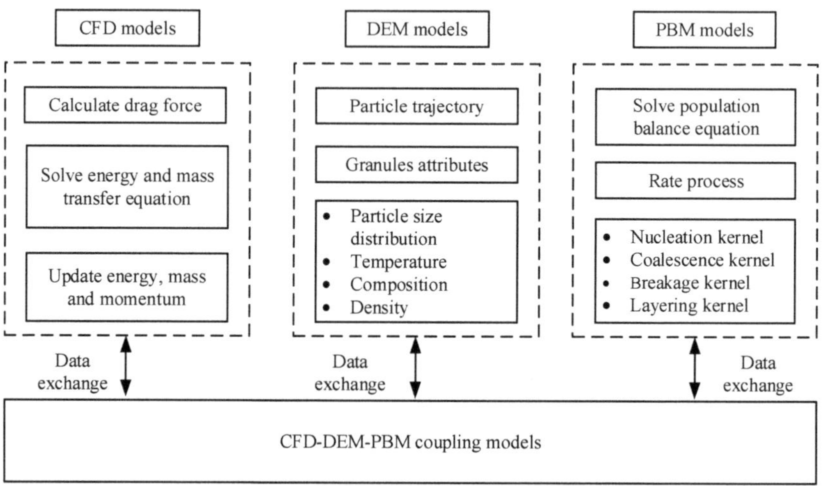

Figure 2. Schematic of the models for granulation.

PBM employs the principle of particle number density conservation and can simulate the granulation process of particles in batch or continuous production mode. Moreover, it can track the particle properties in the granulation process. Different dimensional PBM models have been developed for process simulation, such as 1D PBM [39] and multi-dimensional PBM [40]. The advanced modeling tool can predict granule size, PSD, liquid content distributions, RTD, etc. [41–43]. Based on the PBM framework, the model was also constantly improved. For instance, in improving a 1D PBM [37,44] for process simulation, a high-dimensional stochastic PBM was constructed to estimate the residence times for different screw element geometry [45]. A novel four-dimensional, stochastic PBM for twin-screw granulation was proposed to describe the mechanistic rates and track more complex particle properties and their transformations [46].

Besides the commonly used mechanistic models, a regime map based on the regime theory can present a method to describe the variation and behavior in the granulation process [47]. Moreover, this semi-mechanistic technique can predict the RTD curve [48] and mean residence time (MRT) with decent precision for TSWG.

2.3. Hybrid Model

Compared with the complex numerical simulation models, easier and more efficient modeling methods are an urgent requirement for the simulation application of the TSWG process. The hybrid model combined multiple modeling technique can provide a more comprehensive understanding of the complex pharmaceutical processes, which is an effective approach to improve the limitations of a single kind model. Artificial neural network (ANN) model provides a powerful tool for data analysis and pattern recognition for drug development processes. However, the ANN model has a complex model structure, the network involves input layer, hidden layer(s), and output layer. Many parameters need to be optimized to achieve better model performance. To improve the predictive capability of the artificial neural network (ANN) model, kriging interpolation is applied to get new data to develop an improved ANN model for the mean residence time. The results showed that the hybrid model combined with kriging interpolation could predict the mean residence time with more accuracy for TSWG [49].

In order to obtain a more accurate model, the models combining CFD, DEM, and PBM have also been proposed. A multi-scale, compartmental PBM-DEM model of a continuous TSWG process was presented [50]. The PBM describes the mechanistic rate aggregation, breakage, consolidation, and the DEM was used to obtain residence time information and gather collision and velocity information. The simulated results are consistent with experimental trends, demonstrating the model's qualitative ability to model and predict the effects of screw element configuration and process parameters on the product size distribution, porosity, and liquid distribution.

A new framework based on fuzzy logic is proposed to predict the granule size distribution accurately in the TSG process, outperforming the standard fuzzy logic systems and the ANN [51]. A versatile simulation model with high-performance for TSWG can be further used for defining the design space. A hybrid model combining PBM and ANN was developed, which was utilized to determine particle size distribution and mean residence time, respectively [36]. In addition, the model can simulate and predict the PSD with fairly high accuracy throughout the simulations, as well as define the design space to develop and optimize the TSWG process.

3. Process Monitoring

The purpose of using PAT tools in the TSWG process is to monitor, control, and optimize the process parameters to produce the granule with desired quality. Real-time monitoring of the manufacturing process can help to visualize and provide the product information throughout the process, and with the real-time in-process measurement of CQAs, the off-spec products can be effectively identified and removed from the stream to ensure safety and quality. The common object of granulation technology is to prepare granules with a fine appearance, excellent uniformity, and high yield. Review of the published papers, qualitative and quantitative analysis using PAT tools for granule produced by TSWG have been reported. These analyses include the API content of the granules, RTD, moisture content, PSD, and particle density, etc. The powder properties of particles are complex, the more parameters detected during the process, the more information represented. However, as the number of detection parameters increases, the testing time or the duration of data processing will also increase, ultimately leading to a reduction in detection efficiency. In specific applications, it is necessary to consider comprehensively the experimental conditions and the operability, thus to select the parameters that affect drug quality as much as possible in the TSWG process.

There have been many reports on the application of PAT tools in process monitoring of TSWG, incorporating techniques such as near-infrared (NIR) spectroscopy, Raman spectroscopy, spatial filter velocimetry (SFV), etc. These techniques can contribute to the enhancement of product quality through the development of process knowledge. Moreover, PAT tools are an indispensable element in the implementation of process control strategies [52].

3.1. Spectroscopy

A molecular vibrational spectroscopic technique like NIR spectroscopy and Raman spectroscopy techniques enhanced real-time process monitoring application efficiency. The qualitative and quantitative analysis with non-invasive and non-destructive measurement can be easily used for continuous process flow to obtain multi-dimensional information online.

The NIR spectrum contains both the physical and chemical characteristics of the material with specific absorption and scattering effects. Hence, it is specific and sensitive to the particle properties (e.g., particle size and distribution, surface morphology, density, shape, and surface texture) that affect the path length and light penetration [53]. Therefore, the influence of different installation positions of the NIR spectroscopy probe on in-line measurements was investigated. By characterizing NIR interfaces in different positions of a TSWG process, NIR spectra obtained during the process were analyzed from different aspects by multivariate methods to determine the granules' powder dynamics and API content. The results showed that the linear motion of granules in the barrel produced enough representative measurements for developing a model with a low prediction error [54].

NIR chemical imaging (NIR-CI) is a promising technique that combines traditional optical imaging and NIR spectroscopy. It is especially suitable for the measurement of particles and provides better process visualization and characterization of the TSWG. The NIR-CI image has been used as a qualitative and quantitative analytical method to characterize the mixing and flow of material in the TSWG [55]. The residence time distribution (RTD) was detected and investigated as a function of process parameters (screw rotation speed and material feeding rate) and screw configuration (number and angle of kneading elements). NIR-CI provides the possibility to better characterize and visualize the phenomenon of particle segregation along the TSG barrel and conduct process optimization to ensure granule quality [56]. The mixing behavior and the distribution of liquid and powder during the granulation process can be visualized by NIR-CI. Moisture homogeneity of granules was characterized and verified by the moisture map, which helps analyze the mixing state of liquid and powder material transportation inside the screw chamber [57].

Raman spectra is another representative molecular vibrational spectroscopic method. The application of Raman spectra requires no sample pre-treatment. The fast and non-destructive measurement of the method is suitable and used as the PAT tool. Furthermore, Raman spectra are not sensitive to water, so it is applicable for wet granulation process monitoring. An in-line API quantification method using Raman spectroscopic was developed to determine the API concentration for the TSWG process [58], indicating that Raman spectroscopy is one of the promising PAT tools for the API determination and process monitoring for the TSWG process. It should be noted that the measurement needs to be taken in the dark, as Raman spectra are influenced by light. Thus, the Raman on-line analytical method needs to be optimized to enable measurement in interior light conditions [59]. The results showed that the prediction error could be reduced significantly by optimizing the measurement setup.

3.2. Imaging Technique

The imaging technique showed great potential for process monitoring of the continuous TSWG. It can measure the manufacturing process in a non-contact manner with no sample consumption, saving time and cost. Furthermore, due to the fast response time, it can determine and visualize the agglomeration behavior and the state of the material in static and dynamic modes during the granulation process. Besides, this technique can provide multi-dimensional information such as particle size and distribution, shape and surface morphology, etc.

Eyecon^{TM} 3D imaging system (Innopharma Laboratories, Dublin, Ireland) is a direct imaging particle analyzer generally used for process detection for TSWG. It has a digital camera surrounded by green, red, and blue light sources. To obtain the particle's three-dimensional (3D) features, the illumination direction of the light sources is placed following the rule of photometric stereo. The surface orientation models were calculated by transforming the image intensities on different light illuminants to detect and reconstruct the edges of the particles. An iterative algorithm was adopted to obtain the best fitting ellipse for the granules from the projected two-dimensional (2D) image and compute the equivalent area diameter.

The feasibility of implementing the Eyecon^{TM} 3D imaging system for in-line monitoring of the TSWG process was evaluated. The study showed that Eyecon^{TM} (Innopharma Labs, Dublin, Ireland) provided good in-line images and PSD information despite the granules with a dense moving flow [60]. The capability of the Eyecon^{TM} camera for the in-line size monitoring and controlling of granules in TSWG was evaluated. Eyecon^{TM} could detect the size enlargement and count reduction when the L/S ratio changed [61]. In addition, the Eyecon^{TM} exhibited sensitivity to the perturbations and the variations in the TSWG process [62].

Besides PSD data, imaging techniques can obtain 2D properties distributions, including particle size and its distribution and liquid content of granules, to get better insight into the TSWG [34].

Imaging techniques can also obtain the temperature information of the subjects. For example, the FLIR A655sc infra-red camera, equipped with a 45° lens and a detector, can monitor the temperature of granules. When coupled with the properties of granules produced by TSWG, it can enhance overall understanding of the wetting mechanisms during TSWG [63].

3.3. Acoustic Emissions Technique

The acoustic emissions (AE) technique shares the advantage of NIR spectroscopy in that it is a non-invasive technique that can be utilized in real time monitor of the manufacturing process. However, instead of capturing optical signal, AE methods capture the mechanical information of process events.

The previous research has demonstrated that AE, a non-destructive ultrasonic technique, is a reliable predictor of Gaussian PSD [64–67]. The maximum amplitude of the time-domain AE signal, generated by the collision of a moving particle with a rigid surface, is correlated with particle size to estimate a single descriptor, typically d_{50}. In order to enhance the ability of AE to predict a bimodal PSD for TSWG, H.A. Abdulhussain et al. adopted frequency-domain signal analysis along with artificial intelligence (AI) techniques [68]. Moreover, they developed a novel digital signal filter for the AE signal, which adjusts the signal based on impact mechanics to ensure that all particle sizes are accurately represented in the correlations. When using the filter, AE-based PAT showed exceptional predictive accuracy for near-elastic collisions, as demonstrated in a trial with lactose monohydrate granulation.

3.4. Multi-Technique Integrated Method

Different PAT methods with distinct functioning mechanisms can reflect varied aspects of information from the test subjects. Therefore, combined with the advantages and disadvantages of different methods, the multi-technique integrated method can comprehensively reflect the information on the material properties.

Complementary PAT tools, including Raman, NIR spectrometer, and SFV probe, were used to evaluate the feasibility of solid-state and particle size measurements for the TSWG process [15]. On the premise that the main challenge of probe fouling was solved, it proved that both Raman and NIR were suitable for monitoring the material properties of the model drug (theophylline) throughout the wet granulation process, and the SFV probe showed the ability for the in-line measurement of particle size and the particle size distribution. Furthermore, multivariate data obtained by the three PAT tools were used to monitor the granules' moisture content, particle density, and flowability [53]. The results indicated that the moisture content showed a high degree of correlation with the NIR data, and the imaging data reflected the flowability of the granules. It has been proved that multi-technique integrated for the TSWG process can obtain complementary information.

Meng et al. [69] used three PAT methods, i.e., imaging technique, NIR, and Raman, to monitor the TSWG process. The three probes were mounted over a belt of a conveyor platform that carries on the granules from TSWG to avoid fouling or contamination of probes. For monitoring different properties, Eyecon™ monitored the granule size and shape variation, NIR combined with chemometrics for physical property prediction, and Raman spectroscopy was employed to measure the content of drug components and transformations between materials. The study demonstrated the implementation of different PAT tools for continuously in-line/on-line monitoring of produced granules, which provide a better process design and monitoring of TSWG.

4. Process Control

According to ICH Q10 and Q13 guidelines, the quality of production could be ensured by establishing control strategies to maintain the process under control, in other words, to use process monitoring tools and control methods for the improvement of product quality, assuring the continued applicability and capability of processes [70].

QbD principle is a useful tool for establishing the control strategies for continuous manufacturing, as it provides the basic concepts and scheme for control model construction [71]. To establish variable rate process control strategies for solid oral dosage forms in continuous manufacturing, a comprehensive and systematic methodology was established for improving process understanding and control strategy for TSWG, followed by the QbD principle [72]. Based on risk assessment and DoE results, process optimization could define an appropriate operating parameter range. The empirical control strategy assures product quality consistency through model-based adjustment of process parameters in real-time, increases throughput, and reduces the incident risk of complete shutdown. As a general guideline for the control strategy of continuous manufacturing, it is useable across different unit operation chains.

Multivariate statistical process control (MSPC) strategy based on principal component analysis was a way to develop a control strategy for a continuous pharmaceutical manufacturing line. The barrel temperature, screw speed, and liquid feeding rate were detected during manufacturing for a TSWG [73]. Hotelling's T^2 and the corresponding Q residuals statistics control charts were applied to analyze and evaluate the impact of the fluctuations to get profound knowledge of the process. It was found that the model can monitor the performance of the manufacturing line and can detect process disturbances. In-line measurement of the granule size with the imaging camera, the particle size, and count were then assessed using the Shewhart control charts [61], providing a real-time characterization of product quality attributes for TSWG applications. Fanny Stauffer et al. [74] used the MSPC strategy to display twin-screw granulation lines' process dynamics and deviations. The developed model can identify key points of the product within manufacturing activities. It provides a new way to rationalize the sampling strategy, design the diversion strategy, and continue process verification.

Feedback control strategy as an advanced process control (APC) solution is one way of automation in-process control method. After the monotonic increasing relation between particle size and L/S ratio, the P controller was used to carry out a real-time feedback control strategy for the TSWG process implemented via a camera equipped with a custom image analysis software firstly [75]. The liquid feeding rate was chosen as the process variable (PV) to drive the granule size to the set value. The validation of the developed system showed that the controller could compensate for the impact of interference by adjusting the liquid feeding rate, and the on-line particle size analyzer can provide the granule size with a measurement error of less than 5 µm. The model predictive control method can be implemented for TSWG process control successfully as one of the feedback control strategies. By in-line NIR spectroscopy monitoring, the L/D ratio of the wet granule based on a PLS regression model and an automatic supervisory controller was established for TSWG. When inducing artificial disturbances, it correctly follows the dynamic set point and obtains an RMSE of 0.25% w/w relative to the setpoint [76]. Torque is an important PV in the TSWG process, and the feasibility of the APC approach by adjusting the torque was investigated. The results showed little correlation between the torque and the granule size, which was unsuitable for characterizing the granule size in the APC strategy for the TSWG process. This may be due to the fact that the torque is not only related to the granule size, but also affected by the material properties such as temperature and viscosity, resulting in the change of the shear force [77].

As the utilization of PAT tools becomes more sophisticated and the development of control strategies becomes more scientific, it is feasible to incorporate real-time release testing (RTRT) methods to monitor CQAs in real-time for TSWG process. This ensures the TSWG unit can produce high-quality products that meet the downstream quality requirements with minimized intermediate testing, and improve overall operational efficiencies while ensuring product quality.

5. Conclusions and Future Perspectives

TSWG is a rapidly popular technology for pharmaceutical processes. However, the interaction between the screw configuration (the conveying and mixing elements), the process conditions, and the formulation properties makes the TSWG a complex process. The large diversity in the properties of raw materials, excipients, and binder liquid in different formulations diversify the products manufactured by TSWG. The influence law of these critical process variables on the critical properties of granules for TSWG process remains poorly comprehended. Although the research on TSWG has made great progress in the past few decades, there still exists considerable demand and potential for process understanding and optimization to improve the utilization of this technique. The black box approach such as the DoE method was used to understand the granulation mechanisms; mechanistic modeling tools for mechanism study can help increase the granulation mechanisms compared with the black-box approach, while the mathematical formulations of these numerical methods are extremely complex and slow when applied to practical problems, mainly due to the high computational requirements. Thus, improving the current models is necessary to monitor the continuous production process. The switch from the batch-mode of granulation process to the continuous TSWG process needs to consider how to assess and assure the drug quality and safety.

The application of PAT tools in practical production is facing many challenges. These include the reasonable layout of sensors, the way for large process data processing, multi-sensor integrated control simultaneously, etc., that need to be thoroughly considered for extensive applications. Several research studies have investigated the application of different PAT tools for monitoring of the TSWG process, including NIR spectroscopy, Raman spectroscopy, and acoustic emission techniques as mentioned in the paper. These tools have been shown to provide real-time monitoring of CQAs such as moisture content, granule size, and drug content uniformity. However, the successful application of each PAT tool depends on several factors, including the characteristics of the material, the configuration of the twin-screw granulator, and the operating conditions to ensure the measurement accuracy. Therefore, it is important to carefully assess the applicability of each PAT tool based on the characteristics of the process.

To ensure the quality of the product from CM in meeting the regulatory requirements and guidelines, accurate control strategy for each unit in the CM line including TSWG was needed to regulate the pharmaceutical process, which highlighted the importance of understanding the process. To ensure that the products from CM meet regulatory requirements and guidelines in terms of quality, an accurate control strategy for each unit in the CM line is necessary, which highlighted the importance of process understanding for TSWG. This work summarized the progress in process understanding and development for TSWG to provide a comprehensive reference for future studies.

Author Contributions: Writing—original draft preparation, J.Z.; writing—review and editing, G.T.; supervision, H.Q. All authors have read and agreed to the published version of the manuscript.

Funding: This work was supported by Innovation Team and Talents Cultivation Program of National Administration of Traditional Chinese Medicine (ZYYCXTD-D-2020002), and Postdoctoral Science Foundation of Zhejiang Province (520000-X82101).

Data Availability Statement: Not applicable.

References

1. Fulop, G.; Domokos, A.; Galata, D.; Szabo, E.; Gyurkes, M.; Szabo, B.; Farkas, A.; Madarasz, L.; Demuth, B.; Lender, T.; et al. Integrated twin-screw wet granulation, continuous vibrational fluid drying and milling: A fully continuous powder to granule line. *Int. J. Pharm.* **2021**, *594*, 120126. [CrossRef]
2. Fisher, A.C.; Liu, W.; Schick, A.; Ramanadham, M.; Chatterjee, S.; Brykman, R.; Lee, S.L.; Kozlowski, S.; Boam, A.B.; Tsinontides, S.C.; et al. An audit of pharmaceutical continuous manufacturing regulatory submissions and outcomes in the US. *Int. J. Pharm.* **2022**, *622*, 121778. [CrossRef] [PubMed]
3. Dhenge, R.M.; Fyles, R.S.; Cartwright, J.J.; Doughty, D.G.; Hounslow, M.J.; Salman, A.D. Twin screw wet granulation: Granule properties. *Chem. Eng. J.* **2010**, *164*, 322–329. [CrossRef]
4. Matsui, Y.; Watano, S. Evaluation of properties of granules and tablets prepared by twin-screw continuous granulation and comparison of their properties with those by batch fluidized-bed and high shear granulations. *Powder Technol.* **2018**, *55*, 86–94. [CrossRef]
5. Kytta, K.M.; Lakio, S.; Wikstrom, H.; Sulemanji, A.; Fransson, M.; Ketolainen, J.; Tajarobi, P. Comparison between twin-screw and high-shear granulation—The effect of filler and active pharmaceutical ingredient on the granule and tablet properties. *Powder Technol.* **2020**, *376*, 187–198. [CrossRef]
6. Gorringe, L.J.; Kee, G.S.; Saleh, M.F.; Fa, N.H.; Elkes, R.G. Use of the channel fill level in defining a design space for twin screw wet granulation. *Int. J. Pharm.* **2017**, *519*, 165–177. [CrossRef]
7. Vercruysse, J.; Peeters, E.; Fonteyne, M.; Cappuyns, P.; Delaet, U.; Van Assche, I.; De Beer, T.; Remon, J.P.; Vervaet, C. Use of a continuous twin screw granulation and drying system during formulation development and process optimization. *Eur. J. Pharm. Biopharm.* **2015**, *89*, 239–247. [CrossRef]
8. Chavez, P.-F.; Stauffer, F.; Eeckman, F.; Bostijn, N.; Didion, D.; Schaefer, C.; Yang, H.; El Aalamat, Y.; Lories, X.; Warman, M.; et al. Control strategy definition for a drug product continuous wet granulation process: Industrial case study. *Int. J. Pharm.* **2022**, *624*, 121970. [CrossRef]
9. Beg, S.; Hasnain, M.S.; Rahman, M.; Swain, S. Chapter 1—Introduction to Quality by Design (QbD): Fundamentals, Principles, and Applications. In *Pharmaceutical Quality by Design*; Beg, S., Hasnain, M.S., Eds.; Academic Press: Cambridge, MA, USA, 2019; pp. 1–17.
10. Vargas, J.M.; Nielsen, S.; Cárdenas, V.; Gonzalez, A.; Aymat, E.Y.; Almodovar, E.; Classe, G.; Colón, Y.; Sanchez, E.; Romañach, R.J. Process analytical technology in continuous manufacturing of a commercial pharmaceutical product. *Int. J. Pharm.* **2018**, *538*, 167–178. [CrossRef]
11. Miyai, Y.; Formosa, A.; Armstrong, C.; Marquardt, B.; Rogers, L.; Roper, T. PAT Implementation on a Mobile Continuous Pharmaceutical Manufacturing System: Real-Time Process Monitoring with In-Line FTIR and Raman Spectroscopy. *Organic Process Res. Dev.* **2021**, *25*, 2707–2717. [CrossRef]
12. Portier, C.; Vervaet, C.; Vanhoorne, V. Continuous Twin Screw Granulation: A Review of Recent Progress and Opportunities in Formulation and Equipment Design. *Pharmaceutics* **2021**, *13*, 668. [CrossRef]
13. Morrissey, J.P.; Hanley, K.J.; Ooi, J.Y. Conceptualisation of an Efficient Particle-Based Simulation of a Twin-Screw Granulator. *Pharmaceutics* **2021**, *13*, 2136. [CrossRef]
14. Liu, H.; Ricart, B.; Stanton, C.; Smith-Goettler, B.; Verdi, L.; O'Connor, T.; Lee, S.; Yoon, S. Design space determination and process optimization in at-scale continuous twin screw wet granulation. *Comput. Chem. Eng.* **2019**, *125*, 271–286. [CrossRef]
15. Fonteyne, M.; Vercruysse, J.; Díaz, D.C.; Gildemyn, D.; Vervaet, C.; Remon, J.P.; Beer, T.D. Real-time assessment of critical quality attributes of a continuous granulation process. *Pharm. Dev. Technol.* **2013**, *18*, 85–97. [CrossRef] [PubMed]
16. Peeters, M.; Jiménez, A.A.B.; Matsunami, K.; Stauffer, F.; Nopens, I.; De Beer, T. Evaluation of the influence of material properties and process parameters on granule porosity in twin-screw wet granulation. *Int. J. Pharm.* **2023**, *641*, 123010. [CrossRef]
17. Van de Steene, S.; Van Renterghem, J.; Vanhoorne, V.; Vervaet, C.; Kumar, A.; De Beer, T. Identification of continuous twin-screw melt granulation mechanisms for different screw configurations, process settings and formulation. *Int. J. Pharm.* **2023**, *630*, 122322. [CrossRef] [PubMed]
18. Seem, T.C.; Rowson, N.A.; Ingram, A.; Huang, Z.; Yu, S.; Matas, M.D.; Gabbott, I.; Reynolds, G.K. Twin Screw Granulation—A Literature Review. *Powder Technol.* **2015**, *276*, 89–102. [CrossRef]
19. Kumar, V.; Taylor, M.K.; Mehrotra, A.; Stagner, W.C. Real-Time Particle Size Analysis Using Focused Beam Reflectance Measurement as a Process Analytical Technology Tool for a Continuous Granulation-Drying-Milling Process. *AAPS PharmSciTech* **2013**, *14*, 523–530. [CrossRef]
20. Fonteyne, M.; Correia, A.; De Plecker, S.; Vercruysse, J.; Ilic, I.; Zhou, Q.; Veryaet, C.; Remon, J.P.; Onofre, F.; Bulone, V.; et al. Impact of microcrystalline cellulose material attributes: A case study on continuous twin screw granulation. *Int. J. Pharm.* **2015**, *478*, 705–717. [CrossRef] [PubMed]
21. Kumar, A.; Alakarjula, M.; Vanhoorne, V.; Toiviainen, M.; De Leersnyder, F.; Vercruysse, J.; Juuti, M.; Ketolainen, J.; Vervaet, C.; Remon, J.P.; et al. Linking granulation performance with residence time and granulation liquid distributions in twin-screw granulation: An experimental investigation. *Eur. J. Pharm. Sci.* **2016**, *90*, 25–37. [CrossRef]
22. Meng, W.; Kotamarthy, L.; Panikar, S.; Sen, M.; Pradhan, S.; Marc, M.; Litster, J.D.; Muzzio, F.J.; Ramachandran, R. Statistical analysis and comparison of a continuous high shear granulator with a twin screw granulator: Effect of process parameters on critical granule attributes and granulation mechanisms. *Int. J. Pharm.* **2016**, *513*, 357–375. [CrossRef] [PubMed]

23. Liu, H.L.; Galbraith, S.C.; Ricart, B.; Stanton, C.; Smith-Goettler, B.; Verdi, L.; O'Connor, T.; Lee, S.; Yoon, S. Optimization of critical quality attributes in continuous twin-screw wet granulation via design space validated with pilot scale experimental data. *Int. J. Pharm.* **2017**, *525*, 249–263. [CrossRef] [PubMed]

24. Maniruzzaman, M.; Ross, S.A.; Dey, T.; Nair, A.; Snowden, M.J.; Douroumis, D. A quality by design (QbD) twin-Screw extrusion wet granulation approach for processing water insoluble drugs. *Int. J. Pharm.* **2017**, *526*, 496–505. [CrossRef]

25. Willecke, N.; Szepes, A.; Wunderlich, M.; Remon, J.P.; Vervaet, C.; De Beer, T. A novel approach to support formulation design on twin screw wet granulation technology: Understanding the impact of overarching excipient properties on drug product quality attributes. *Int. J. Pharm.* **2018**, *545*, 128–143. [CrossRef] [PubMed]

26. Stauffer, F.; Vanhoorne, V.; Piker, G.; Chavez, P.F.; Vervaet, C.; De Beer, T. Managing API raw material variability during continuous twin-screw wet granulation. *Int. J. Pharm.* **2019**, *561*, 265–273. [CrossRef]

27. Meng, W.; Rao, K.S.; Snee, R.D.; Ramachandran, R.; Muzzio, F.J. A comprehensive analysis and optimization of continuous twin-screw granulation processes via sequential experimentation strategy. *Int. J. Pharm.* **2019**, *556*, 349–362. [CrossRef]

28. Portier, C.; De Vriendt, C.; Vigh, T.; Di Pretoro, G.; De Beer, T.; Vervaet, C.; Vanhoorne, V. Continuous twin screw granulation: Robustness of lactose/MCC-based formulations. *Int. J. Pharm.* **2020**, *588*, 119756. [CrossRef]

29. Portier, C.; Pandelaere, K.; Delaet, U.; Vigh, T.; Di Pretoro, G.; De Beer, T.; Vervaet, C.; Vanhoorne, V. Continuous twin screw granulation: A complex interplay between formulation properties, process settings and screw design. *Int. J. Pharm.* **2020**, *576*, 119004. [CrossRef]

30. Miyazaki, Y.; Lenhart, V.; Kleinebudde, P. Switch of tablet manufacturing from high shear granulation to twin-screw granulation using quality by design approach. *Int. J. Pharm.* **2020**, *579*, 119139. [CrossRef]

31. Portier, C.; Pandelaere, K.; Delaet, U.; Vigh, T.; Kumar, A.; Di Pretoro, G.; De Beer, T.; Vervaet, C.; Vanhoorne, V. Continuous twin screw granulation: Influence of process and formulation variables on granule quality attributes of model formulations. *Int. J. Pharm.* **2020**, *576*, 118981. [CrossRef]

32. Portier, C.; Vigh, T.; Di Pretoro, G.; Leys, J.; Klingeleers, D.; De Beer, T.; Vervaet, C.; Vanhoorne, V. Continuous twin screw granulation: Impact of microcrystalline cellulose batch-to-batch variability during granulation and drying—A QbD approach. *Int. J. Pharm. X* **2021**, *3*, 100077. [CrossRef] [PubMed]

33. Plath, T.; Korte, C.; Sivanesapillai, R.; Weinhart, T. Parametric Study of Residence Time Distributions and Granulation Kinetics as a Basis for Process Modeling of Twin-Screw Wet Granulation. *Pharmaceutics* **2021**, *13*, 645. [CrossRef] [PubMed]

34. Ismail, H.Y.; Albadarin, A.B.; Iqbal, J.; Walker, G.M. Image processing for detecting complete two dimensional properties' distribution of granules produced in twin screw granulation. *Int. J. Pharm.* **2021**, *600*, 120472. [CrossRef]

35. Koster, C.; Pohl, S.; Kleinebudde, P. Evaluation of Binders in Twin-Screw Wet Granulation. *Pharmaceutics* **2021**, *13*, 241. [CrossRef] [PubMed]

36. Ismail, H.Y.; Singh, M.; Shirazian, S.; Albadarin, A.B.; Walker, G.M. Development of high-performance hybrid ANN-finite volume scheme (ANN-FVS) for simulation of pharmaceutical continuous granulation. *Chem. Eng. Res. Des.* **2020**, *163*, 320–326. [CrossRef]

37. Jimenez, A.A.B.; Van Hauwermeiren, D.; Peeters, M.; De Beer, T.; Nopens, I. Improvement of a 1D Population Balance Model for Twin-Screw Wet Granulation by Using Identifiability Analysis. *Pharmaceutics* **2021**, *13*, 692. [CrossRef]

38. Zheng, C.; Zhang, L.; Govender, N.; Wu, C.Y. DEM analysis of residence time distribution during twin screw granulation. *Powder Technol.* **2021**, *377*, 924–938. [CrossRef]

39. Kumar, A.; Vercruysse, J.; Mortier, S.T.F.C.; Vervaet, C.; Remon, J.P.; Gernaey, K.V.; De Beer, T.; Nopens, I. Model-based analysis of a twin-screw wet granulation system for continuous solid dosage manufacturing. *Comput. Chem. Eng.* **2016**, *89*, 62–70. [CrossRef]

40. Shirazian, S.; Ismail, H.Y.; Singh, M.; Shaikh, R.; Croker, D.M.; Walker, G.M. Multi-dimensional population balance modelling of pharmaceutical formulations for continuous twin-screw wet granulation: Determination of liquid distribution. *Int. J. Pharm.* **2019**, *566*, 352–360. [CrossRef]

41. Ismail, H.Y.; Shirazian, S.; Singh, M.; Whitaker, D.; Albadarin, A.B.; Walker, G.M. Compartmental approach for modelling twin-screw granulation using population balances. *Int. J. Pharm.* **2020**, *576*, 118737. [CrossRef]

42. Ismail, H.Y.; Singh, M.; Albadarin, A.B.; Walker, G.M. Complete two dimensional population balance modelling of wet granulation in twin screw. *Int. J. Pharm.* **2020**, *591*, 118737. [CrossRef] [PubMed]

43. Muddu, S.V.; Ramachandran, R. A Population Balance Methodology incorporating Semi-Mechanistic residence time metrics for twin screw granulation. *Processes* **2022**, *10*, 292. [CrossRef]

44. Wang, L.G.; Pradhan, S.U.; Wassgren, C.; Barrasso, D.; Slade, D.; Litster, J.D. A breakage kernel for use in population balance modelling of twin screw granulation. *Powder Technol.* **2020**, *363*, 525–540. [CrossRef]

45. McGuire, A.D.; Mosbach, S.; Reynolds, G.K.; Patterson, R.I.A.; Bringley, E.; Eaves, N.; Dreyer, J.A.H.; Kraft, M. Analysing the effect of screw configuration using a stochastic twin-screw granulation model. *Chem. Eng. Sci.* **2019**, *203*, 358–379. [CrossRef]

46. McGuire, A.D.; Mosbach, S.; Lee, K.F.; Reynolds, G.; Kraft, M. A high-dimensional, stochastic model for twin-screw granulation–Part 1: Model description. *Chem. Eng. Sci.* **2018**, *188*, 221–237. [CrossRef]

47. Kumar, A.; Dhondt, J.; Vercruysse, J.; De Leersnyder, F.; Vanhoorne, V.; Vervaet, C.; Remon, J.P.; Gernaey, K.V.; De Beer, T.; Nopens, I. Development of a process map: A step towards a regime map for steady-state high shear wet twin screw granulation. *Powder Technol.* **2016**, *300*, 73–82. [CrossRef]

48. Kotamarthy, L.; Ramachandran, R.J.P.T. Mechanistic understanding of the effects of process and design parameters on the mixing dynamics in continuous twin-screw granulation. *Powder Technol.* **2021**, *390*, 73–85. [CrossRef]

49. Ismail, H.Y.; Singh, M.; Darwish, S.; Kuhs, M.; Shirazian, S.; Croker, D.M.; Khraisheh, M.; Albadarin, A.B.; Walker, G.M. Developing ANN-Kriging hybrid model based on process parameters for prediction of mean residence time distribution in twin-screw updates wet granulation. *Powder Technol.* **2019**, *343*, 568–577. [CrossRef]

50. Barrasso, D.; Eppinger, T.; Pereira, F.E.; Aglave, R.; Debus, K.; Bermingham, S.K.; Ramachandran, R. A multi-scale, mechanistic model of a wet granulation process using a novel bi-directional PBM–DEM coupling algorithm. *Chem. Eng. Sci.* **2015**, *123*, 500–513. [CrossRef]

51. Wafa'H, A.; Khorsheed, B.; Mahfouf, M.; Reynolds, G.K.; Salman, A.D. An interpretable fuzzy logic based data-driven model for the twin screw granulation process. *Powder Technol.* **2020**, *364*, 135–144.

52. Kim, E.J.; Kim, J.H.; Kim, M.S.; Jeong, S.H.; Choi, D.H. Process Analytical Technology Tools for Monitoring Pharmaceutical Unit Operations: A Control Strategy for Continuous Process Verification. *Pharmaceutics* **2021**, *13*, 919. [CrossRef]

53. Rosas, J.G.; Blanco, M.; Gonzalez, J.M.; Alcala, M. Real-time determination of critical quality attributes using near-infrared spectroscopy: A contribution for Process Analytical Technology (PAT). *Talanta* **2012**, *97*, 163–170. [CrossRef] [PubMed]

54. Roman-Ospino, A.D.; Tamrakar, A.; Igne, B.; Dimaso, E.T.; Airiau, C.; Clancy, D.J.; Pereira, G.; Muzzio, F.J.; Singh, R.; Ramachandran, R. Characterization of NIR interfaces for the feeding and in-line monitoring of a continuous granulation process. *Int. J. Pharm.* **2020**, *574*, 118848. [CrossRef] [PubMed]

55. Kumar, A.; Vercruysse, J.; Toiviainen, M.; Panouillot, P.E.; Juuti, M.; Vanhoorne, V.; Vervaet, C.; Remon, J.P.; Gernaey, K.V.; De Beer, T.; et al. Mixing and transport during pharmaceutical twin-screw wet granulation: Experimental analysis via chemical imaging. *Eur. J. Pharm. Biopharm.* **2014**, *87*, 279–289. [CrossRef] [PubMed]

56. Mundozah, A.L.; Yang, J.K.; Tridon, C.C.; Cartwright, J.J.; Omar, C.S.; Salman, A.D. Assessing Particle Segregation Using Near-Infrared Chemical Imaging in Twin Screw Granulation. *Int. J. Pharm.* **2019**, *568*, 118541. [CrossRef]

57. Vercruysse, J.; Toiviainen, M.; Fonteyne, M.; Helkimo, N.; Ketolainen, J.; Juuti, M.; Delaet, U.; Van Assche, I.; Remon, J.P.; Vervaet, C.; et al. Visualization and understanding of the granulation liquid mixing and distribution during continuous twin screw granulation using NIR chemical imaging. *Eur. J. Pharm. Biopharm.* **2014**, *86*, 383–392. [CrossRef]

58. Harting, J.; Kleinebudde, P. Development of an in-line Raman spectroscopic method for continuous API quantification during twin-screw wet granulation. *Eur. J. Pharm. Biopharm.* **2018**, *125*, 169–181. [CrossRef]

59. Harting, J.; Kleinebudde, P. Optimisation of an in-line Raman spectroscopic method for continuous API quantification during twin-screw wet granulation and its application for process characterisation. *Eur. J. Pharm. Biopharm.* **2019**, *137*, 77–85. [CrossRef]

60. Kumar, A.; Dhondt, J.; De Leersnyder, F.; Vercruysse, J.; Vanhoorne, V.; Vervaet, C.; Remon, J.P.; Gernaey, K.V.; De Beer, T.; Nopens, I. Evaluation of an in-line particle imaging tool for monitoring twin-screw granulation performance. *Powder Technol.* **2015**, *285*, 80–87. [CrossRef]

61. Sayin, R.; Martinez-Marcos, L.; Osorio, J.G.; Cruise, P.; Jones, I.; Halbert, G.W.; Lamprou, D.A.; Litster, J.D. Investigation of an 11 mm diameter twin screw granulator: Screw element performance and in-line monitoring via image analysis. *Int. J. Pharm.* **2015**, *496*, 24–32. [CrossRef]

62. El Hagrasy, A.S.; Cruise, P.; Jones, I.; Litster, J.D. In-line Size Monitoring of a Twin Screw Granulation Process Using High-Speed Imaging. *J. Pharm. Innov.* **2013**, *8*, 90–98. [CrossRef]

63. Stauffer, F.; Ryckaert, A.; Vanhoorne, V.; Van Hauwermeiren, D.; Funke, A.; Djuric, D.; Vervaet, C.; Nopens, I.; De Beer, T. In-line temperature measurement to improve the understanding of the wetting phase in twin-screw wet granulation and its use in process development. *Int. J. Pharm.* **2020**, *584*, 119451. [CrossRef] [PubMed]

64. Hu, Y.; Huang, X.; Qian, X.; Gao, L.; Yan, Y. Online particle size measurement through acoustic emission detection and signal analysis. In Proceedings of the 2014 IEEE International Instrumentation and Measurement Technology Conference (I2MTC) Proceedings, Montevideo, Uruguay, 12–15 May 2014; pp. 949–953.

65. Hu, Y.; Qian, X.; Huang, X.; Gao, L.; Yan, Y. Online continuous measurement of the size distribution of pneumatically conveyed particles by acoustic emission methods. *Flow Meas. Instrum.* **2014**, *40*, 163–168. [CrossRef]

66. Hu, Y.; Wang, L.; Huang, X.; Qian, X.; Gao, L.; Yan, Y. On-line Sizing of Pneumatically Conveyed Particles Through Acoustic Emission Detection and Signal Analysis. *IEEE Trans. Instrum. Meas.* **2015**, *64*, 1100–1109. [CrossRef]

67. Uher, M.; Beneš, P. Measurement of particle size distribution by the use of acoustic emission method. In Proceedings of the 2012 IEEE International Instrumentation and Measurement Technology Conference Proceedings, Graz, Austria, 13–16 May 2012; pp. 1194–1198.

68. Abdulhussain, H.A.; Thompson, M.R. Considering inelasticity in the real-time monitoring of particle size for twin-screw granulation via acoustic emissions. *Int. J. Pharm.* **2023**, *639*, 122949. [CrossRef] [PubMed]

69. Meng, W.; Roman-Ospino, A.D.; Panikar, S.S.; O'Callaghan, C.; Gilliam, S.J.; Ramachandran, R.; Muzzio, F.J. Advanced process design and understanding of continuous twin-screw granulation via implementation of in-line process analytical technologies. *Adv. Powder Technol.* **2019**, *30*, 879–894. [CrossRef]

70. Dahlgren, G.; Tajarobi, P.; Simone, E.; Ricart, B.; Melnick, J.; Puri, V.; Stanton, C.; Bajwa, G. Continuous Twin Screw Wet Granulation and Drying-Control Strategy for Drug Product Manufacturing. *J. Pharm. Sci.* **2019**, *108*, 3502–3514. [CrossRef]

71. Ishimoto, H.; Kano, M.; Sugiyama, H.; Takeuchi, H.; Terada, K.; Aoyama, A.; Shoda, T.; Demizu, Y.; Shimamura, J.; Yokoyama, R.; et al. Approach to Establishment of Control Strategy for Oral Solid Dosage Forms Using Continuous Manufacturing. *Chem. Pharm. Bull.* **2021**, *69*, 211–217. [CrossRef]

72. Pauli, V.; Elbaz, F.; Kleinebudde, P.; Krumme, M. Methodology for a Variable Rate Control Strategy Development in Continuous Manufacturing Applied to Twin-screw Wet-Granulation and Continuous Fluid-bed Drying. *J. Pharm. Innov.* **2018**, *13*, 247–260. [CrossRef]

73. Silva, A.F.; Sarraguca, M.C.; Fonteyne, M.; Vercruysse, J.; De Leersnyder, F.; Vanhoorne, V.; Bostijn, N.; Verstraeten, M.; Vervaet, C.; Remon, J.P.; et al. Multivariate statistical process control of a continuous pharmaceutical twin-screw granulation and fluid bed drying process. *Int. J. Pharm.* **2017**, *528*, 242–252. [CrossRef]

74. Stauffer, F.; Boulanger, E.; Pilcer, G. Sampling and diversion strategy for twin-screw granulation lines using batch statistical process monitoring. *Eur. J. Pharm. Sci.* **2022**, *171*. [CrossRef] [PubMed]

75. Madarasz, L.; Nagy, Z.K.; Hoffer, I.; Szabo, B.; Csontos, I.; Pataki, H.; Demuth, B.; Szabo, B.; Csorba, K.; Marosi, G. Real-time feedback control of twin-screw wet granulation based on image analysis. *Int. J. Pharm.* **2018**, *547*, 360–367. [CrossRef]

76. Nicolai, N.; De Leersnyder, F.; Copot, D.; Stock, M.; Ionescu, C.M.; Gernaey, K.V.; Nopens, I.; De Beer, T. Liquid-to-solid ratio control as an advanced process control solution for continuous twin-screw wet granulation. *Aiche J.* **2018**, *64*, 2500–2514. [CrossRef]

77. Ryckaert, A.; Stauffer, F.; Funke, A.; Djuric, D.; Vanhoorne, V.; Vervaet, C.; De Beer, T. Evaluation of torque as an in-process control for granule size during twin-screw wet granulation. *Int. J. Pharm.* **2021**, *602*, 120642. [CrossRef] [PubMed]

 *biomedicines*

Review

Tumor Microenvironment-Responsive Drug Delivery Based on Polymeric Micelles for Precision Cancer Therapy: Strategies and Prospects

Zhu Jin *,†, Majdi Al Amili † and Shengrong Guo *

Shanghai Frontiers Science Center of Drug Target Identification and Delivery, School of Pharmaceutical Sciences, Shanghai Jiao Tong University, Shanghai 200240, China; majdi92@sjtu.edu.cn
* Correspondence: jinzhude@sjtu.edu.cn (Z.J.); srguo@sjtu.edu.cn (S.G.)
† These authors contributed equally to this work.

Abstract: In clinical practice, drug therapy for cancer is still limited by its inefficiency and high toxicity. For precision therapy, various drug delivery systems, including polymeric micelles self-assembled from amphiphilic polymeric materials, have been developed to achieve tumor-targeting drug delivery. Considering the characteristics of the pathophysiological environment at the drug target site, the design, synthesis, or modification of environmentally responsive polymeric materials has become a crucial strategy for drug-targeted delivery. In comparison to the normal physiological environment, tumors possess a unique microenvironment, characterized by a low pH, high reactive oxygen species concentration, hypoxia, and distinct enzyme systems, providing various stimuli for the environmentally responsive design of polymeric micelles. Polymeric micelles with tumor microenvironment (TME)-responsive characteristics have shown significant improvement in precision therapy for cancer treatment. This review mainly outlines the most promising strategies available for exploiting the tumor microenvironment to construct internal stimulus-responsive drug delivery micelles that target tumors and achieve enhanced antitumor efficacy. In addition, the prospects of TME-responsive polymeric micelles for gene therapy and immunotherapy, the most popular current cancer treatments, are also discussed. TME-responsive drug delivery via polymeric micelles will be an efficient and robust approach for developing clinical cancer therapies in the future.

Keywords: tumor microenvironment; cancer therapy; stimuli-responsiveness; drug delivery; precision

Citation: Jin, Z.; Al Amili, M.; Guo, S. Tumor Microenvironment-Responsive Drug Delivery Based on Polymeric Micelles for Precision Cancer Therapy: Strategies and Prospects. *Biomedicines* **2024**, *12*, 417. https://doi.org/10.3390/biomedicines12020417

Academic Editor: Letizia Polito

Received: 30 December 2023
Revised: 7 February 2024
Accepted: 8 February 2024
Published: 11 February 2024

1. Introduction

Cancer encompasses a range of conditions marked by the unregulated proliferation of abnormal cells with the potential to infiltrate surrounding tissues. Globally, it is the primary contributor to mortality, accounting for approximately 7.6 million deaths in 2008, equivalent to nearly 13% of the total fatalities. More recent statistics from 2020 indicate an alarming increase, with approximately 10 million deaths attributed to cancer. Current projections suggest a potential surge in cancer cases, reaching an unprecedented 22.2 million by 2030 [1].

Surgery is the primary treatment modality for cancer, but it is effective only for eradicating tumors in the early stages of the disease [2]. In most cases, patients are required to undergo drug therapy. The therapeutic agents used in cancer treatment include small-molecule chemotherapeutics, peptides, monoclonal antibodies, and others, which need to be delivered to the target site through specific mechanisms to exert their therapeutic effects. However, practical applications often face challenges such as low solubility, short half-life, and instability of these drugs in vivo. Moreover, a lack of target selectivity affects both cancerous and normal cells, leading to severe side effects on tissues such as bone marrow and the gastrointestinal tract. The resulting challenges, including multidrug resistance and a narrow therapeutic index, limit its efficacy. Dose reductions due to side effects further compromise therapeutic outcomes and may contribute to potential metastasis [3,4].

Nanotechnology advancements paved the way for cancer treatment via nanodrug delivery systems [5]. For instance, mesoporous silica nanoparticles [6], gold nanorods [7], liposomes [8], and micelles [9] are extensively utilized for drug delivery and cancer treatment. Among these nanoparticles, micelles, which are made from the self-assembly of amphiphilic polymers, not only increase the solubility of hydrophobic drugs but also increase their biocompatibility, pharmacokinetics, and cellular uptake; extend in vivo circulation; prevent drugs from being quickly decomposed; and protect against enzymatic degradation [10]. Micelles have an outstanding small particle size due to the number of monomers forming a micelle, which is controlled in a thermodynamically dependent manner and formed within a narrow space. The particle size of micelles is crucial since it can impact biodistribution [11,12]. The micellar particle size can be controlled by tuning the structure of the amphiphiles, the aggregation number of the amphiphiles, the molecular weight of the amphiphiles, the synthesis process, and the hydrophilic/hydrophobic segment ratio. The structure of amphiphiles endows micelles with specific properties essential for drug delivery, such as the ability to solubilize hydrophobic drugs in the hydrophobic core, self-assembly, and drug encapsulation by simple physical mixing. The corona of micelles can be tailored to actively target drug molecules at the site of interest by conjugating ligands specific to target tissues or cells, facilitating molecular recognition and interaction between micelles and target cells. Interestingly, altering amphiphilic copolymer components can easily modify several physiological properties, including surface charge, surface properties, and particle size. Other essential properties, such as biodegradation, biocompatibility, and elimination, can also be utilized in amphiphiles [13–15].

Solid tumors are characterized by poorly developed blood vessels and hypervascularization containing gaps in the endothelial lining. In addition, the hyperpermeability of the tumor microvasculature allows polymeric micelles to passively diffuse and reach tumor tissues through the secretion of materials and factors from tumors, such as primary fibroblast growth factor, nitric oxide, vascular permeability factors, prostaglandins, vascular endothelial growth factor, and bradykinin [16]. To maximize their tumor-targeting capacity, smart micelles have been developed to control drug release; these micelles remain stable under physiological conditions and in healthy tissue while releasing the drug upon exposure to certain conditions in the cancer niche and unavailable in normal tissue [17]. Solid tumors also exhibit a unique microenvironment, including an acidic pH, a reducing environment, the overexpression of certain enzymes, hypoxia, reactive oxygen species (ROS), and increased adenosine-5′-triphosphate (ATP) levels, which can be harnessed for designing smart stimulus-responsive micelles.

In addition to traditional chemotherapy, gene therapy and immunotherapy have garnered widespread attention for their application in cancer treatment in recent years. The efficient delivery of nucleic acid drugs such as DNA and RNA and immunomodulators, including antibodies, peptides, and small molecules, has become a focal point of research. In synthesizing polymer micelles, the hydrophobic core and hydrophilic shell structure of the micelles can be tailored by selecting appropriate materials, holding significant promise for enhancing both in vitro and in vivo gene delivery efficacy. Cationic polymer-based micelles, specifically those composed of polyethyleneimine (PEI), are being explored as advantageous carriers for gene encapsulation. These micelles, which feature stimulus responsiveness and active targeting, enhance the stability and permeability of genes through cell membranes and specificity. Notably, polymeric micelles exhibit promise in cancer immunotherapy by enhancing the delivery of immunostimulatory agents and improving the pharmacokinetics and biodistribution of immune-modulating drugs. The dual potential of polymeric micelles in gene delivery and cancer immunotherapy highlights their significance in advancing cancer treatment strategies [18,19]. This review summarizes various strategies for tumor microenvironment (TME)-responsive drug delivery using micelles, focusing on pH, redox reaction, enzyme, ROS, and hypoxia responsiveness, and outlines the prospects of polymeric micelles in gene therapy and cancer immunotherapy.

2. Tumor Microenvironment

Tumor cells induce substantial molecular, cellular, and structural alterations within their host tissues. The evolving TME is intricate and continuously changing. Although the composition of the TME varies among different types of tumors, common features include immune cells, stromal cells, blood vessels, and the extracellular matrix. It is widely recognized that the TME is not a passive bystander but rather an active facilitator of cancer progression. During the early stages of tumor growth, dynamic and reciprocal interactions occur between cancer cells and TME components, supporting cancer cell survival, local invasion, and metastasis [20]. To counteract hypoxic and acidic conditions, the TME orchestrates an angiogenic program to restore oxygen and nutrient supplies while eliminating metabolic waste. A diverse array of adaptive and innate immune cells infiltrates tumors, exhibiting both pro- and antitumorigenic effects [21].

The TME possesses distinctive attributes that can be harnessed for TME-targeted nanoparticles. Notably, the extracellular pH in the TME tends to be more acidic (pH 6.5 to pH 6.9) than the physiological pH of normal tissue (7.2 to 7.5). This acidity arises from the heightened glycolysis rate in cancer cells, which converts glucose into lactic acid to meet their energy demands. This pH variance in tumor cells offers the potential for designing pH-responsive cancer-targeting systems [22]. Another unique feature is hypoxia, where deep-seated tumor cells suffer from oxygen deprivation due to irregular vascular networks within solid tumors. These slowly proliferating cells in hypoxic regions display reduced susceptibility to conventional antiproliferative drugs [23].

Additionally, the TME exhibits altered expression of specific enzymes, often from the protease family, such as matrix metalloproteinases (MMPs), or from the lipase family, such as phospholipase A2. Enzyme–substrate specificity has spurred the development of enzyme-responsive nanomaterials for targeted drug delivery [24]. Furthermore, tumor cells in the TME face heightened oxidative stress, which is attributed to elevated levels of superoxide anion radicals, hydroxyl radicals, and hydrogen peroxide. To combat this, tumor cells increase their redox potential by expressing redox species such as superoxide dismutase and reduced glutathione (GSH). This imbalance in oxidation and reduction potentials within the TME presents an excellent opportunity for designing TME-targeted nanoparticles that recognize elevated levels of ROS compared to those of normal cells due to their aerobic metabolism resulting from oncogenic transformation [25–27]. These inherent TME stimuli offer promising prospects for the development of TME-responsive nanoparticles. While targeting the TME for cancer treatment holds significant promise, current FDA-approved treatments have limited effectiveness. As our understanding of how the TME contributes to tumorigenesis continues to evolve, new therapeutic targets and strategies will undoubtedly emerge.

3. Strategies for TME-Responsive Drug Delivery

The amphiphiles of micelles can be specifically designed to respond to internal stimuli such as pH, redox conditions, enzymes, and temperature for tumor-specific drug delivery. The nanoparticles should interact with the tumor site and prevent interaction with healthy tissues, which ideal targeted nanoparticles can offer. Stimulus-responsive micelles are highly preferred for this approach due to their desirable features. They can interact well with the site of action cells and respond to the surrounding tumor microenvironments, such as through changes in pH, enzymes, or redox [28]. The mainstream strategies for TME-responsive drug delivery are summarized as shown in Figure 1. Some typical stimulus-sensitive cleavage linkers that can be exploited to construct TME-responsive amphiphilic polymers are listed in Figure 2 [29].

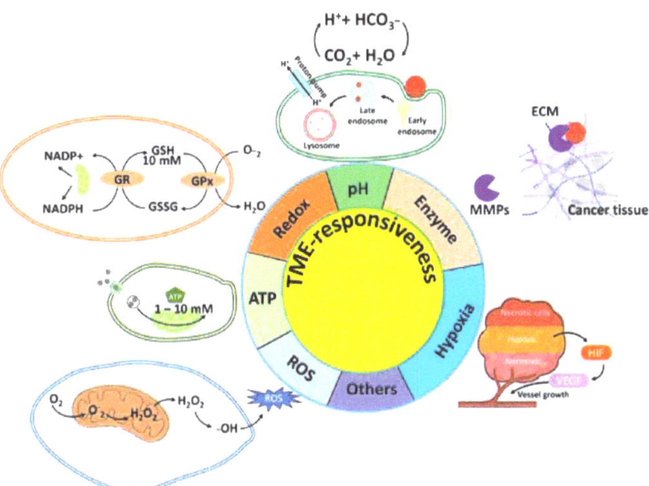

Figure 1. Schematic illustration of TME-responsiveness for drug delivery based on polymeric micelles.

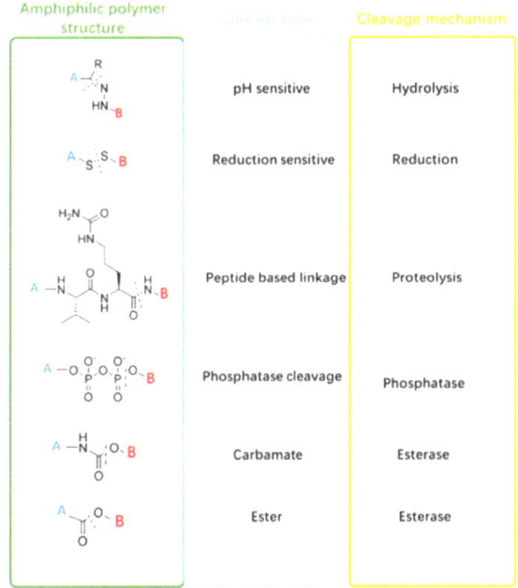

Figure 2. Schematic illustrating amphiphilic polymer structures, linkage types, and cleavage mechanisms of some typical TME-responsive linkers. A and B stand for hydrophilic and hydrophobic segments, respectively.

3.1. pH Responsiveness

The human body sustains a steady pH of 7.4 in healthy physiological tissues [30]. Moreover, tumor disease sites have a pH of 5~6 due to the accumulation of lactic acid that results from the rapid division of cancer cells [31]. This pH variation can be exploited to release the cargo of micelles in a controlled manner exclusively at the tumor site in response to its acidic pH (Figure 3) [32]. Cargo release can be accomplished by either breaking the labile bond, which forms the amphiphile of micelles, or destabilizing micelles via alteration of the size, shape, or hydrophilic–hydrophobic balance. Diverse pH-responsive micelles have been investigated and developed for cancer treatment. The strategy for selecting polymers relies on the existence of ionizable chemical groups. These polymers can be classified into anionic pH-responsive micelles and cationic pH-responsive micelles depending on the content of the ionizable chemical group of polycarboxylic or polyamine groups. To select a suitable polymer that is usually a weak acid or base, it should be considered whether its pKa is suitable for the desired pH at the site of action [33]. A negatively charged polymer containing carboxylic groups is used to design anionic pH-responsive micelles. The carboxylic acid groups are protonated (nonionized) at basic pH with hydrophobic properties. They can be deprotonated under acidic conditions (ionic form) while being hydrophilic at acidic pH. This change from hydrophobic to hydrophilic allows destabilization of the micellar system. This results in cargo release at tumor sites. Polymers with these characteristics usually contain carboxylic acid groups such as polymethacrylic acid, poly(acrylic acid) (PAAc), polyglycolic acid, and polyglutamic acid. In addition, polymers consisting of sulfonic acid groups such as poly(4-styrene sulfonic acid) and poly(2-acrylamido-2-methylpropane sulfonic acid) are also utilized as negatively charged polymers [34].

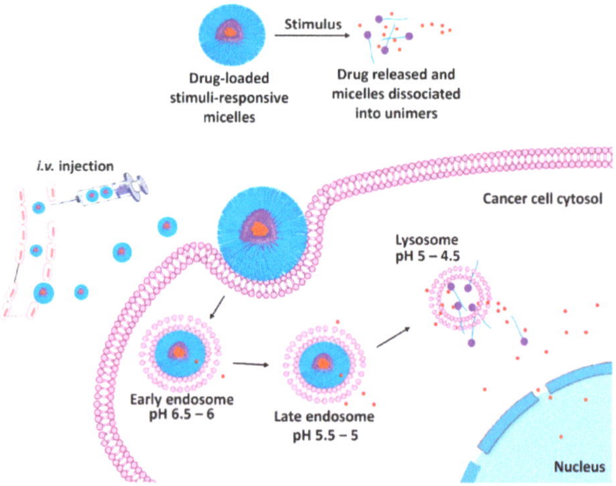

Figure 3. Schematic illustration showing drug release via micelle dissociation under different internal acidic conditions. Red dots are the released drug. Following micelles' endocytosis, they are entrapped within the early endosome, late endosome, and then the lysosome. Red dots are the released drug, concentrated in the lysosomal area due to the highest acidity level.

These polymers are usually desired for the synthesis of hydrogels capable of shrinking and swelling in a pH-dependent manner to achieve cargo release. A positively charged polymer, which contains polybases such as PEI and polyamines, is used to synthesize these types of micelles. The amine group in the polymeric chain can accept protons at acidic pH and donate protons at basic pH. Cationic pH-responsive micelles can improve cellular uptake due to their positive surface charge resulting from ionizable polyamines, for instance, poly(N,N'-dimethylamino ethyl methacrylate) and PEI. Tuning the pKa of the cationic amine groups allows the polymers to be protonated at acidic pH and deprotonated at basic pH. This charge alteration destabilizes the micellar complex and triggers cargo release. Unfortunately, cationic polymers are considered more cytotoxic than anionic polymers due to their ability to interact with negatively charged proteins during blood circulation non-specifically. Moreover, positively charged micelles cause serum inhibition, rapid clearance, and instability with opsonin. Anionic polymers are impeded for efficient cellular uptake due to cellular repulsion by negatively charged plasmalemma. Charge-reversal pH-responsive micelles, which can be transformed from negatively to positively charged micelles and vice versa in a pH-dependent manner, have recently been developed. This approach's advantage is achieving active tumor targeting without a specific targeting ligand [35].

Moreover, it increases the circulation half-life of micelles in the blood, enhances cellular uptake, and accomplishes efficient drug release in target cells [36]. Peng et al.

designed charge-reversal micelles consisting of positively charged micelles composed of an amphiphilic copolymer core with a mitochondrial active targeting moiety (triphenylphosphonium) (TPP) and denoted as Ce6@TPPM. Furthermore, the pH-responsive outer layer consisted of anionic 2,3-dimethyl maleic anhydride (DMA)-conjugated biotin-polyethylene glycol (PEG)-NH$_2$ for active tumor targeting and Ce6 delivery, denoted as Ce6@TPPM-BioPEG-DMA. The system was stable at pH 7.4 in physiological environments with a negatively charged surface (Ce6@TPPM-BioPEG-DMA). Upon reaching the acidic tumor microenvironment, the system efficiently accumulated due to ligand–receptor-mediated active targeting; subsequently, the system converted to a positively charged layer (Ce6@TPPM) at pH~6.5, which further accelerated the accumulation of micelles in the tumor tissue and allowed the TPP to be re-exposed inside tumor cells to actively target the mitochondria (Figure 4) [37]. This strategy overcomes the cytotoxicity and rapid clearance of positively charged micelles and enhances the cellular uptake of negatively charged micelles.

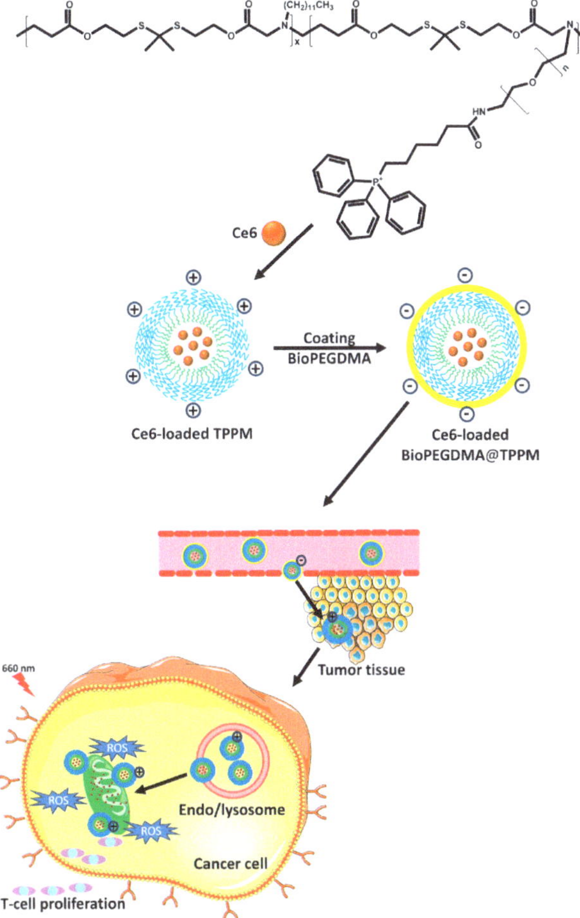

Figure 4. Schematic illustration showing a pH-responsive micelle of BioPEGDMA@TPPM micelles for enhanced PDT. The self-assembly and Ce6 loading were followed by coating with BioPEGDMA via electrostatic interaction, resulting in tumor-targeted delivery and charge reversal in TME. The endo/lysosomal escape, mitochondria targeting, generation of ROS under laser irradiation, and stimulated immune responses based on BioPEGDMA@TPPM.

3.2. Redox Responsiveness

The difference in the concentrations of the tumoral reductants, which are represented mainly by GSH, was approximately $2–10 \times 10^{-3}$ M, especially in multidrug-resistant tumors. Moreover, the GSH concentration in the extracellular fluid ranged from 2 to 10×10^{-6} M. There are three prominent redox couples, GSH/GSSG [38], NAD(P)H/NAD(P)$^+$ [39], and Tex(SH)$_2$/TrxSS [40], which exist in the TME. Redox-responsive micelles containing redox-sensitive moieties have been designed to deliver and release cargoes exclusively at tumor sites in response to tumor-reductive environments.

The most commonly employed redox-responsive polymer is a disulfide bond-containing polymer that can rapidly respond to and be cleaved by redox components such as GSH. Disulfide bonds can be readily cleaved by GSH to form a sulfhydryl group, leading to destabilization of the micellar system and cargo release. Rapid cleavage of disulfide bonds (within minutes to hours) is more favorable than long cleavage periods (from weeks to months), represented by polycarbonates and aliphatic polyesters, due to rapid intracellular drug release, which is advantageous for inhibiting cancer cell growth during the first hours after injection [41]. Diselenide (Se-Se) and carbon–selenide (C-Se) bonds have also attracted attention from many studies due to their lower bond energies (Se-Se 172 kJ/mol and C-Se 244 kJ/mol) [42,43], which require less energy for bond cleavage than disulfide bonds (S-S 268 kJ/mol) [44]. Moreover, the maleimide–thioester bond (C-S 272 kJ/mol) exhibited increased blood stability and decreased cargo release [43].

Sahoo et al. fabricated the most promising redox-responsive micellar system by self-assembly of poly(N,N'-dimethylaminoethylmethacrylate)-b-(poly(2-(methacryloyl)-oxyethyl-2′-hydroxyethyl disulfidecholate)-r-2-(methacryloyloxy)ethyl-1-pyrenebutyrate). The anticancer agent doxorubicin (DOX) was encapsulated in the micellar core, and DNA was complexed with the outer layer of micelles to form micelleplexes. In the absence of GSH, 8~10% of the DOX was released within 48 h. Moreover, in the presence of 10 mM GSH, 90% of the DOX was released, confirming the cleavage of disulfide bonds that conjugate the hydrophobic cholate group with the polymeric backbone, allowing DOX molecules to be released into the tumor medium due to the disassembly of micelles (Figure 5) [45]. An earlier study in 2018 showed the least drug leakage in a formulation based on redox-responsive drug release [46]. The authors achieved less than 5% leakage of paclitaxel (PTX) within 48 h by conjugating the drug to the hydrophobic part of the amphiphilic block copolymer to form PEG-b-poly(5-methyl-5-propargyl-1,3-dioxan-2-one)-g-PTX, which self-assembled into micelles for the treatment of HeLa tumor-bearing mice. However, in the presence of 10 mM dithiothreitol (DTT), the release reached 70%. Briefly, the advantage of redox responsiveness is stability in healthy tissues, which results in fewer side effects and cytotoxicity.

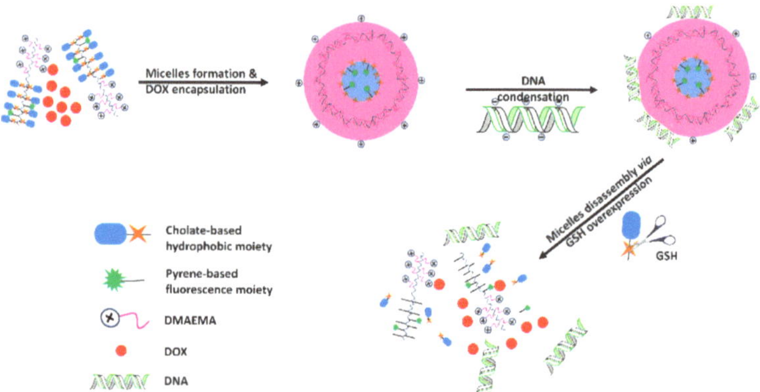

Figure 5. Schematic illustration showing the micelles' self-assembly and DOX loading followed by DNA condensation. DOX and DNA are released upon GSH overexpression in the TME.

Disulfide crosslinking chemistry was extensively used for not only small-molecule drugs but also macromolecular therapeutics such as nucleic acids. It was extensively utilized in advancing the development of polyplex-based carriers to deliver a wide range of cargo molecules, such as plasmid DNA, siRNA, and mRNA. Interestingly, the introduction of charge-preserved disulfide crosslinking via 1-amidine-3-mercaptopropyl groups presented high protection to packaged nucleic acids against enzymatic degradation compared to charge-compensated crosslinking via 3-mercaptopropionyl groups to polycation segment of PEG, thereby facilitating maximized intracellular delivery of nucleic acids [47,48].

3.3. Enzyme Responsiveness

Enzymes play essential roles in metabolic and biological processes in the human body due to their catalytic properties and high specificity [49]. In tumors, several enzymes are dysregulated and overexpressed in cancer cells. Exploiting these overexpressed enzymes as triggers for cargoes loaded in micelles can be achieved by incorporating enzyme-responsive moieties into the side chain or main chain of the micelles that can be degraded by these enzymes in either the intracellular or extracellular tumor microenvironments to release the cargo [50,51]. There are two main types of enzyme-responsive micelles: oxidoreductases and hydrolyzed micelles.

Oxidoreductases, such as oxygenases (oxygen transfer from molecular oxygen), oxidases (electron transfer to molecular oxygen), peroxidases (electron transfer to peroxidases), and dehydrogenases (hydride transfer), function as catalysts for oxidation–reduction reactions. Oxidoreductases have been exploited for enzyme-responsive drug release due to the oxidative environments they can produce in many diseases, including cancer. The ability of oxidoreductases to catalyze the transfer of electrons between biological molecules requires the presence of an enzyme cofactor, which can function as an electron carrier, such as NAD^+ or $NADP^+$; accordingly, the electron donor is a reductant. In contrast, the electron acceptor substrate is an oxidant [52].

On the other hand, some hydrolysis enzymes are also overexpressed in many stages of human cancers and are involved in cancer initiation, progression, angiogenesis regulation, and metastasis [53]. Generally, MMPs are the most utilized enzymes for stimuli-responsive drug delivery. Chen et al. developed a promising micellar nanoplatform formed by the self-assembly of the biotin-PEG-block-poly(*L*-lysine)(Mal)-peptide-DOX (biotin-PEG-*b*-PLL(Mal)-peptide-DOX) amphiphilic copolymer for the treatment of mouse squamous cell carcinoma and African green monkey kidney fibroblast cells (Figure 6). The peptide is an MMP-2-sensitive linker that can be cleaved in the presence of the MMP-2 enzyme to release DOX in the tumor milieu. The presence of the MMP-2 enzyme induced 46.2% DOX release within 6 h, and almost no drug leakage occurred in the absence of MMP-2 or MMP-2 in combination with the inhibitor [54]. This system has precise enzyme responsiveness and active targeting properties via biotin ligands and can be considered among the most promising nanoplatforms for safe and precise drug delivery. Despite newer studies reporting that 30% [55], 40% [56], and 62.5% [57] of the loaded drugs leaked out in the absence of MMP-2, these drugs cannot be considered to have precise responsiveness or safety for clinical use compared with the biotin-PEG-*b*-PLL(Mal)-peptide-DOX formulation. Cathepsin B is one of the most widely overexpressed cysteine cathepsins in various cancers, and it is involved in the degradation of fibronectin, type IV collagen, and laminin, which leads to cell migration, invasion, and angiogenesis [58].

A micellar nanoplatform was formed by the self-assembly of the [(DEAMEMA)-c-(BMA)]-*b*-[(PEGMA$_{300}$)-c-(peptide)] amphiphile, which is responsive to cathepsin B. The BIM peptide was conjugated to the amphiphile through the FKFL peptide linker, which can be cleaved in the presence of cathepsin B to release BIM specifically in the endolysosome/lysosomes of tumor cells. To confirm the cathepsin B-triggered cleavage of FKFL, the authors performed a cathepsin B cleavage assay; then, the cleavage was quantified via RP-HPLC and mass spectrometry. As a result, cathepsin B rapidly and specifically cleaved the FKFL linker to release BIM in the endosome of SKOV3 ovarian cancer cells and induced apoptosis to kill cancer cells [59]. Phospholipase A2 (PLA2) is overexpressed in several types of tumors at 22-fold higher concentrations than those in normal tissues, especially in prostate cancer [60]. Gao et al. designed PLA_2-responsive phospholipid micelles loaded with superparamagnetic iron oxide nanoparticles (SPIONs) to successfully release drugs in response to PLA_2 and via noninvasive magnetic resonance imaging (MRI) [61]. A later study reported biocompatible upconversion nanoparticle (UCNP)-loaded phosphate micelles for bioimaging prostate cancer cells. The release of UCNPs via PLA-2-responsive cleavage was achieved exclusively at tumor sites rather than in healthy cells [62].

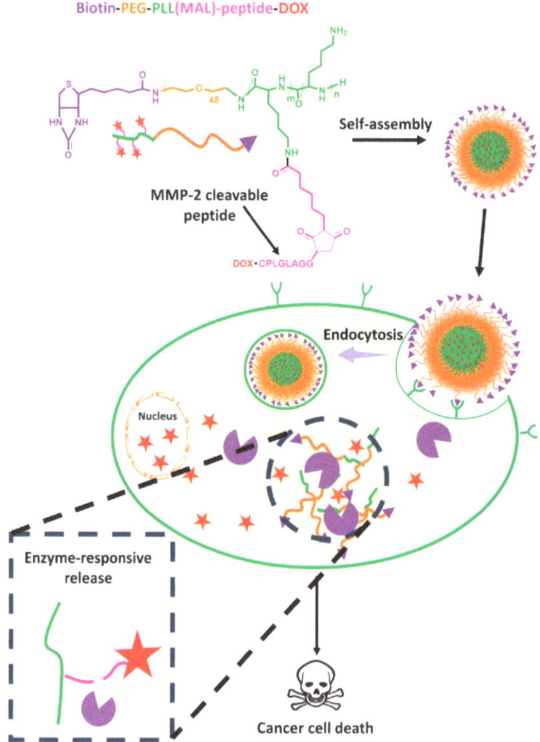

Figure 6. Schematic illustration of the formation of MMP-2-responsive polymeric micelle-based biotin-PEG-*b*-PLL(Mal)-peptide-DOX amphiphilic and intracellular delivery. The red star stands for DOX.

3.4. ROS Responsiveness

ROS, such as hydrogen peroxide (H_2O_2), singlet oxygen, superoxides, hypochlorite ions, peroxynitrites, superoxide anions, and hydroxyl radicals, are byproducts produced by electron transport reactions in the mitochondria of healthy cells [63,64]. They are important for metabolism and intercellular signal functioning and exist at low concentrations of approximately 20×10^{-9} M. In cancer cells, these levels increase 1000-fold due to abnormal metabolism, mitochondrial malfunction, and oncogene expression, which result in abnormal metabolism, proliferation, and survival [65]. This difference can be exploited to trigger the release of drug-loaded micelles in the tumor microenvironment specifically. ROS-responsive micelles can be developed by employing various ROS-sensitive materials, such as thioketals, thioethers, ferrocene groups, boronic esters, and sulfides [66]. These materials can undergo certain reactions, such as hydrophobic-to-hydrophilic or hydrophilic-to-hydrophobic transitions, upon elevated ROS levels in tumor microenvironments, leading to micellar system destabilization and drug release. Thioketal is widely used as a ROS-sensitive linker due to the ease of cleaving this linkage [67]. For example, Wang et al. developed micelles for prostate-specific membrane antigen-negative (PISMA (-)) prostate cancer treatment. The micellar system was made by the self-assembly of two copolymers. First, the DUP-1 peptide was conjugated with PEG-1,2-distearoyl-sn-glycero-3-phosphoethanolamine (DUP-PEG-DSPE). Second, DOX and TK linkers were decorated with the side chain of PEG-*b*-PLL to produce a ROS-responsive polymer prodrug (P(L-TK-DOX)) to form a micellar system loaded with a ROS generation agent (α-tocopherol succinate, α-TOS). The micellar system was shown to accumulate and internalize cancer cells via the DUP-1-targeting peptide specific for PSMA (-). α-TOS release upon exposure to elevated concentrations of ROS in the tumor microenvironment could further elevate ROS levels to induce toxicity and enhance ROS responsiveness. Interestingly, less than 5% of the DOX leaked out in the absence of ROS. Moreover, in the presence of different ROS concentrations (20 nM, 0.1 mM, and 1 mM H_2O_2), the DOX release increased to 18%, 57%, and 79%, respectively, within 48 h. The study conclusion revealed that the combination of ROS-sensitive drug release behavior and active targeting of tumors is an effective treatment for human PSMA(-) prostate cancer [68].

3.5. Hypoxia Responsiveness

Hypoxia is involved in many diseases, such as cardiovascular disorders, rheumatoid arthritis, anemia, and cancer [69]. The oxygen level decreases due to rapid cancer cell growth and insufficient blood supply, which results in tumor angiogenesis and metastasis. An abnormal vascular network is incapable of providing adequate oxygen to cancer cells, especially in the center of the tumor region, which results in an acute hypoxic or oxygen deficiency gradient, which increases from tumor terminals or blood vessels to the tumor center. Hypoxia is a significant stimulus that triggers drug release due to its rarity in normoxic cells [70]. However, the distance between hypoxic region blood vessels and increased efflux transporters prevents the influx of nanoparticles to these hypoxic regions [71]. To overcome this dilemma, several strategies exist for developing drug delivery systems that can provide deep penetration into the hypoxic region. The accumulation of nanoparticles in hypoxic regions could be facilitated by developing nanoparticles upon size-, shape-, and charge-dependent uptake via passive diffusion [72]. Active targeting of nanoparticles is a robust and efficient method to deeply penetrate tumor sites by conjugating targeting ligands such as the cyclic peptide internalizing RGD (iRGD) on the outer surface of the nanosystem or therapeutic drug to increase their ability to penetrate the neuropilin-1 receptor, which is overexpressed in tumor cells and angiogenetic blood vessels [73].

Hypoxia is also involved in the induction of radioresistance, chemoresistance, and cancer recurrence through hypoxia evasion of apoptosis, inactivity of stem cells, and dysregulation of the cell cycle [74]. Hypoxia-bioreductive prodrugs or hypoxia-activated prodrugs, also known as hypoxia-selective cytotoxins (for instance, quinone derivatives such as mitomycin C, N-oxide derivatives such as banoxantrone dihydrochloride and tirapazamine, and nitroimidazole derivatives such as 2-nitroimidazole), are inactive compounds that are spontaneously converted to cytotoxic substances upon specific metabolic pathways that exist in the hypoxic microenvironment [75]. This conversion can be employed directly to kill cancer cells. In addition, upon the prodrug's conversion from one property to another, a micellar system containing these prodrugs is constructed to trigger drug release, specifically in hypoxic tumor regions, resulting in micelle destabilization and drug release.

3.6. Other Stimulus Responsiveness

Cancer cells tend to absorb large amounts of glucose to promote tumor growth, metastasis, and survival [76,77]. Therefore, glucose-responsive micelles can be designed by incorporating glucose oxidase (GOx) within the polymeric chains of the micellar system. Upon reaching the tumor microenvironment, a competitive combination of GOx and glucose destabilizes the micellar system and triggers drug release in cancerous tissues [78].

Most of the applied strategies involve treating diabetes. However, several cancers are involved in diabetes and glucose metabolism. A glucose-responsive nanosystem can indirectly trigger drug release via the conversion of glucose to gluconic acid by the catalysis of GOx, which decreases the pH of the medium and triggers release. Another strategy for killing cancer cells via glucose-responsive treatment is carried out by competing with cancer cells for glucose consumption and converting glucose to gluconic acid and H_2O_2; this approach is known as cancer starvation therapy. H_2O_2 is essential for physiological processes, such as cell growth and the immune response. Moreover, increasing concentrations of H_2O_2 resulted in increased cytotoxicity and cancer cell death [79].

ATP is present in all organisms, is involved in the production and degradation of many cellular compounds, and is the primary source of cellular energy for signaling and metabolism. A substantial concentration of ATP was observed in the intracellular compartment ($\sim$3 mM) compared to the extracellular environment ($\sim$0.4 mM) [80]. Utilizing this difference in ATP concentration gradient between extracellular and intracellular spaces, ATP-responsive drug delivery systems were engineered, which stably encapsulate the therapeutic cargoes in the extracellular medium, smoothly releasing those cargoes after exposure to the high-ATP concentration milieu of the cytosol [81]. For example, a polyplex micelle was formulated with reversible phenyl-boronate ester linkages with phenylboronic acid moieties in the block copolymers and polyol moieties of oligoRNAs hybridized with mRNA in the polymeric micelle (PM) core. This design substantially protected the mRNA from enzymatic degradation in the extracellular space and efficiently released entrapped mRNA to the cytosol for efficient translation.

Polymers that undergo phase transitions upon exposure to certain salt concentrations exhibit reduced electrostatic strength due to the high salt concentration, making these polymers ionic strength- or salt-responsive materials. An increase in salt concentration decreases the electrostatic repulsion between the copolymers, leading to precipitation and drug release [82]. Salt-responsive materials exhibit unusual rheological behavior due to the attractive Coulombic interactions between oppositely charged species, which cause alterations in the solubility, size, length, and surface charge of the polymer [83,84]. These materials respond to the ionic strength of PAAc and methacrylic acid, which undergo viscosity changes and shrinkage upon exposure to high salt concentrations due to the attraction pairs of the ions [85]. Compared with those in normal lactating breast epithelium, salt concentrations in breast cancer tissues were significantly greater. Brain cancers also exhibit an influx of intracellular sodium ions to promote tumor cell proliferation. The epithelial sodium channel (ENaC) regulates the entry of sodium into the intracellular compartment. Abnormalities in ENaC function correlated with tumors result in antiapoptotic effects, uncontrolled tumor growth, cell migration, and angiogenesis [86]. Exploiting the difference in salt concentration between the tumor microenvironment and healthy tissue to construct ionic strength-responsive drug delivery systems has not been widely investigated.

3.7. Multistimulus Responsiveness

Polymeric micelles that are responsive to multiple stimuli are gaining prominence and demonstrate significant promise for targeted drug delivery and cancer therapy. The integration of various sensitivities into a single polymeric micellar system allows more precise control of drug delivery and release, leading to enhanced anticancer activity both in vitro and in vivo. These combined sensitivities to different stimuli can occur simultaneously or sequentially, offering versatility in therapeutic applications. Luo et al. designed pH- and redox-responsive PMs to deliver the anticancer drug DOX (Figure 7). The present study investigated the pH sensitivity of the PMs by determining the pKb values, which were 6.45, 6.57, and 6.72 for PM-1, PM-2, and PM-3, respectively. The low critical micelle concentration (CMC) values (3.1 mg/L, 4.2 mg/L, and 6.4 mg/L) indicated the thermodynamic stability of the polymeric micelles, which made them efficient drug carriers. The redox responsiveness of the PMs was evaluated through size and zeta potential changes in the presence of DTT, which demonstrated increased particle sizes due to the cleavage of disulfide bonds. The in vitro drug release profiles of DOX-loaded PMs were examined at different pH values and in the presence of DTT. The results indicated that controlled drug release was triggered by specific microenvironmental cues, such as pH and reducing agents. Cytotoxicity assays revealed that blank PMs had negligible cytotoxicity, while DOX-loaded PMs exhibited greater cytotoxicity against HepG2 cells than free DOX [87]. Zhang et al. designed a multifunctional polymeric system for dual-enzyme- and redox-triggered intracellular drug release to improve cancer treatment efficacy. The key components of their system were the enzyme-responsive polymer PBA-PEG-Azo-PCL and the redox-responsive prodrug mPEG-ss-CPT. Azo bonds in PBA-PEG-Azo-PCL were shown to be cleaved by azoreductase and the coenzyme NADPH, mimicking the tumor tissue microenvironment. The micelles exhibited highly sensitive tumor microenvironment responsiveness, with changes in size indicating successful cleavage of the azo bonds. Additionally, the disulfide bonds in mPEG-ss-CPT were cleaved in the presence of GSH, increasing the micelle size. The dual-responsive behaviors were explained by a series of chemical reactions, ensuring controlled drug release. In vitro drug release studies demonstrated that the micelles exhibited good stability in blood circulation but rapidly released their cargo inside tumor cells, particularly under conditions mimicking the tumor microenvironment. In vivo and ex vivo fluorescence imaging confirmed the selective accumulation of the micelles at the tumor site. The dual-responsive micelles exhibited significant anticancer activity with minimal side effects on normal tissues, as demonstrated by tumor volume changes, survival rates, and histological analyses [88].

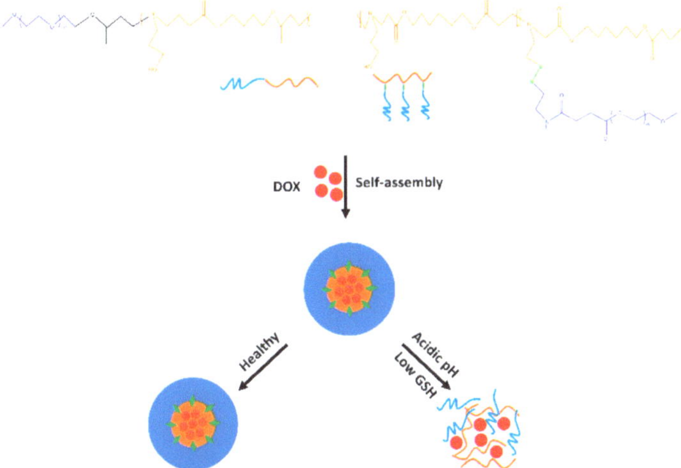

Figure 7. Schematic illustration indicating the co-micellization of pH/redox dual-responsive diblock copolymers for drug delivery and controlled release triggered by acidic pH and overexpression of GSH in TME.

4. Prospects in Cancer Therapy

4.1. Gene Therapy

Gene delivery requires appropriate carriers with high gene transfer efficiency, good biocompatibility, and low cytotoxicity. Genetic materials such as plasmid DNA (pDNA), siRNA, and RNA demand cationic polymers to successfully complex these negatively charged genetic agents with cationic polymer-based micelles [89]. Naked genetic materials suffer rapid elimination from the body, instability in the blood circulation, and inability to diffuse through the cell membrane, ascribed to their large anionic charge. Consequently, a nanocarrier is required to entrap these genes with a high buffering capacity for improved transfection and the capability to efficiently target diseased cells and release the genes in the intracellular compartment, allowing the siRNA to target the cytoplasm and

DNA to target the nucleus. Employing viral carriers to encapsulate genes is unfavorable due to their potential genotoxicity and chance of producing replication-competent viruses [90]. Cationic polymers have been extensively used to deliver nucleic acids through their electrostatic polyionic self-assembly with nucleic acids, allowing the polyion complex formation, named "polyplex" [91]. Polyplex micelles are stealth-polymer shielded nucleic acid delivery systems constructed through electrostatic self-assembly between nucleic acids and polycationic block copolymers of PEG or poly(oxazoline), where the nucleic acid is packaged with polycations as the core compartment and the stealth polymer chains surrounding the core as the protective shell compartment. This characteristic core–shell architecture of polyplex micelles protects the nucleic acid cargoes from hydrolytic and enzymatic degradation, which increases the stability and permeability of the nucleic acids through the cell membrane. Surface modification of PEG (PEGylation) onto nucleic acid delivery carriers is a well-known strategy for extending blood retention and improving therapeutic outcomes in vivo. However, PEG shells often present a trade-off between prolonged blood retention and promoted transfection because high-stealth shielding is advantageous in prolonging blood circulation, whereas it is disadvantageous in obtaining efficient transfection due to low cellular uptake and inefficient endosomal escape (PEG dilemma) by minimizing the interactions with the plasma membrane before internalization and endolysosomal membranes after internalization of the targeted cells. To overcome this PEG dilemma, Nishiyama et al. developed stepwise acidic pH-responsive plasmid DNA delivery nanocarriers with a surface covered by ethylenediamine-based polycarboxybetaines. These nanocarriers switched their surface charge potential from a neutral charge at pH 7.4 to a positive charge at tumoral pH 6.5 and endolysosomal pH 5.5, thereby promoting cellular uptake and increasing the endosomal escape toward efficient gene transfection [92]. In another strategy, cyclic RGD (Arg-Gly-Asp) (cRGD) peptide as a ligand was introduced to the distal end of the PEG chains to improve the specific integrin-mediated uptake of disulfide core-crosslinked polyplex micelle-based gene carriers [93,94].

Utilizing smart micelles with stimulus responsiveness and active targeting can lead to robust, highly specific, and optimal development of gene delivery strategies. Polycations made of PEI for gene delivery are among the most common polycations because they exhibit high transfection ability, attributed to the high buffering capacity of the polycation due to the proton sponge effect, by which the polycation buffers endosomal acidification and escalates endosomal ion osmotic pressure due to protonation of the amine groups, resulting in rupture of the membrane of the endosome and the subsequent release of the captured system into the cytoplasm. Accordingly, polycations also demand a high N/P ratio (molar ratio of amino groups (N) in polymer to phosphate groups (P) in nucleic acid) to form a secured complex, resulting in high stability and transfection ability [95,96].

A novel dual-responsive PEI-based polymeric micelle system has been developed for the targeted delivery of therapeutic agents in cancer treatment. Overcoming challenges such as mucosal barriers, nonspecific uptake, and intracellular drug resistance is crucial for achieving high therapeutic efficiency. The key feature of this system is the presence of a sheddable PEI shell, which responds to variations in extracellular pH and intracellular GSH levels. The micelle system exhibited ultrasensitive negative-to-positive charge reversal in response to the extracellular pH. When exposed to the acidic environment commonly found at tumor sites, the surface charge of the micelles changes from negative to positive. This transformation enhances electrostatic interactions, significantly improving the uptake of micelles by cancer cells. Upon internalization by cancer cells, the micelles encounter another layer of responsiveness. The disulfide linkages within the system can be cleaved by the presence of GSH in the cytoplasm. Importantly, GSH concentrations are often greater in cancer cells than in normal cells. The cleavage of disulfide linkages triggers the deshielding of the hydrophilic PEI shell, leading to the rapid release of the encapsulated therapeutic agent. The mechanism of this dual-responsive polymer micelle involves several stages. Initially, micelles, with their originally negatively charged surface, exhibit prolonged circulation time in the bloodstream. At the tumor site, they take advantage of the enhanced permeability and retention effect, accumulating at higher concentrations. The ultrasensitive negative-to-positive charge reversal that occurs in response to an acidic pH facilitates efficient internalization by cancer cells through electronic interactions and folate receptor (FR)-mediated endocytosis. Furthermore, the micelles escape from lysosomes, a cellular organelle, via a proton sponge effect. Excess intracellular GSH triggers the cleavage of disulfide linkages, resulting in the deshielding of the PEI shell and rapid release of the therapeutic agent into the nucleus [97].

Gao et al. developed pH/redox dual-responsive polyplexes demonstrating promising characteristics for codelivering siRNA and DOX. The polyplex exhibited efficient encapsulation of DOX and siRNA, along with pH-/redox-triggered payload release, facilitated by protonation of PHis and disulfide bond cleavage. Specifically, at an N/P ratio of 7, the polyplex displayed superior payload delivery efficiency, MDR1 gene silencing, cytotoxicity against MCF-7/ADR cells, and more potent inhibition of MCF-7/ADR tumor growth than at higher N/P ratios. This enhanced performance at N/P 7 was attributed to the increased electrostatic attraction between the siRNA and oligoethylenimines (OEIs), which suppressed the release of MDR1 siRNA and OEIs. A stronger electrostatic interaction was crucial for overcoming payload endolysosomal sequestration by OEI-induced membrane permeabilization [98]. Pan et al. synthesized dendrimer micelles self-assembled from two copolymers. First, PEG_{2k}-1,2-dioleoyl-sn-glycero-3-phosphoethanolamine (PEG_{2k}-DOPE)-conjugated generation 4 polyamidoamine dendrimer (G4-PAMAM-D)-incorporated MDR-1 siRNA (siMDR-1) was used. Second, the PEG_{5k}-DOPE-conjugated tumor-specific monoclonal antibody 2C5 (mAb 2C5) was used for chemotherapeutic DOX and gene codelivery. The results revealed significant specific binding between cell surface-attached nucleosome tumor cells and the mAb 2C5, which enhanced cellular uptake and increased cytotoxicity in MDR cancer cell lines [99]. An exciting pH-responsive crosslinked polyplex micelle was engineered for mRNA delivery based on cis-aconitic anhydride-modified poly(ethylene glycol)-poly(L-lysine). This polyplex micelle was stable at pH 7.4, whereas it released the packaged mRNA when the pH was decreased below 6.5 (tumoral pH), thus providing high protein expression in the tumor compared to the commercial transfection reagent PEI [100].

4.2. Immunotherapy

Cancer immunotherapy has emerged as a successful treatment strategy following conventional surgical, chemical, and radiotherapeutic approaches [101]. Biological therapy harnesses the body's immune system to induce an attack on tumor cells, resulting in an antitumor effect. The immune system is trained to identify and target specific cancer cells, enhancing the effectiveness of immune cells in eliminating cancer [102]. Notably, cancer immunotherapy addresses both primary tumor and secondary tumor metastasis by triggering a systemic immune response [103]. Additionally, it can impede tumor recurrence by fostering a cancer-specific memory immune response that becomes activated upon encountering tumor-associated antigens [104]. Among the various immunotherapy modalities explored, immune checkpoint inhibitors (ICIs), chimeric antigen receptor T cells, and oncolytic viruses have been extensively studied for their notable achievements in clinical trials. High-dose interleukin-2 has been among the earliest immunotherapies used to activate T cells. Both ICIs and adoptive cell transfer therapy have been demonstrated to be effective against various malignancies [105,106]. Cancer immunotherapy has been proven to mitigate metastasis, prevent tumor recurrence, and reverse multidrug resistance in tumor cells. Notably, its efficacy has been established for treating head and neck cancer, lung carcinoma, leukemia, breast carcinoma, ovarian carcinoma, renal carcinoma, and bladder tumors [107], as polymeric micelles show promise in the realm of cancer immunotherapy by serving various purposes, such as improving the delivery of immunostimulatory agents and enhancing the pharmacokinetics and biodistribution of immune-modulating drugs [108].

These immunotherapeutics include but are not limited to antibodies, small molecules, peptides, and cytokines. A study developed pH- and enzyme-responsive micelles for PD-1 and PTX codelivery, resulting in synergistic cancer chemoimmunotherapy via antitumor immunity by PTX-induced immunogenic cell death (ICD), while aPD-1 blocks the PD-1/PD-L1 axis to suppress immune escape due to PTX-induced PD-L1 upregulation [109]. Different types of cytokines, such as interleukins, interferons, and colony-stimulating factor (CSF), are employed for immunotherapy. In two studies, pIL-12 was codelivered with DOX via pH-/enzyme-responsive micelles to synergistically enhance NK cells and tumor-infiltrated cytotoxic T lymphocytes to achieve synergistic antitumor immune responses through cancer immunity cycle (CIC) cascade activation and amplification, providing therapeutic antitumor and antimetastatic efficacy [110]; this combination of nanosystems was subsequently used to develop a complete CIC-boosted combinatory strategy for developing immunotherapies against cancer and modulating the polarization of protumor M2 macrophages to activated antitumor M1 macrophages [111]. Interferon cytokine-based Mn and ABZI codelivery activate the STING pathway, mature dendritic cells (DCs), and eventually kill tumor cells via cytotoxic CD8+ T cells and NK cells [112]. Mao et al. loaded M-CSF in pH-responsive micelles. The results showed significant inhibition of tumor growth by promoting T-cell tumor infiltration and reversing the M1/M2 polarization balance within the tumor microenvironment [113]. Li et al. designed polymeric micelles based on the amphiphilic diblock copolymer poly(2-ethyl-2-oxazoline)-poly-(D, L-lactide) (PEOz-PLA) in combination with carboxyl-terminated Pluronic F127, for the codelivery of the antigen ovalbumin (OVA) and the Toll-like receptor-7 agonist CL264 (carboxylated-NPs/OVA/CL264) to lymph node-resident DCs. Surface modification with carboxylic groups endows micelles with endocytic receptor-targeting ability, promoting internalization by DCs through the scavenger receptor-mediated pathway. Adjusting the mass ratio of PEOz-PLA to carboxylated Pluronic F127 in the mixed micelles enabled the release of encapsulated CL264 to the early endosome. This resulted in increased expression of costimulatory molecules and secretion of cytokines stimulated by DCs, contributing to an enhanced immune response. Moreover, incorporating PEOz outside the micellar shell facilitated major histocompatibility complex I antigen presentation by facilitating endosome escape and cytosolic release of antigens. This study validated the efficacy of the system in inducing potent immune responses in vivo. Immunization with this codelivery system in tumor-bearing mice not only significantly inhibited tumor growth but also prolonged survival. The findings highlighted the potential clinical applications of this system as an effective antitumor vaccine for cancer immunotherapy. The study also emphasized the importance of particle surface characteristics in enhancing immune responses and demonstrated the advantages of carboxylated NPs in comparison to other formulations [114].

5. Conclusions and Outlooks

This review examines recent advancements in TME-responsive polymeric micelles, which are increasingly utilized for delivering chemotherapeutics and biologics in cancer treatment. We categorize TME-responsive strategies based on the endogenous characteristics of the tumor environment. Currently, genes and immunotherapies are at the forefront of cancer treatment, with TME-responsive micelles contributing significantly to this area and showing potential for further development. Numerous preclinical studies have demonstrated the efficacy of these micelles in targeted drug delivery for precise cancer therapy. However, comprehensive safety assessments, including long-term toxicity, biodistribution, pharmacokinetic, metabolic, and excretion studies, are essential before proceeding to clinical trials.

One challenge in this field is the complexity of the manufacturing and quality control processes required for sophisticated micellar systems, which limits industrial scaling. Several micellar formulations, such as Genexol-PM, have reached clinical trials and received FDA approval for breast cancer treatment. However, achieving the desired therapeutic effectiveness and clinical translation remains challenging. A deeper understanding of micelle–biological component interactions is necessary, particularly concerning biodistribution and microenvironmental responses. Issues such as insufficient stimulus sensitivity and nonspecific distribution can lead to off-target effects. Identifying the most effective stimuli for targeted delivery with minimal off-target effects remains a critical research area.

The development of site-specific, patient-tailored micelles is essential for advancing precision cancer treatment. Designing micelles with targeted therapeutic agents and moieties that align with an individual's genetic profile can enhance delivery efficiency and ensure treatment safety and efficacy. These micelles, integral to precision medicine, deliver precise therapeutic doses to cancer sites and facilitate disease monitoring through advanced imaging techniques. We anticipate that polymeric micelles will emerge as a robust platform for clinical cancer therapy in the near future.

Author Contributions: Z.J.: conception and design, manuscript revision; M.A.A.: manuscript draft and revision; S.G.: manuscript discussion and revision. All authors have read and agreed to the published version of the manuscript.

Funding: This work was supported by the Shanghai Jiao Tong University Foundation (grant number YG2021QN33), the National Natural Science Foundation of China (grant number 82104075), and the Shanghai Sailing Program (grant number 21YF1420400).

Conflicts of Interest: The authors declare no conflicts of interest.

References

1. Gyanani, V.; Haley, J.C.; Goswami, R. Challenges of current anticancer treatment approaches with focus on liposomal drug delivery systems. *Pharmaceuticals* **2021**, *14*, 835. [CrossRef]
2. Tohme, S.; Simmons, R.L.; Tsung, A. Surgery for cancer: A trigger for metastases. *Cancer Res.* **2017**, *77*, 1548–1552. [CrossRef]
3. Chivere, V.T.; Kondiah, P.P.D.; Choonara, Y.E.; Pillay, V. Nanotechnology-based biopolymeric oral delivery platforms for advanced cancer treatment. *Cancers* **2020**, *12*, 522. [CrossRef]
4. Wang, J.; Li, S.; Han, Y.; Guan, J.; Chung, S.; Wang, C.; Li, D. Poly (ethylene glycol)–polylactide micelles for cancer therapy. *Front. Pharmacol.* **2018**, *9*, 202. [CrossRef]
5. Borgheti-Cardoso, L.N.; Viegas, J.S.R.; Silvestrini, A.V.P.; Caron, A.L.; Praca, F.G.; Kravicz, M.; Bentley, M.V.L.B. Nanotechnology approaches in the current therapy of skin cancer. *Adv. Drug Deliv. Rev.* **2020**, *153*, 109–136. [CrossRef]
6. Pallares, R.M.; Agbo, P.; Liu, X.; An, D.D.; Gauny, S.S.; Zeltmann, S.E.; Minor, A.M.; Abergel, R.J. Engineering mesoporous silica nanoparticles for targeted alpha therapy against breast cancer. *ACS Appl. Mater. Interfaces* **2020**, *12*, 40078–40084. [CrossRef]
7. Shukla, N.; Singh, B.; Kim, H.J.; Park, M.H.; Kim, K. Combinational chemotherapy and photothermal therapy using a gold nanorod platform for cancer treatment. *Part. Part. Syst. Charact.* **2020**, *37*, 2000099. [CrossRef]
8. Hu, Y.; Zhang, Y.; Wang, X.; Jiang, K.; Wang, H.; Yao, S.; Liu, Y.; Lin, Y.-Z.; Wei, G.; Lu, W. Treatment of lung cancer by peptide-modified liposomal irinotecan endowed with tumor penetration and NF-κB inhibitory activities. *Mol. Pharm.* **2020**, *17*, 3685–3695. [CrossRef]
9. Xu, C.; Xu, J.; Zheng, Y.; Fang, Q.; Lv, X.; Wang, X.; Tang, R. Active-targeting and acid-sensitive pluronic prodrug micelles for efficiently overcoming MDR in breast cancer. *J. Mater. Chem. B* **2020**, *8*, 2726–2737. [CrossRef]
10. Gallego-Arranz, T.; Pérez-Cantero, A.; Torrado-Salmerón, C.; Guarnizo-Herrero, V.; Capilla, J.; Torrado-Durán, S. Improvement of the pharmacokinetic/pharmacodynamic relationship in the treatment of invasive aspergillosis with voriconazole. Reduced drug toxicity through novel rapid release formulations. *Colloids Surf. B Biointerfaces* **2020**, *193*, 111119. [CrossRef]
11. Kabanov, A.V.; Batrakova, E.V.; Alakhov, V.Y. Pluronic® block copolymers as novel polymer therapeutics for drug and gene delivery. *J. Control. Release* **2002**, *82*, 189–212. [CrossRef] [PubMed]
12. Yue, J.; Liu, S.; Xie, Z.; Xing, Y.; Jing, X. Size-dependent biodistribution and antitumor efficacy of polymer micelle drug delivery systems. *J. Mater. Chem. B* **2013**, *1*, 4273–4280. [CrossRef]
13. Perin, F.; Motta, A.; Maniglio, D. Amphiphilic copolymers in biomedical applications: Synthesis routes and property control. *Mater. Sci. Eng. C* **2021**, *123*, 111952. [CrossRef]
14. Lu, Y.; Zhang, E.; Yang, J.; Cao, Z. Strategies to improve micelle stability for drug delivery. *Nano Res.* **2018**, *11*, 4985–4998. [CrossRef]
15. Perumal, S.; Atchudan, R.; Lee, W. A review of polymeric micelles and their applications. *Polymers* **2022**, *14*, 2510. [CrossRef]
16. Kedar, U.; Phutane, P.; Shidhaye, S.; Kadam, V. Advances in polymeric micelles for drug delivery and tumor targeting. *Nanomedicine* **2010**, *6*, 714–729. [CrossRef]
17. Yang, F.; Xu, J.; Fu, M.; Ji, J.; Chi, L.; Zhai, G. Development of stimuli-responsive intelligent polymer micelles for the delivery of doxorubicin. *J. Drug Target.* **2020**, *28*, 993–1011. [CrossRef]
18. Wan, Z.; Zheng, R.; Moharil, P.; Liu, Y.; Chen, J.; Sun, R.; Song, X.; Ao, Q. Polymeric micelles in cancer immunotherapy. *Molecules* **2021**, *26*, 1220. [CrossRef]

19. Cheng, L.; Yu, J.; Hao, T.; Wang, W.; Wei, M.; Li, G. Advances in Polymeric Micelles: Responsive and Targeting Approaches for Cancer Immunotherapy in the Tumor Microenvironment. *Pharmaceutics* **2023**, *15*, 2622. [CrossRef]
20. Baghban, R.; Roshangar, L.; Jahanban-Esfahlan, R.; Seidi, K.; Ebrahimi-Kalan, A.; Jaymand, M.; Kolahian, S.; Javaheri, T.; Zare, P. Tumor microenvironment complexity and therapeutic implications at a glance. *Cell Commun. Signal.* **2020**, *18*, 59. [CrossRef]
21. Anderson, N.M.; Simon, M.C. The tumor microenvironment. *Curr. Biol.* **2020**, *30*, R921–R925. [CrossRef]
22. Uthaman, S.; Huh, K.M.; Park, I.-K. Tumor microenvironment-responsive nanoparticles for cancer theragnostic applications. *Biomater. Res.* **2018**, *22*, 22. [CrossRef]
23. Li, X.; Kwon, N.; Guo, T.; Liu, Z.; Yoon, J. Innovative strategies for hypoxic-tumor photodynamic therapy. *Angew. Chem. Int. Ed. Engl.* **2018**, *57*, 11522–11531. [CrossRef] [PubMed]
24. Hu, Q.; Katti, P.S.; Gu, Z. Enzyme-responsive nanomaterials for controlled drug delivery. *Nanoscale* **2014**, *6*, 12273–12286. [CrossRef] [PubMed]
25. Yang, P.; Tao, J.; Chen, F.; Chen, Y.; He, J.; Shen, K.; Zhao, P.; Li, Y. Multienzyme-mimic ultrafine alloyed nanoparticles in metal organic frameworks for enhanced chemodynamic therapy. *Small* **2021**, *17*, 2005865. [CrossRef]
26. Jiang, X.; Gray, P.; Patel, M.; Zheng, J.; Yin, J.-J. Crossover between anti-and pro-oxidant activities of different manganese oxide nanoparticles and their biological implications. *J. Mater. Chem. B* **2020**, *8*, 1191–1201. [CrossRef] [PubMed]
27. Ding, X.; Wang, Z.; Yu, Q.; Michał, N.; Roman, S.; Liu, Y.; Peng, N. Superoxide Dismutase-Like Regulated Fe/Ppa@ PDA/B for Synergistically Targeting Ferroptosis/Apoptosis to Enhance Anti-Tumor Efficacy. *Adv. Healthc. Mater.* **2023**, *12*, 2301824. [CrossRef]
28. Trimaille, T.; Verrier, B. Copolymer Micelles: A Focus on Recent Advances for Stimulus-Responsive Delivery of Proteins and Peptides. *Pharmaceutics* **2023**, *15*, 2481. [CrossRef]
29. Sheyi, R.; de la Torre, B.G.; Albericio, F. Linkers: An Assurance for Controlled Delivery of Antibody-Drug Conjugate. *Pharmaceutics* **2022**, *14*, 396. [CrossRef]
30. Zhou, X.X.; Jin, L.; Qi, R.Q.; Ma, T. pH-responsive polymeric micelles self-assembled from amphiphilic copolymer modified with lipid used as doxorubicin delivery carriers. *R. Soc. Open Sci.* **2018**, *5*, 171654. [CrossRef]
31. Augustine, R.; Kalva, N.; Kim, H.A.; Zhang, Y.; Kim, I. pH-responsive polypeptide-based smart nano-carriers for theranostic applications. *Molecules* **2019**, *24*, 2961. [CrossRef]
32. Feng, J.; Wen, W.; Jia, Y.-G.; Liu, S.; Guo, J. pH-responsive micelles assembled by three-armed degradable block copolymers with a cholic acid core for drug controlled-release. *Polymers* **2019**, *11*, 511. [CrossRef]
33. Ratemi, E. pH-responsive polymers for drug delivery applications. In *Stimuli Responsive Polymeric Nanocarriers for Drug Delivery Applications*; Woodhead Publishing: Sawston, UK, 2018; Volume 1, pp. 121–141.
34. Kocak, G.; Tuncer, C.; Bütün, V. pH-Responsive polymers. *Polym. Chem.* **2017**, *8*, 144–176. [CrossRef]
35. Zhang, H.; Kong, X.; Tang, Y.; Lin, W. Hydrogen sulfide triggered charge-reversal micelles for cancer-targeted drug delivery and imaging. *ACS Appl. Mater. Interfaces* **2016**, *8*, 16227–16239. [CrossRef]
36. He, W.; Zheng, X.; Zhao, Q.; Duan, L.; Lv, Q.; Gao, G.H.; Yu, S. pH-triggered charge-reversal polyurethane micelles for controlled release of doxorubicin. *Macromol. Biosci.* **2016**, *16*, 925–935. [CrossRef]
37. Peng, N.; Yu, H.; Yu, W.; Yang, M.; Chen, H.; Zou, T.; Deng, K.; Huang, S.; Liu, Y. Sequential-targeting nanocarriers with pH-controlled charge reversal for enhanced mitochondria-located photodynamic-immunotherapy of cancer. *Acta Biomater.* **2020**, *105*, 223–238. [CrossRef]
38. Lv, H.; Zhen, C.; Liu, J.; Yang, P.; Hu, L.; Shang, P. Unraveling the potential role of glutathione in multiple forms of cell death in cancer therapy. *Oxidative Med. Cell. Longev.* **2019**, *2019*, 3150145. [CrossRef] [PubMed]
39. Ju, H.-Q.; Lin, J.-F.; Tian, T.; Xie, D.; Xu, R.-H. NADPH homeostasis in cancer: Functions, mechanisms and therapeutic implications. *Signal Transduct. Target. Ther.* **2020**, *5*, 231. [CrossRef] [PubMed]
40. Monteiro, H.P.; Ogata, F.T.; Stern, A. Thioredoxin promotes survival signaling events under nitrosative/oxidative stress associated with cancer development. *Biomed. J.* **2017**, *40*, 189–199. [CrossRef] [PubMed]
41. Xu, X.; Wu, J.; Liu, S.; Saw, P.E.; Tao, W.; Li, Y.; Krygsman, L.; Yegnasubramanian, S.; De Marzo, A.M.; Shi, J. Redox-responsive nanoparticle-mediated systemic RNAi for effective cancer therapy. *Small* **2018**, *14*, 1802565. [CrossRef] [PubMed]
42. Wang, L.; Wang, W.; Cao, W.; Xu, H. Multi-hierarchical responsive polymers: Stepwise oxidation of a selenium-and tellurium-containing block copolymer with sensitivity to both chemical and electrochemical stimuli. *Polym. Chem.* **2017**, *8*, 4520–4527. [CrossRef]
43. Martin, J.R.; Duvall, C.L. Oxidation state as a bioresponsive trigger. In *Oxidative Stress and Biomaterials*; Elsevier: Amsterdam, The Netherlands, 2016; pp. 225–250.
44. Vijeth, S.; Heggannavar, G.B.; Kariduraganavar, M.Y. Encapsulating wall materials for micro-/nanocapsules. In *Microencapsulation: Processes, Technologies and Industrial Applications*; BoD–Books on Demand: Norderstedt, Germany, 2019; pp. 1–19. [CrossRef]
45. Sahoo, S.; Kayal, S.; Poddar, P.; Dhara, D. Redox-responsive efficient DNA and drug co-release from micelleplexes formed from a fluorescent cationic amphiphilic polymer. *Langmuir* **2019**, *35*, 14616–14627. [CrossRef]
46. Yi, X.; Dai, J.; Han, Y.; Xu, M.; Zhang, X.; Zhen, S.; Zhao, Z.; Lou, X.; Xia, F. A high therapeutic efficacy of polymeric prodrug nano-assembly for a combination of photodynamic therapy and chemotherapy. *Commun. Biol.* **2018**, *1*, 202. [CrossRef] [PubMed]

47. Dirisala, A.; Uchida, S.; Tockary, T.A.; Yoshinaga, N.; Li, J.; Osawa, S.; Gorantla, L.; Fukushima, S.; Osada, K.; Kataoka, K. Precise tuning of disulphide crosslinking in mRNA polyplex micelles for optimising extracellular and intracellular nuclease tolerability. *J. Drug Target.* **2019**, *27*, 670–680. [CrossRef] [PubMed]
48. Christie, R.J.; Matsumoto, Y.; Miyata, K.; Nomoto, T.; Fukushima, S.; Osada, K.; Halnaut, J.; Pittella, F.; Kim, H.J.; Nishiyama, N. Targeted polymeric micelles for siRNA treatment of experimental cancer by intravenous injection. *Acs Nano* **2012**, *6*, 5174–5189. [CrossRef]
49. Robinson, P.K. Enzymes: Principles and biotechnological applications. *Essays Biochem.* **2015**, *59*, 1. [CrossRef]
50. Li, Y.; Zhang, C.; Li, G.; Deng, G.; Zhang, H.; Sun, Y.; An, F. Protease-triggered bioresponsive drug delivery for the targeted theranostics of malignancy. *Acta Pharm. Sin. B* **2021**, *11*, 2220–2242. [CrossRef]
51. Shahriari, M.; Zahiri, M.; Abnous, K.; Taghdisi, S.M.; Ramezani, M.; Alibolandi, M. Enzyme responsive drug delivery systems in cancer treatment. *J. Control. Release* **2019**, *308*, 172–189. [CrossRef] [PubMed]
52. Beilke, M.C.; Klotzbach, T.L.; Treu, B.L.; Sokic-Lazic, D.; Wildrick, J.; Amend, E.R.; Gebhart, L.M.; Arechederra, R.L.; Germain, M.N.; Moehlenbrock, M.J. Enzymatic biofuel cells. In *Micro Fuel Cells*; Elsevier: Amsterdam, The Netherlands, 2009; pp. 179–241.
53. Yao, Q.; Kou, L.; Tu, Y.; Zhu, L. MMP-responsive 'smart' drug delivery and tumor targeting. *Trends Pharmacol. Sci.* **2018**, *39*, 766–781. [CrossRef] [PubMed]
54. Chen, W.-H.; Luo, G.-F.; Lei, Q.; Jia, H.-Z.; Hong, S.; Wang, Q.-R.; Zhuo, R.-X.; Zhang, X.-Z. MMP-2 responsive polymeric micelles for cancer-targeted intracellular drug delivery. *Chem. Commun.* **2015**, *51*, 465–468. [CrossRef]
55. Li, W.; Tao, C.; Wang, J.; Le, Y.; Zhang, J. MMP-responsive in situ forming hydrogel loaded with doxorubicin-encapsulated biodegradable micelles for local chemotherapy of oral squamous cell carcinoma. *RSC Adv.* **2019**, *9*, 31264–31273. [CrossRef]
56. Zhang, X.; Wang, X.; Zhong, W.; Ren, X.; Sha, X.; Fang, X. Matrix metalloproteinases-2/9-sensitive peptide-conjugated polymer micelles for site-specific release of drugs and enhancing tumor accumulation: Preparation and in vitro and in vivo evaluation. *Int. J. Nanomed.* **2016**, *11*, 1643–1661.
57. Wang, X.; Chen, Q.; Zhang, X.; Ren, X.; Zhang, X.; Meng, L.; Liang, H.; Sha, X.; Fang, X. Matrix metalloproteinase 2/9-triggered-release micelles for inhaled drug delivery to treat lung cancer: Preparation and in vitro/in vivo studies. *Int. J. Nanomed.* **2018**, *13*, 4641–4659. [CrossRef] [PubMed]
58. Vidak, E.; Javoršek, U.; Vizovišek, M.; Turk, B. Cysteine cathepsins and their extracellular roles: Shaping the microenvironment. *Cells* **2019**, *8*, 264. [CrossRef] [PubMed]
59. Kern, H.B.; Srinivasan, S.; Convertine, A.J.; Hockenbery, D.; Press, O.W.; Stayton, P.S. Enzyme-cleavable polymeric micelles for the intracellular delivery of proapoptotic peptides. *Mol. Pharm.* **2017**, *14*, 1450–1459. [CrossRef]
60. Alnaim, A.S.; Eggert, M.W.; Nie, B.; Quanch, N.D.; Jasper, S.L.; Davis, J.; Barnett, G.S.; Cummings, B.S.; Panizzi, P.R.; Arnold, R.D. Identifying uptake and biodistribution of liposome nanoparticles associated with secretes phospholipase A2 proteins and PLA2 receptors within a prostate cancer. *Cancer Res.* **2019**, *79* (Suppl. S13), 2985. [CrossRef]
61. áJames Delikatny, E. PLA 2-responsive and SPIO-loaded phospholipid micelles. *Chem. Commun.* **2015**, *51*, 12313–12315.
62. Sharipov, M.; Tawfik, S.M.; Gerelkhuu, Z.; Huy, B.T.; Lee, Y.-I. Phospholipase A2-responsive phosphate micelle-loaded UCNPs for bioimaging of prostate cancer cells. *Sci. Rep.* **2017**, *7*, 16073. [CrossRef]
63. Phaniendra, A.; Jestadi, D.B.; Periyasamy, L. Free radicals: Properties, sources, targets, and their implication in various diseases. *Indian J. Clin. Biochem.* **2015**, *30*, 11–26. [CrossRef] [PubMed]
64. Liou, G.-Y.; Storz, P. Reactive oxygen species in cancer. *Free Radic. Res.* **2010**, *44*, 479–496. [CrossRef]
65. Na, Y.; Lee, J.S.; Woo, J.; Ahn, S.; Lee, E.; Choi, W.I.; Sung, D. Reactive oxygen species (ROS)-responsive ferrocene-polymer-based nanoparticles for controlled release of drugs. *J. Mater. Chem. B* **2020**, *8*, 1906–1913. [CrossRef]
66. Li, R.; Peng, F.; Cai, J.; Yang, D.; Zhang, P. Redox dual-stimuli responsive drug delivery systems for improving tumor-targeting ability and reducing adverse side effects. *Asian J. Pharm. Sci.* **2020**, *15*, 311–325. [CrossRef]
67. Mohammed, F.; Ke, W.; Mukerabigwi, J.F.M.; Japir, A.A.-W.M.; Ibrahim, A.; Wang, Y.; Zha, Z.; Lu, N.; Zhou, M.; Ge, Z. ROS-responsive polymeric nanocarriers with photoinduced exposure of cell-penetrating moieties for specific intracellular drug delivery. *ACS Appl. Mater. Interfaces* **2019**, *11*, 31681–31692. [CrossRef]
68. Wang, Y.; Zhang, Y.; Ru, Z.; Song, W.; Chen, L.; Ma, H.; Sun, L. A ROS-responsive polymeric prodrug nanosystem with self-amplified drug release for PSMA (−) prostate cancer specific therapy. *J. Nanobiotechnol.* **2019**, *17*, 91. [CrossRef]
69. Deng, Y.; Yuan, H.; Yuan, W. Hypoxia-responsive micelles self-assembled from amphiphilic block copolymers for the controlled release of anticancer drugs. *J. Mater. Chem. B* **2019**, *7*, 286–295. [CrossRef] [PubMed]
70. Im, S.; Lee, J.; Park, D.; Park, A.; Kim, Y.-M.; Kim, W.J. Hypoxia-triggered transforming immunomodulator for cancer immunotherapy via photodynamically enhanced antigen presentation of dendritic cell. *ACS Nano* **2018**, *13*, 476–488. [CrossRef] [PubMed]
71. He, H.; Zhu, R.; Sun, W.; Cai, K.; Chen, Y.; Yin, L. Selective cancer treatment via photodynamic sensitization of hypoxia-responsive drug delivery. *Nanoscale* **2018**, *10*, 2856–2865. [CrossRef] [PubMed]
72. Yang, G.; Phua, S.Z.F.; Lim, W.Q.; Zhang, R.; Feng, L.; Liu, G.; Wu, H.; Bindra, A.K.; Jana, D.; Liu, Z. A hypoxia-responsive albumin-based nanosystem for deep tumor penetration and excellent therapeutic efficacy. *Adv. Mater.* **2019**, *31*, 1901513. [CrossRef] [PubMed]
73. Sahu, A.; Choi, W.I.; Tae, G. Recent progress in the design of hypoxia-specific nano drug delivery systems for cancer therapy. *Adv. Ther.* **2018**, *1*, 1800026. [CrossRef]

74. Wang, Y.; Shang, W.; Niu, M.; Tian, J.; Xu, K. Hypoxia-active nanoparticles used in tumor theranostic. *Int. J. Nanomed.* **2019**, *14*, 3705–3722. [CrossRef] [PubMed]
75. Sharma, A.; Arambula, J.F.; Koo, S.; Kumar, R.; Singh, H.; Sessler, J.L.; Kim, J.S. Hypoxia-targeted drug delivery. *Chem. Soc. Rev.* **2019**, *48*, 771–813. [CrossRef]
76. Manna, U.; Patil, S. Glucose-triggered drug delivery from borate mediated layer-by-layer self-assembly. *ACS Appl. Mater. Interfaces* **2010**, *2*, 1521–1527. [CrossRef] [PubMed]
77. Sun, S.; Sun, Y.; Rong, X.; Bai, L. High glucose promotes breast cancer proliferation and metastasis by impairing angiotensinogen expression. *Biosci. Rep.* **2019**, *39*, BSR20190436. [CrossRef] [PubMed]
78. Xiong, Y.; Qi, L.; Niu, Y.; Lin, Y.; Xue, Q.; Zhao, Y. Autonomous Drug Release Systems with Disease Symptom-Associated Triggers. *Adv. Intell. Syst.* **2020**, *2*, 1900124. [CrossRef]
79. Yu, S.; Chen, Z.; Zeng, X.; Chen, X.; Gu, Z. Advances in nanomedicine for cancer starvation therapy. *Theranostics* **2019**, *9*, 8026. [CrossRef] [PubMed]
80. Gorman, M.W.; Feigl, E.O.; Buffington, C.W. Human plasma ATP concentration. *Clin. Chem.* **2007**, *53*, 318–325. [CrossRef] [PubMed]
81. Sun, W.; Gu, Z. ATP-responsive drug delivery systems. *Expert Opin. Drug Deliv.* **2016**, *3*, 311–314. [CrossRef]
82. Gil, E.S.; Hudson, S.M. Stimuli-reponsive polymers and their bioconjugates. *Prog. Polym. Sci.* **2004**, *29*, 1173–1222. [CrossRef]
83. Kathmann, E.E.; White, L.A.; McCormick, C.L. Water-Soluble Polymers. 73. Electrolyte- and pH-Responsive Zwitterionic Copolymers of 4-[(2-Acrylamido-2-methylpropyl)-dimethylammonio]butanoate with 3-[(2-Acrylamido-2-methylpropyl)dimethylammonio]propanesulfonate. *Macromolecules* **1997**, *30*, 5297–5304. [CrossRef]
84. Fu, J.; Schlenoff, J.B. Driving Forces for Oppositely Charged Polyion Association in Aqueous Solutions: Enthalpic, Entropic, but Not Electrostatic. *J. Am. Chem. Soc.* **2016**, *138*, 980–990. [CrossRef]
85. Bediako, J.K.; Mouele, E.S.; Ouardi, Y.; Repo, E. Saloplastics and the polyelectrolyte complex continuum: Advances, challenges and prospects. *Chem. Eng. J.* **2023**, *462*, 142322. [CrossRef]
86. Amara, S.; Tiriveedhi, V. Inflammatory role of high salt level in tumor microenvironment. *Int. J. Oncol.* **2017**, *50*, 1477–1481. [CrossRef]
87. Luo, Y.; Yin, X.; Yin, X.; Chen, A.; Zhao, L.; Zhang, G.; Liao, W.; Huang, X.; Li, J.; Zhang, C.Y. Dual pH/redox-responsive mixed polymeric micelles for anticancer drug delivery and controlled release. *Pharmaceutics* **2019**, *11*, 176. [CrossRef]
88. Zhang, L.; Wang, Y.; Zhang, X.; Wei, X.; Xiong, X.; Zhou, S. Enzyme and redox dual-triggered intracellular release from actively targeted polymeric micelles. *ACS Appl. Mater. Interfaces* **2017**, *9*, 3388–3399. [CrossRef]
89. Wen, J.; Mao, H.-Q.; Li, W.; Lin, K.Y.; Leong, K.W. Biodegradable polyphosphoester micelles for gene delivery. *J. Pharm. Sci.* **2004**, *93*, 2142–2157. [CrossRef]
90. Nishiyama, N.; Bae, Y.; Miyata, K.; Fukushima, S.; Kataoka, K. Smart polymeric micelles for gene and drug delivery. *Drug Discov. Today Technol.* **2005**, *2*, 21–26. [CrossRef]
91. Dirisala, A.; Uchida, S.; Li, J.; Van Guyse, J.F.R.; Hayashi, K.; Vummaleti, S.V.C.; Kaur, S.; Mochida, Y.; Fukushima, S.; Kataoka, K. Effective mRNA Protection by Poly (l-ornithine) Synergizes with Endosomal Escape Functionality of a Charge-Conversion Polymer toward Maximizing mRNA Introduction Efficiency. *Macromol. Rapid Commun.* **2022**, *43*, 2100754. [CrossRef] [PubMed]
92. Shen, X.; Dirisala, A.; Toyoda, M.; Xiao, Y.; Guo, H.; Honda, Y.; Nomoto, T.; Takemoto, H.; Miura, Y.; Nishiyama, N. pH-responsive polyzwitterion covered nanocarriers for DNA delivery. *J. Control. Release* **2023**, *360*, 928–939. [CrossRef]
93. Dirisala, A.; Osada, K.; Chen, Q.; Tockary, T.A.; Machitani, K.; Osawa, S.; Liu, X.; Ishii, T.; Miyata, K.; Oba, M. Optimized rod length of polyplex micelles for maximizing transfection efficiency and their performance in systemic gene therapy against stroma-rich pancreatic tumors. *Biomaterials* **2014**, *35*, 5359–5368. [CrossRef] [PubMed]
94. Kagaya, H.; Oba, M.; Miura, Y.; Koyama, H.; Ishii, T.; Shimada, T.; Takato, T.; Kataoka, K.; Miyata, T. Impact of polyplex micelles installed with cyclic RGD peptide as ligand on gene delivery to vascular lesions. *Gene Ther.* **2012**, *19*, 61–69. [CrossRef]
95. Kim, K.; Chen, W.C.W.; Heo, Y.; Wang, Y. Polycations and their biomedical applications. *Prog. Polym. Sci.* **2016**, *60*, 18–50. [CrossRef]
96. Kondinskaia, D.A.; Gurtovenko, A.A. Supramolecular complexes of DNA with cationic polymers: The effect of polymer concentration. *Polymer* **2018**, *142*, 277–284. [CrossRef]
97. Guo, X.; Shi, C.; Yang, G.; Wang, J.; Cai, Z.; Zhou, S. Dual-responsive polymer micelles for target-cell-specific anticancer drug delivery. *Chem. Mater.* **2014**, *26*, 4405–4418. [CrossRef]
98. Gao, Y.; Jia, L.; Wang, Q.; Hu, H.; Zhao, X.; Chen, D.; Qiao, M. pH/Redox dual-responsive polyplex with effective endosomal escape for codelivery of siRNA and doxorubicin against drug-resistant cancer cells. *ACS Appl. Mater. Interfaces* **2019**, *11*, 16296–16310. [CrossRef] [PubMed]
99. Pan, J.; Attia, S.A.; Subhan, M.A.; Filipczak, N.; Mendes, L.P.; Li, X.; Kishan Yalamarty, S.S.; Torchilin, V.P. Monoclonal antibody 2C5-modified mixed dendrimer micelles for tumor-targeted codelivery of chemotherapeutics and siRNA. *Mol. Pharm.* **2020**, *17*, 1638–1647. [CrossRef] [PubMed]
100. Yang, W.; Chen, P.; Boonstra, E.; Hong, T.; Cabral, H. Polymeric micelles with pH-responsive cross-linked core enhance in vivo mRNA delivery. *Pharmaceutics* **2022**, *14*, 1205. [CrossRef]
101. Debela, D.T.; Muzazu, S.G.Y.; Heraro, K.D.; Ndalama, M.T.; Mesele, B.W.; Haile, D.C.; Kitui, S.K.; Manyazewal, T. New approaches and procedures for cancer treatment: Current perspectives. *SAGE Open Med.* **2021**, *9*, 20503121211034366. [CrossRef]

102. Papaioannou, N.E.; Beniata, O.V.; Vitsos, P.; Tsitsilonis, O.; Samara, P. Harnessing the immune system to improve cancer therapy. *Ann. Transl. Med.* **2016**, *4*, 261. [CrossRef]
103. Janssen, L.M.E.; Ramsay, E.E.; Logsdon, C.D.; Overwijk, W.W. The immune system in cancer metastasis: Friend or foe? *J. Immunother. Cancer* **2017**, *5*, 79. [CrossRef]
104. Gao, S.; Yang, X.; Xu, J.; Qiu, N.; Zhai, G. Nanotechnology for boosting cancer immunotherapy and remodeling tumor microenvironment: The horizons in cancer treatment. *ACS Nano* **2021**, *15*, 12567–12603. [CrossRef]
105. Mishra, A.K.; Ali, A.; Dutta, S.; Banday, S.; Malonia, S.K. Emerging trends in immunotherapy for cancer. *Diseases* **2022**, *10*, 60. [CrossRef]
106. Kirtane, K.; Elmariah, H.; Chung, C.H.; Abate-Daga, D. Adoptive cellular therapy in solid tumor malignancies: Review of the literature and challenges ahead. *J. Immunother. Cancer* **2021**, *9*, e002723. [CrossRef]
107. Edwards, S.C.; Hoevenaar, W.H.M.; Coffelt, S.B. Emerging immunotherapies for metastasis. *Br. J. Cancer* **2021**, *124*, 37–48. [CrossRef] [PubMed]
108. Hari, S.K.; Gauba, A.; Shrivastava, N.; Tripathi, R.M.; Jain, S.K.; Pandey, A.K. Polymeric micelles and cancer therapy: An ingenious multimodal tumor-targeted drug delivery system. *Drug Deliv. Transl. Res.* **2023**, *13*, 135–163. [CrossRef] [PubMed]
109. Su, Z.; Xiao, Z.; Wang, Y.; Huang, J.; An, Y.; Wang, X.; Shuai, X. Codelivery of anti-PD-1 antibody and paclitaxel with matrix metalloproteinase and pH dual-sensitive micelles for enhanced tumor chemoimmunotherapy. *Small* **2020**, *16*, 1906832. [CrossRef]
110. Sun, Y.; Liu, L.; Zhou, L.; Yu, S.; Lan, Y.; Liang, Q.; Liu, J.; Cao, A.; Liu, Y. Tumor microenvironment-triggered charge reversal polymetformin-based nanosystem co-delivered doxorubicin and IL-12 cytokine gene for chemo–gene combination therapy on metastatic breast cancer. *ACS Appl. Mater. Interfaces* **2020**, *12*, 45873–45890. [CrossRef]
111. Cao, Y.; Li, J.; Liang, Q.; Yang, J.; Zhang, X.; Zhang, J.; An, M.; Bi, J.; Liu, Y. Tumor Microenvironment Sequential Drug/Gene Delivery Nanosystem for Realizing Multistage Boosting of Cancer–Immunity Cycle on Cancer Immunotherapy. *ACS Appl. Mater. Interfaces* **2023**, *15*, 54898–54914. [CrossRef]
112. Li, J.; Ren, H.; Qiu, Q.; Yang, X.; Zhang, J.; Zhang, C.; Sun, B.; Lovell, J.F.; Zhang, Y. Manganese coordination micelles that activate stimulator of interferon genes and capture in situ tumor antigens for cancer metalloimmunotherapy. *ACS Nano* **2022**, *16*, 16909–16923. [CrossRef]
113. Mao, K.; Cong, X.; Feng, L.; Chen, H.; Wang, J.; Wu, C.; Liu, K.; Xiao, C.; Yang, Y.-G.; Sun, T. Intratumoral delivery of M-CSF by calcium crosslinked polymer micelles enhances cancer immunotherapy. *Biomater. Sci.* **2019**, *7*, 2769–2776. [CrossRef] [PubMed]
114. Li, C.; Zhang, X.; Chen, Q.; Zhang, J.; Li, W.; Hu, H.; Zhao, X.; Qiao, M.; Chen, D. Synthetic polymeric mixed micelles targeting lymph nodes trigger enhanced cellular and humoral immune responses. *ACS Appl. Mater. Interfaces* **2018**, *10*, 2874–2889. [CrossRef]

Review

Advances and Prospects for Hydrogel-Forming Microneedles in Transdermal Drug Delivery

Xiaolin Hou [1,†], Jiaqi Li [1], Yongyu Hong [2,†], Hang Ruan [1], Meng Long [1], Nianping Feng [1,*] and Yongtai Zhang [1,*]

[1] Department of Pharmaceutics, Shanghai University of Traditional Chinese Medicine, No. 1200 Cailun Road, Pudong New Area, Shanghai 201203, China; houxiao_lin2008@163.com (X.H.); alijiaqi0@163.com (J.L.); 18221817885@163.com (H.R.); longmeng0608@126.com (M.L.)

[2] Xiamen Hospital of Chinese Medicine, No. 1739 Xiangyue Road, Huli District, Xiamen 361015, China; 13906055433@126.com

[*] Correspondence: npfeng@shutcm.edu.cn (N.F.); zyt@shutcm.edu.cn (Y.Z.)

[†] These authors contributed equally to this work.

Abstract: Transdermal drug delivery (TDD) is one of the key approaches for treating diseases, avoiding first-pass effects, reducing systemic adverse drug reactions and improving patient compliance. Microneedling, iontophoresis, electroporation, laser ablation and ultrasound facilitation are often used to improve the efficiency of TDD. Among them, microneedling is a relatively simple and efficient means of drug delivery. Microneedles usually consist of micron-sized needles (50–900 μm in length) in arrays that can successfully penetrate the stratum corneum and deliver drugs in a minimally invasive manner below the stratum corneum without touching the blood vessels and nerves in the dermis, improving patient compliance. Hydrogel-forming microneedles (HFMs) are safe and non-toxic, with no residual matrix material, high drug loading capacity, and controlled drug release, and they are suitable for long-term, multiple drug delivery. This work reviewed the characteristics of the skin structure and TDD, introduced TDD strategies based on HFMs, and summarized the characteristics of HFM TDD systems and the evaluation methods of HFMs as well as the application of HFM drug delivery systems in disease treatment. The HFM drug delivery system has a wide scope for development, but the translation to clinical application still has more challenges.

Keywords: hydrogel-forming microneedles; transdermal drug delivery; controlled release; permeation pathway; environmental response

Citation: Hou, X.; Li, J.; Hong, Y.; Ruan, H.; Long, M.; Feng, N.; Zhang, Y. Advances and Prospects for Hydrogel-Forming Microneedles in Transdermal Drug Delivery. *Biomedicines* **2023**, *11*, 2119. https://doi.org/10.3390/biomedicines11082119

Academic Editors: Ali Nokhodchi and Rowan S. Hardy

Received: 31 May 2023
Revised: 12 July 2023
Accepted: 21 July 2023
Published: 27 July 2023

1. Introduction

Transdermal drug delivery (TDD) is a route of drug delivery for treating or preventing disease by absorbing drugs through the skin, permeating into the skin and further into the blood circulation [1]. TDD avoids first-pass effects, prolongs the action of drugs with short half-lives through slow release and avoids fluctuations in blood levels, reduces side effects and improves patient compliance [2]. The stratum corneum barrier plays a key role in TDD, and many methods have been used to improve the efficiency of TDD, including the use of chemical penetration enhancers and different physical enhancement approaches, such as microneedling [2,3], iontophoresis [4], electroporation [5], laser ablation [6] and ultrasound facilitation [7–9].

In recent years, microneedles have gained widespread interest in TDD and have shown brilliant achievements in delivering both chemical small molecules and biomacromolecules whilst being minimally invasive and painless [10–17]. Microneedles usually consist of micrometer-sized needles (50–900 μm in length) in the form of microneedle arrays that can successfully penetrate the stratum corneum and deliver drugs in a minimally invasive manner below the stratum corneum without damaging blood vessels and nerves in the dermis [18,19], improving patient compliance and allowing drugs exposed in the epidermis or dermis to be rapidly absorbed by surrounding capillaries and lymph nodes [20–22].

Some of the microneedle products approved for marketing by the FDA and undergoing clinical trials are shown in Table 1 and **??**, respectively. As shown in Table 1, microneedling devices are often intended for aesthetic use rather than medical purposes. Additionally, clinical trials (Table **??**) of microneedles for disease treatment, such as influenza, psoriasis, and diabetes, have gradually increased in recent years.

Microneedles can be fabricated with different materials and can be classified into five main types (Figure 1), namely solid microneedles, coated microneedles, hollow microneedles, dissolving microneedles and hydrogel-forming microneedles (HFMs) [23–25]. Among these, HFMs, an attractive type of microneedles first reported in 2012, consist of a swellable polymer (cross-linked hydrogel) that enables the sustained delivery of drugs for long periods of time by either incorporating the drug into the polymer structure during preparation or by loading the drug into a separate reservoir and attaching it to the HFMs [26]. In the following, the application and evaluation methods of HFMs in TDD are analyzed and discussed in detail.

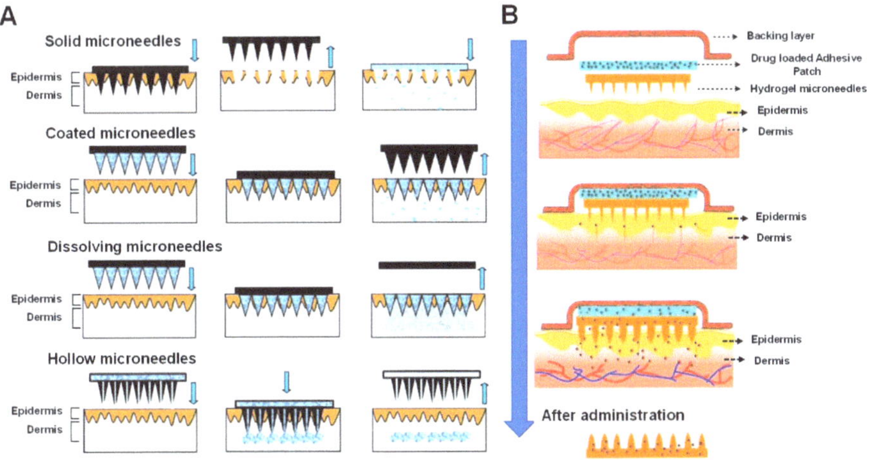

Figure 1. Schematic representation of methods of traditional (**A**) and hydrogel (**B**) microneedles mediated drug delivery across skin (arrows point to the order of operations). The figure was adopted from ref. [11] with permission from WILEY-VCH VERLAG GMBH & CO. KGAA.

Table 1. Microneedle products approved by the FDA.

Products Name	Applicant	De Novo or 510 (k) Number	Date Approved	Design of the Product	Use
DP4 Micro needling device	Equipmed USA LLC, Newport Beach, CA, USA	K221070	20 December 2022	Powered MN device with 16 stainless steel microneedle plate adapts to the skin's surface and has a maximal needle length of 3 mm.	Aesthetic use
SkinPen Precision system	Crown Aesthetics, Dallas, TX, USA	K220506 K202243	7 March 2022 2 April 2021	The microneedling pen handpiece comes with a sterile needle cartridge and includes 14 total solids medical-grade stainless steel, maximum needle length of 1.5 mm.	Aesthetic use
INTRAcel RF Microneedle System	Jeisys Medical Inc., Seoul, Republic of Korea	K183284	15 January 2020	Fractional radio frequency combined with insulated microneedling	General and plastic surgery
Exceed Microneedling device	MT. DERM GmbH, Berlin, Germany	K182407 K180778	19 July 2019 7 September 2018	Powered microneedling device with 6 stainless steel microneedle plate adapts to the skin's surface and has a maximal needle length of 1.5 mm.	Aesthetic use
SkinPen precision system	Bellus Medical, LLC, Lindon, UT, USA	DEN160029	1 March 2018	Microneedling pen handpiece with a sterile needle cartridge	Aesthetic use
MICRONJET 600	NANOPASS TECHNOLOGIES LTD., Ness Ziona, Israel	K092746	3 February 2010	Hollow microneedles that consist of a needle holder, a needle tube, a needle tube liner and a protective sleeve	Intradermal injection

Table 1. *Cont.*

Number of Clinical Trials	Study Title	Conditions	Interventions	Study Phase	Status
00837512	Insulin Delivery Using Microneedles in Type 1 Diabetes	Type 1 Diabetes Mellitus	Device: Microneedle; Device: Subcutaneous insulin catheter	Phase 2 Phase 3	Completed
03855397	Pain and Safety of Microneedles in Oral Cavity	Oral Cavity Disease	Other: Microneedle Other: Hypodermic needle Other: Flat patch	Not Applicable	Completed
04583852	Evaluate the Efficacy and Safety of Brightening Micro-needle Patch on Facial Solar Lentigines	Solar Lentigines	Other: AIVÍA, Ultra-Brightening Spot Microneedle Patch Other: Placebo Micro-needle Patch	Not Applicable	Completed
05108714	Intradermal Lidocaine Via MicronJet600 Microneedle Device	Local Anaesthesia	Device: Intravenous cannulation after intradermal injection of lidocaine via MicronJet600 microneedle device (1) Device: Intravenous cannulation after intradermal injection of saline via MicronJet600 microneedle device Device: Intravenous cannulation after intradermal injection of lidocaine via MicronJet600 microneedle device (2) Procedure: Intravenous cannulation after without prior interventions	Not Applicable	Completed
05267938	Microneedle Pretreatment as a Strategy to Improve the Effectiveness of Topical Anesthetics Formulations	Oral Cavity Disease	Drug: Topical Anesthetic Drug: Local anesthetic	Phase 1	Completed
04989361	Soluble Hyaluronic Acid Microneedle VS. Non-ablative Fractional Laser on Infraorbital Wrinkles	Wrinkle	Procedure: Soluble Hyaluronic Acid Microneedle Procedure: Nonfractional laser	Not Applicable	Recruiting
05377905	Microneedle Array Plus Doxorubicin in Cutaneous Squamous Cell Cancer (cSCC)	Cutaneous Squamous Cell Carcinoma Skin Cancers-Squamous Cell Carcinoma	Drug: Microneedle Array Doxorubicin (MNA-D)	Phase 1 Phase 2	Not yet recruiting

Table 1. *Cont.*

Number of Clinical Trials	Study Title	Conditions	Interventions	Study Phase	Status
02966067	A Split Mouth Trial to Compare Microneedles vs. Standard Needles in Dental Anaesthetic Delivery	Dental Pain Anesthesia, Loca	Device: Microneedle Device (Experimental) Device: 30-gauge Short Hypodermic Needle	Not Applicable	Completed
03054480	Fractional Micro-Needle Radiofrequency and I Botulinum Toxin A for Primary Axillary Hyperhidrosis	Primary Axillary Hyperhidrosis	Device: Fractional Micro-Needle Radiofrequency Drug: Botulinum toxin type A	Not Applicable	Completed
03207763	Microneedle Patch Study in Healthy Infants/Young Children	Vaccination Skin Absorption	Device: Microneedle Formulation 1 Device: Microneedle Formulation 2	Not Applicable	Completed
02682056	Glucose Measurement Using Microneedle Patches	Diabetes	Device: Microneedle patch Device: Intravenous (IV) catheter Device: Lancet	Not Applicable	Completed
04732195	Pilocarpine Microneedles for Sweat Induction (PMN-SI)	Cystic Fibrosis	Device: Pilocarpine microneedle patch Device: Pilocarpine Iontophoresis	Not Applicable	Completed
05694858	Transdermal Microneedle Lignocaine Delivery Versus EMLA Patch for Topical Analgesia Before Venepuncture Procedure To Adults in Clinical Setting	Glaucoma Cataract	Combination Product: Lignocaine loaded maltose microneedle array patch Drug: EMLA 5% patch	Phase 1 Phase 2	Not yet recruiting
02995057	Safety Demonstration of Microneedle Insertion	Allergic Reaction to Nickel	Device: Gold- or silver-coated, or uncoated nickel microneedles	Not Applicable	Completed
04552015	Microneedles for Diagnosis of LTBI	Latent Tuberculosis	Diagnostic Test: TST vs. PPD microneedle test	Not Applicable	Terminated
03203174	The Use of Microneedles With Topical Botulinum Toxin for Treatment of Palmar Hyperhidrosis	Hyperhidrosis	Device: Microneedle Device: Sham Microneedle Drug: Botulinum Toxin Type A Other: Saline	Phase 1	Completed
03332628	Racial/Ethnic Differences in Microneedle Response	Healthy	Device: Microneedle patch	Not Applicable	Completed

Table 1. *Cont.*

Number of Clinical Trials	Study Title	Conditions	Interventions	Study Phase	Status
05078463	Efficacy of Transdermal Microneedle Patch for Topical Anesthesia Enhancement in Paediatric Thalassemia Patients	Thalassemia in Children	Device: Microneedle Drug: 1 Finger Tip Unit (FTU) EMLA Cream (30-min application time) Drug: 1 Finger Tip Unit (FTU) EMLA (15-min application time) Drug: 0.5 Finger Tip Unit (FTU) EMLA (30-min application time) Device: Sham Patch	Phase 2	Completed
00539084	A Study to Assess the Safety and Efficacy of a Microneedle Device for Local Anesthesia	Local Anesthesia Intradermal Injections	Device: MicronJet	Not Applicable	Completed
02596750	The Effect of Microneedle Pretreatment on Topical Anesthesia	Pain	Device: Microneedle Roller Device: Sham microneedle Roller	Not Applicable	Completed
01812837	The Use of Microneedles in Photodynamic Therapy	Actinic Keratosis	Device: Microneedle Drug: Aminolevulinic Acid Radiation: Blue light	Not Applicable	Completed
05710068	Effects of RF Microneedle on Photoaging Skin	Pigmentation Pigmentation Disorder	Device: RF Microneedle Drug: Combination cream	Not Applicable	Completed
03629041	A Study of the Use of Microneedle Patches to Deliver Topical Lidocaine in the Oral Cavity	Topical Anaesthesia	Device: Microneedle Patch Device: Patch with no microneedles	Phase 1	Completed
03795402	Analysis of Non-Invasively Collected Microneedle Device Samples From Mild Plaque Psoriasis for Use in Transcriptomics Profiling	Psoriasis Vulgaris	Device: Microneedle Device	Not Applicable	Completed
02594644	The Use of Microneedles to Expedite Treatment Time in Photodynamic Therapy	Keratosis, Actinic	Device: Microneedle Roller Drug: Aminolevulinic Acid Radiation: Blue Light	Not Applicable	Completed
03415373	Clinical Evaluation of Healthy Subjects Receiving Intradermal Saline Using the Microneedle Adapter (Model UAR-2S)	Intradermal Injection	Device: Microneedle Adapter (Model UAR-2S) Device: Hypodermic needle + syringe	Not Applicable	Completed

Table 1. *Cont.*

Number of Clinical Trials	Study Title	Conditions	Interventions	Study Phase	Status
03607903	Adalimumab Microneedles in Healthy Volunteers	Pain Injection Site	Biological: Adalimumab ID Biological: Adalimumab SC Other: Saline ID Other: Saline SC	Phase 1 Phase 2	Completed
04394689	Measles and Rubella Vaccine Microneedle Patch Phase 1–2 Age De-escalation Trial	Measles Rubella Vaccination Healthy	Biological: Measles Rubella Vaccine (MRV-SC) Biological: MRVMNP Other: PLA-MNP Other: PLA-SC	Phase 1 Phase 2	Completed
03739398	A Study on the Effectiveness and Safety Evaluation of Combination Therapy With 1927 nm Thulium Laser and Fractional Microneedle Radiofrequency Equipment for Improvement of Skin Aging	Wrinkle	Device: LUTRONIC GENUS laser (Fractional Microneedle Radiofrequency (FMR)) Device: LASEMED laser (The Thulium laser with 1927 nm wavelength)	Not Applicable	Completed
02438423	Inactivated Influenza Vaccine Delivered by Microneedle Patch or by Hypodermic Needle	Influenza	Biological: Inactivated influenza vaccine Other: Placebo	Phase 1	Completed
04928222	Placebo Microneedles in Healthy Volunteers (Part I) and Efficacy/Safety of Doxorubicin Microneedles in Basal Cell Cancer Subjects (Part II)	Basal Cell Carcinoma	Combination Product: Doxorubicin containing MNA Drug: Placebo containing MNA	Phase 1 Phase 2	Active, not recruiting
02632110	Microneedle Lesion Preparation Prior to Aminolevulinic Acid Photodynamic Therapy (ALAPDT) for AK on Face	Actinic Keratosis	Drug: ALA Drug: Topical Solution Vehicle Device: IBL 10 mW Procedure: Microneedle lesion preparation Device: IBL 20 mW	Phase 2	Completed
02745392	Safety and Efficacy of ZPZolmitriptan Intracutaneous Microneedle Systems for the Acute Treatment of Migraine	Acute Migraine	Drug: ZPZolmitriptan Drug: Placebo	Phase 2 Phase 3	Completed
01789320	Safety Study of Suprachoroidal Triamcinolone Acetonide Via Microneedle to Treat Uveitis	Uveitis Intermediate Uveitis Posterior Uveitis Panuveitis Noninfectious Uveitis	Drug: triamcinolone acetonide (Triesence®)	Phase 1 Phase 2	Completed

2. Characteristics of HFMs as TDD System

HFMs are safe, have no residual matrix material and are suitable for long-term, multiple-drug delivery. HFMs have the advantages of resisting the closure of skin tissue pores after puncture into the skin, and no matrix material remains when the HFM patch is removed due to the inherent swelling insolubility and viscoelastic properties of the matrix material [26,27]. HFMs also bring benefits from avoiding drug deposition after microneedle tip penetration [28].

HFMs are characterized by water absorption and swelling, and sustainable and controlled drug release. Drugs can be loaded in HFMs in two ways, either by incorporating the drug into the microneedle matrix during preparation or by loading the drug into a separate reservoir and then attaching it to the hydrogel microneedle as a substrate [29]. Both methods of preparation allow for the continuous delivery of the drug over a long period of time. Materials used to prepare HFMs are non-toxic, degradable and biocompatible [30], and commonly used materials include natural compounds such as gelatin and polymer copolymers such as poly (methyl vinyl ether-co-maleic acid) cross-linked with polyethylene glycol (PMVE/MAPEG) [10,31]. Among these, PMVE/MAPEG has an excellent water absorption capacity and allows the preparation of super-swollen HFMs that can absorb fluids and swell up to 20 times their original size [11,32]. HFMs pierce the skin and rapidly absorb interstitial fluid, causing the hydrogel to swell, creating a continuous, unobstructed hydrogel conduit for the drug permeating into the skin [33].

HFMs are able to control the drug release behavior through the crosslinking density of the hydrogel microneedle matrix material, thus achieving controlled drug delivery kinetics [34]. For example, the degree of swelling of PMVE/MAPEG decreases with increasing cross-linkage. When the PMVE/MAPEG ratio was 2:1 and 4:3, respectively, the degree of swelling increased by 294% and 250%, respectively [35]. In addition, increasing the concentration of the cross-linking agent ethylene glycol dimethacrylate (EGDMA) decreases the release of the loaded drug, and the $t_{1/2}$ of the drug increased from 2.64 h to 45.67 h when EGDMA was added from 1% to 8%. It indicates that by increasing the cross-linking agent, the cross-linking degree can be increased and the swelling degree can be decreased, which results in lower and more sustained drug release [36]. Some main characteristics of different HFM formulations have been summarized in Table 2.

Table 2. Characteristics of different HFM formulations.

Compounds	Polymer	Characteristics	Ref.
Fluorescein, FITC-Dextran, Doxorubicin	Methacrylated hyaluronic acid	Methacrylated hyaluronic acid microneedles fully swelled within 1 min, with swelling ratio of ~2.74; >80% of fluorescein, FITC-Dextran, and >50% of doxorubicin were released from the microneedle patches within 30 min.	[37]
HRP	Silk	The beta sheet content in the microneedle devices was increased from 14% to 15% and 21% as the water vapor annealing time increased from 0 h to 2 h and 8 h, and the HRP release reduced to 37% and 18%.	[38]
FITC-dextran	Silk fibroin, urea, N-dimethylformamidee, glycine and 2-ethoxyethanol	The swelling-modified silk fibroin microneedles with different microscopic pore size attains 250–650% swelling ratio after PBS immersion; the swelling-modified silk fibroin microneedles display significantly enhanced transdermal drug release kinetics compared with the controlled silk fibroin films, with 2–10-times larger accumulative release ratio than the corresponding control groups during the entire release process in vitro.	[2]

Table 2. *Cont.*

Compounds	Polymer	Characteristics	Ref.
Light-responsive ibuprofen conjugates	Crosslinked 2-hydroxyethyl methacrylate	The crosslinked 2-hydroxyethyl methacrylate hydrogel shows maximum swelling degrees of around 50% after 24 h; the system allows the release of ibuprofen during prolonged periods of time (up to 160 h).	[39]
Sildenafil citrate	Polyvinyl alcohol and polyvinylpyrrolidone crosslinked by tartaric acid	The hydrogel's swelling percentage was 348.07–72,897% with different formulations.	[40]
Doxorubicin	Methacrylated hyaluronic acid	The swelling ratio increased rapidly and reached a maximum of 337% at 10 min; the release profile of doxorubicin/SMNs dramatically turned into a slow rate after 90 min and less than 90% doxorubicin was released at the end of the detection point (12 h). The maximum doxorubicin concentration that appeared at 1 h was 0.58 ± 0.35 µg/mL for the doxorubicin/DMNs group, and was 1.28 ± 0.32 µg/mL for the doxorubicin/SMNs group at 2 h, respectively. The crosslinking network of SMNs significantly retarded the diffusion of small molecule drugs within the needle matrix, and extended the drug release duration, increasing the drug transdermal efficacy.	[41]
Nicotinamide mononucleotide	Polyvinyl alcohol, carboxymethyl cellulose, DMSO	Formulations of microneedles containing 2.9% carboxymethyl cellulose have a higher swelling ratio (186%) in comparison with the 0% carboxymethyl cellulose composite (48%) and a higher nicotinamide mononucleotide release of $91.94 \pm 4.03\%$ at 18 h compared with the carboxymethyl cellulose-free polyvinyl alcohol matrix of $50.48 \pm 3.73\%$ at 18 h.	[42]

3. Materials for Forming HFMs

Similar to other types of microneedles, the common preparation method for HFMs is mainly micromolding [43,44], where a microneedle matrix in its flowing hydrogel state is injected into the mold using centrifugal and decompression methods and then dried. The release and permeation behavior of the active ingredients is mainly controlled by the nature of the polymers that make up the microneedle matrix, independent of the microneedle preparation process.

Preferred microneedle materials should be biocompatible and non-immunogenic, have fine-mechanical strength, and be able to carry potentially large and complex drugs without damage. The commonly used materials for HFMs include Gantrez S-97, a co-polymer of poly(methylvinylether co. maleic acid) (PMVE/MA) [12,45], methacrylated hyaluronic acid (MeHA) [46], Gantrez AN-139, a co-polymer of poly(methylvinylether co. maleic anhydride) (PMVE/MAH) [47,48], polyvinyl alcohol (PVA)/polymer blends [49,50] and crosslinked PVA [51], crosslinked 2-hydroxyethyl methacrylate (pHEMA) [39], poly(styrene-b-acrylic acid) (PS-*b*-PAA) [52], modified silk [2] and clay/polymer blends [53].

Some cross-linked polymers, such as those prepared through esterification reactions, have different degrees of polymerization, resulting in widely varying structures and therefore different properties of water absorption and swelling. This results in different rates of swelling after the microneedle tips are inserted into the skin, leading to different drug release profiles, such as burst and sustained drug release [11]. For example, HFMs based on the 'super-swelling' polymer PMVE/MAPEG consisted of drug reservoirs and microneedle tips which did not contain the drug [11]. When the tips were inserted into the skin, the microneedle tips rapidly absorbed the interstitial fluid, creating continuous conduits between the dermal microcirculation and the attached patch-type drug reservoirs, which sustainedly released the drug. In another report, HFMs (two layers) assembled from a lyophilized drug reservoir layer and a microneedle layer consisting of 20% (*w/w*) poly(methyl vinyl-maleic acid) crosslinked by esterification with 7.5% (*w/w*) poly(ethylene glycol) (Mw 10,000 Da) significantly enhanced the penetration of metformin hydrochloride in subcutaneous neonatal pig skin in vitro. The combined HFMs delivered 27.6-fold and 71.2-fold more drug over 6 h and 24 h, respectively, than controls using only drug reservoirs [12]. In addition, on-demand drug release can be achieved using HFMs through stimulation control. Hardy et al. prepared HFMs using the light-responsive materials 2-hydroxyethyl methacrylate (HEMA) and ethylene glycol dimethacrylate (EGDMA) to achieve on-demand delivery of ibuprofen [39]. In another report, the glucose-sensitive molecule 4-(2-acrylamidoethylcarbamoyl)-3-fluorophenylboronic acid (AFPBA) was chosen for integration with the hydrogel scaffold for its suitable equilibrium association constant with glucose (Figure 2C). After immersion in a glucose solution, the pore size of Gel-AFPBA-ins hydrogel (a hydrogel in which GelMA and AFPBA are cross-linked and then copolymerized with insulin) (Figure 2B) increased and insulin was released faster and more with an increasing glucose solution concentration (Figure 3) [54].

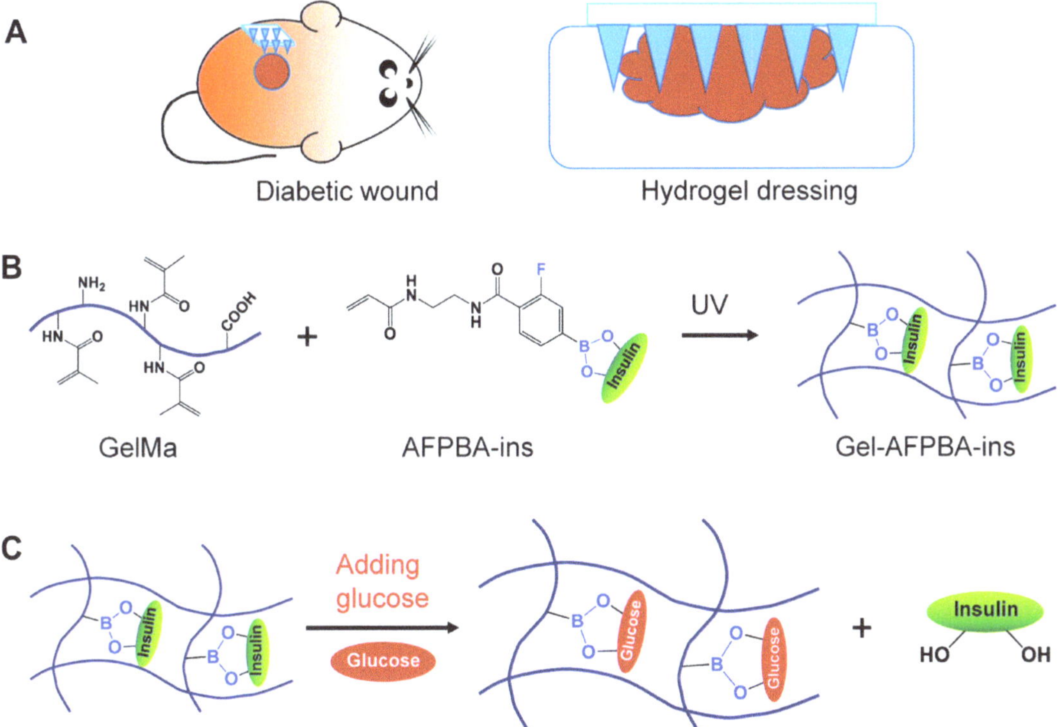

Figure 2. Schematic representation of responsive microneedle dressing for diabetic wound healing. (**A**) Diabetic wounds in mice treated with the hydrogel-based microneedle dressing. (**B**) Preparation of glucose-responsive insulin-releasing Gel–AFPBA–ins hydrogels. (**C**) Mechanism of glucose-responsive insulin release from the prepared hydrogels. The figure was cited from ref. [54] with permission from the Royal Society of Chemistry.

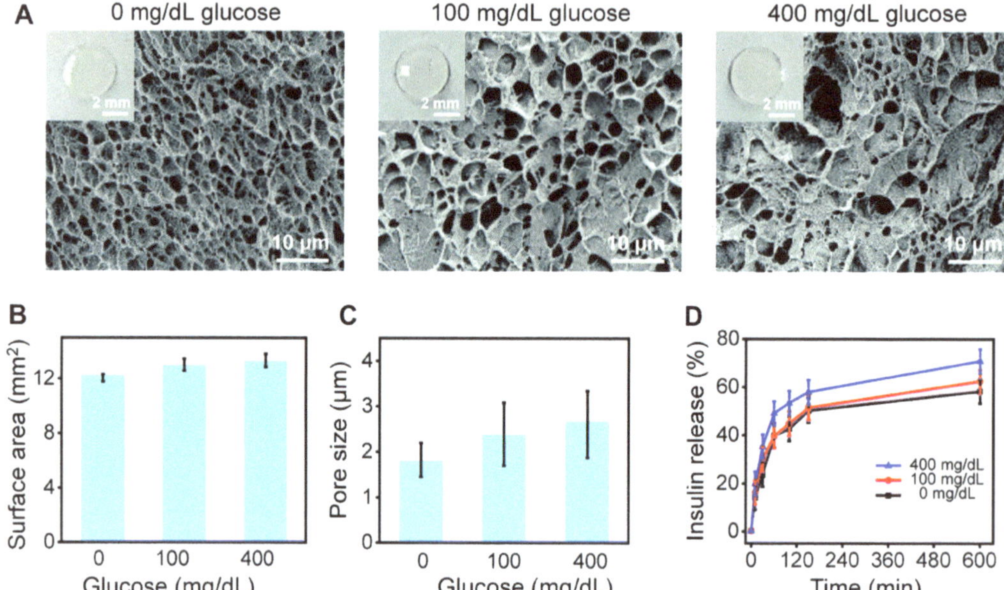

Figure 3. Schematic representation of glucose responsiveness and insulin release of the Gel–AFPBA–ins hydrogels. (**A**) Scanning electron microscope images of the cross-sectional morphology of the hydrogels after reaction with different glucose solutions. The insets show the appearance of the corresponding hydrogels. (**B**) Surface area and (**C**) average pore size of the hydrogels after reaction with different glucose solutions. (**D**) Insulin release kinetics from the hydrogels in different glucose solutions. Mean ± s.d. (n = 5 for 3 repetitions). The figure was cited from ref. [54] with permission from the Royal Society of Chemistry.

4. Evaluation Methods for HFMs

The materials, design, and preparation process of HFMs are important parameters in determining the properties of microneedles, while effective drug delivery also depends on the mechanical strength, skin penetration and release kinetics of HFMs.

4.1. Appearance and Morphology

The morphology and dimensions of HFMs (including the tip radius, height, width, length and spacing) can be characterized by optical microscopy, scanning electron microscopy (SEM) or optical coherence tomography (OCT), confocal laser scanning microscopy (CLSM), and multiphoton microscopy (MPM).

Optical microscopy and SEM are commonly used to image and measure the morphology of HFM arrays and the height, width and spacing of microneedles [55]. OCT imaging is highly accurate, has a certain imaging depth and imaging speed, and is often used to observe, in situ, the penetration depth of microneedle patches after puncturing into isolated or in vivo skin or to record the process of microneedle changes within the skin [56,57]. By loading the microneedles with fluorescent dyes similar to the physicochemical properties of the drug, the distribution of the drug in the microneedles can be assessed by CLSM [58].

4.2. Swellability and Water Insolubility

The swelling properties of HFMs were determined by placing the microneedle array in distilled water or PBS and removing and weighing at specific time intervals to calculate the percentage of swelling [59]. Ex vivo skin such as porcine skin was also used, with the subcutaneous tissue layer carefully removed and the skin placed on tissue paper equilibrated with PBS (pH 7.4). HFM patches were punctured into the isolated skin and then removed at specific time intervals and their base-width swelling capacity was measured using digital microscopy [60].

Water insolubility is an important property of HFMs. The solubility of HFMs was calculated by swelling them sufficiently and then placing them at 90 °C to dry completely to a constant weight and comparing the weight before swelling with the weight after swelling and drying to a constant weight [59].

4.3. Mechanical Strength

The shape of the microneedle determines how much force can be applied to the microneedle before the needle breaks. The diameter and angle of the needle tip, as well as the height and basal measurement of the microneedle, determine whether the microneedle can be safely and reliably inserted into the skin [61]. In general, smaller tip diameters, smaller tip angles and higher tip height-to-width ratios facilitate successful skin penetration. It was found that the average depth of penetration, as determined for the nine microneedles, was significantly higher for the triangular and square base geometries, (340 μm and 343 μm, respectively) than for microneedle arrays with a hexagonal base (197 μm). Accordingly, the average distance between the microneedle base plate and the stratum corneum was estimated at 660 μm, 657 μm and 803 μm for the triangular (34% penetration), square (34% penetration) and hexagonal (20% penetration) base geometries, respectively [62]. Mechanical strength is generally tested using a texturizer or a motorized force-measuring table [63,64]. For fracture testing, arrays of microneedles are microscopically observed before and after testing to determine height differences.

4.4. Skin Piercing and Transdermal Permeation Properties

Microneedles act on the skin surface, puncturing the epidermis and creating microscopic pores through which the drug diffuses into the dermal microcirculation. The success of microneedle puncture can be assessed using a paraffin membrane or porcine skin. The porcine skin has similar physical properties to human skin and can be used as a simulated human skin model [63,65,66]. When conducting relevant experiments, the skin was first washed with PBS (pH 7.4), and then the skin was placed dermally downwards on a wax sheet [67]. The HFMs were then pressed into the skin with the thumb for 30 s. The microneedle arrays were removed from the skin and stained with 150 μL of 1% methylene blue solution for 5 min to assess the position of the stained microneedle pinholes. Excess staining solution was gently washed away with PBS. The stained skin was imaged with a digital microscope and the percentage of stained blue microneedles was calculated to assess the skin puncture performance of the microneedle arrays. The 100% success rate indicated that all microneedle arrays would be observed in the skin [68]. In general, parameters such as the microneedle tip diameter, basal width, length, type of microneedle and its mechanical strength play a crucial role in forming the size of the microchannel in the skin [69].

OCT can be used for in situ observation of the depth of microneedle puncture into the skin in vitro. Kaiyue et al. inserted microneedle patches into rat skin in vitro and imaged the microneedle patches together with the treated skin with OCT; the microneedle tips reach a depth of about 300 μm into the skin and do not break during the insertion process [58].

Fluorescence microscopy can be used to examine the distribution and accumulation of the drug in the skin. Using fluorescence imaging, Aljuffali et al. observed that after transdermal administration, fluorescence was only detected on the skin surface in the free fluorescent probe group and only a weak fluorescent signal was present in the hair follicles, whereas fluorescence was significantly enhanced in the skin of the fluorescently labelled nanocarrier group, suggesting a pro-permeation effect of the nanocarriers [70].

When the drug itself is fluorescent or the drug delivery system is labelled with fluorescence, CLSM is often used to observe fluorescence at different skin depths, allowing verification of the depth of penetration of the agent into the skin tissue and visualization of the accumulation of the agent in the skin tissue. Alvarez-Roman et al. used CLSM to determine the penetration, distribution and accumulation of polymeric nanoparticles in isolated porcine skin [71]. Moreover, by loading coumarin 6 and rhodamine B into the inter-layer and tip-layer of the microneedles, respectively, and inserting into the skin of the knee joint of hairless rats for 30 min, the depth and distribution of the microneedles in the skin were evaluated by tracing the fluorescence of coumarin 6 and rhodamine B by using CLSM and performing 3D reconstruction [58].

MPM is suitable for the characterization of human skin and allows the assessment of skin morphology and layers at a subcellular level. The two-photon excitation principle overcomes the limitations of fluorescence imaging and allows for in vivo non-toxic manipulation. Excitation occurs almost exclusively at the target inspection site without damaging surrounding tissue [72]. MPM also extends the applicability of fluorescence lifetime imaging microscopy (FLIM), and MPM-FLIM allows non-invasive, high-resolution examination of human skin for in vitro, ex vivo, and even clinical in vivo applications [73]. MPM has been used to evaluate the pathophysiological features of inflamed skin, skin permeation and delivery of drugs [74–76].

4.5. In Vitro Release and Transdermal Behaviour

The in vitro TDD can be assessed by Franz diffusion in the donor compartment of the diffusion cell, with the stratum corneum of porcine skin fixed face up to the receiving cell, with PBS (pH 7.4) kept constantly at 37 °C as the receiving medium [77]. The microneedle arrays were applied to the isolated skin and samples were taken from the receiving cell at set intervals. For measuring in vitro drug release, microneedles are placed in PBS (pH 7.4, 37 °C), and samples are taken at set intervals to determine drug concentrations. Skin permeation of drugs can also be evaluated by in vivo animal models, often in suitable rats or mice. The hair of the anaesthetized animal is removed and the skin is then punctured using a microneedle patch, whilst other parameters associated with drug efficacy can be assessed, such as the microneedle strength, permeation efficiency and irritation [78].

It has been noted that skin structure and immune responses in animal models differ significantly from those in humans. In addition, the biochemical properties of ex vivo human skin are different compared to in vivo human skin [79]. Therefore, human trials need to be included in the study when conducting pharmacodynamic studies [80].

4.6. Biosafety and Stability

Biosafety and stability are two important issues that limit the widespread use of HFMs. One of the safety aspects of HFM systems for clinical use is biocompatibility. To ensure that HFM products are acceptable for human exposure, several tests are required to assess their biocompatibility, based on exposure times of less than 24 h, 24 to 30 h, and more than 30 h [81]. Cytotoxicity, sensitization, irritation, intracutaneous reactivity tests, genotoxicity and subacute/subchronic systematic toxicity tests are recommended for the different periods of HFM use [81].

The hemolytic assay is one of the early ways to assess toxicity [82]. Elim et al. used red blood cells from rats to evaluate the hemolytic of HFMs, and no hemolytic was observed, indicating that the materials used were haemocompatible [40]. Vicente-Perez et al. investigated the effect of repeated application HFMs arrays prepared by Gantrez® S-97 BF and polyethylene glycol in the mouse skin, which displayed mild erythema, but did not stimulate the humoral immune system or cause infection or trigger an inflammatory response cascade [83]. Al-Kasasbeh et al. demonstrated for the first time in human volunteers that repeat HFM application and wear did not induce prolonged skin reactions or prolonged disruption of skin barrier function. Importantly, concentrations of specific systemic biomarkers of inflammation (C-reactive protein (CRP); tumor necrosis factor-α (TNF-α)), infection (interleukin-1β (IL-1β), allergy (immunoglobulin E (IgE)) and immunity (immunoglobulin G (IgG))) were all recorded over the course of this fixed study period. No biomarker concentrations above the normal, documented adult ranges were recorded over the course of the study, indicating that no systemic reactions were initiated in volunteers [84].

The stability of HFMs can be evaluated to ensure that active ingredients are protected during storage. This is usually done by storing HFMs and their cargo at various temperatures, including −25 °C, 4 °C, 20 °C, 40 °C and 60 °C, followed by analytical assessments. Generally, the protein cargo of HFMs has better storage stability and a longer shelf-life due to the rigid glassy microneedle matrices restraining molecular mobility and limiting access to atmospheric oxygen. Water should be particularly focused when non-vacuum storage conditions are present, as they can not only destroy the stability of cargo but also the mechanical properties of the HFMs themselves [85].

5. Application of HFMs in Disease Treatment

HFMs have been widely used for the treatment of various diseases, such as cardiovascular diseases, metabolism-related diseases and cancer, due to their outstanding advantages mentioned above.

5.1. Anticancer

Using HFMs for transdermal anticancer drug delivery can overcome the disadvantages of low bioavailability and side effects of oral administration and can also be used for the local administration of drugs for the treatment of superficial tumors such as melanoma, improving bioavailability while avoiding systemic exposure of the drug. Taking advantage of the abundant immune cells including antigen-presenting cells and Langerhans cells in the epidermis and dermis, the activation of the skin's immune microenvironment can act synergistically with the drugs delivered by HFMs.

Chen et al. prepared HFMs with cross-linking polyvinylpyrrolidone (PVP) and PVA as a matrix, loaded with 1-methyltryptophan and indocyanine green-encapsulated nanoparticles for the treatment of melanoma [86]. This system successfully induced immunogenic cell death, enhanced immune response and provided a promising melanoma treatment (Figure 4).

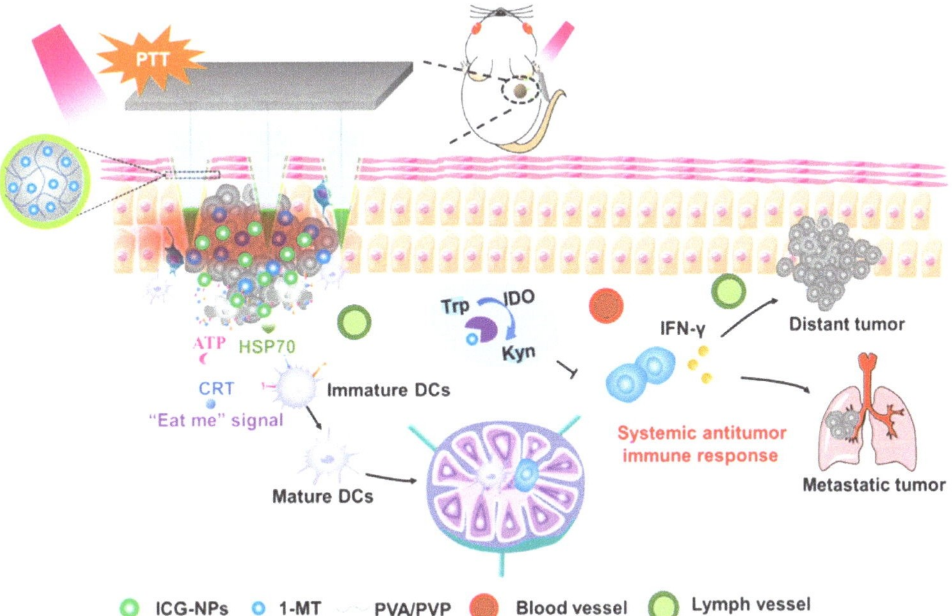

Figure 4. Schematic illustration of the mechanism of antitumor immunity. The figure was cited from ref. [61] with permission from the American Chemical Society.

Huang et al. prepared HFMs loaded with doxorubicin (DOX) and trametinib (Tra) using photo-cross-linked dextrose methacrylate (DexMA) as the microneedle matrix and successfully achieved the slow release of the drugs and exploited the synergistic effect of DOX and Tra (Figure 5) [87].

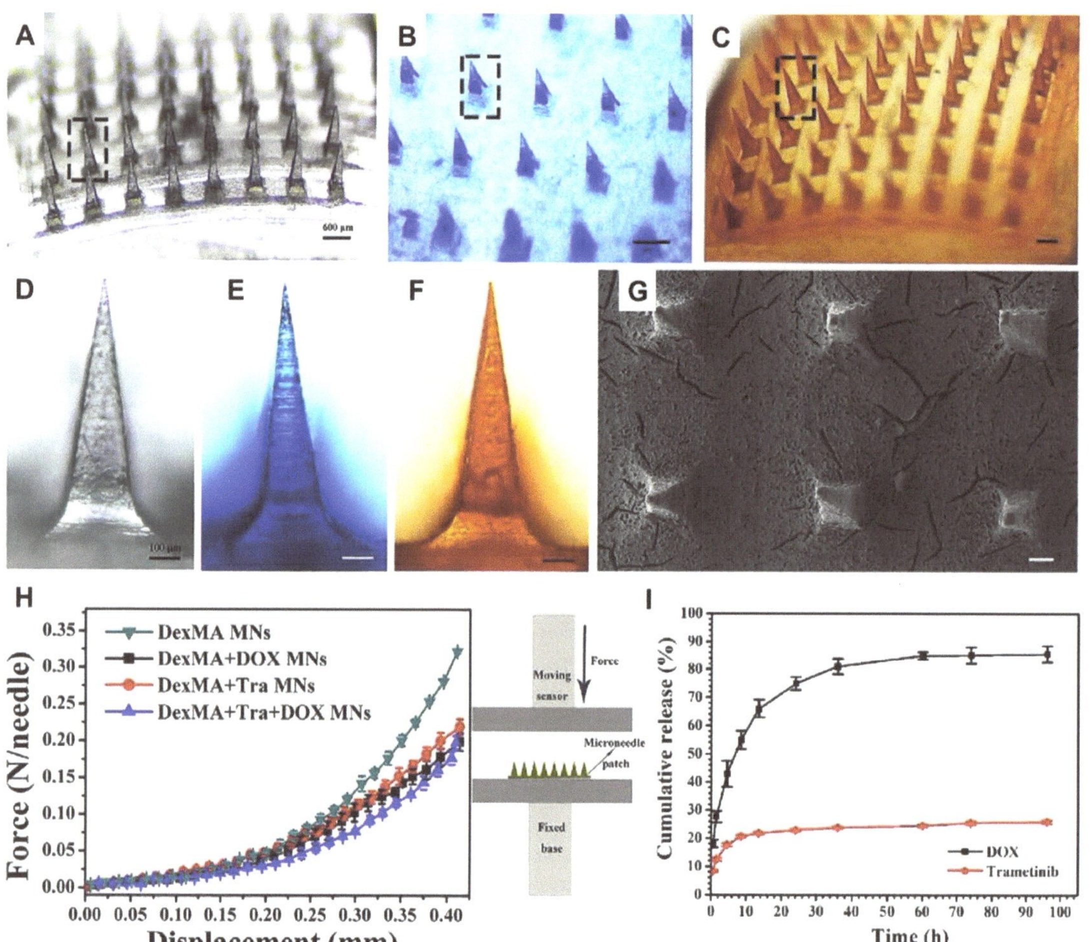

Figure 5. Characterization of DexMA hydrogel microneedles (MNs). (**A–C**) Images of blank MNs, methylene blue-loaded MNs, and DOX-loaded MNs, respectively. (**D–F**) Images of a single microneedle corresponding to ((**A–C**), marked with a dotted box), respectively. (**G**) SEM image of MNs. (**H**) MNs mechanical property. (**I**) In vitro drug release from MNs. The figure was adopted from ref. [87] with permission from Elsevier, Limited.

5.2. Treating Diabetes

HFMs are a promising drug delivery system for the treatment of diabetes because they are minimally invasive, painless, have no microneedle matrix residue and can be repeatedly administered multiple times.

Chen et al. used silk protein and phenylboronic acid/acrylamide as a microneedle matrix loaded with insulin to prepare glucose-responsive smart HFMs [88]. After the microneedles penetrated the skin, insulin was released autonomously to control the blood glucose concentration when the glucose concentration in the skin tissue increased. The

HFMs also retain their original needle shape after a week in water, offering the potential for safe, residue-free and sustained drug release.

Wang et al. fabricated HFMs by using PVA as a microneedle matrix, loaded with glucose oxidase (core) and catalase (shell) and loaded 4-nitrophenyl 4-(4,4,5,5-tetramethyl-1,3,2-dioxol-2yl) benzyl carbonate (insulin NBC) modified insulin into PVA, and PVA was further gelated by an H_2O_2-labile linker: N^1-(4-boronobenzyl)-N^3-(4-boronophenyl)-N^1,N^1,N^3,N^3-tetramethylpropane-1,3-diaminium (TSPBA) [89]. When microneedles were exposed to high glucose concentrations, local high levels of H_2O_2 were produced and insulin NBC was oxidized and hydrolyzed, leading to the rapid release of free insulin and control of blood glucose concentrations (Figure 6).

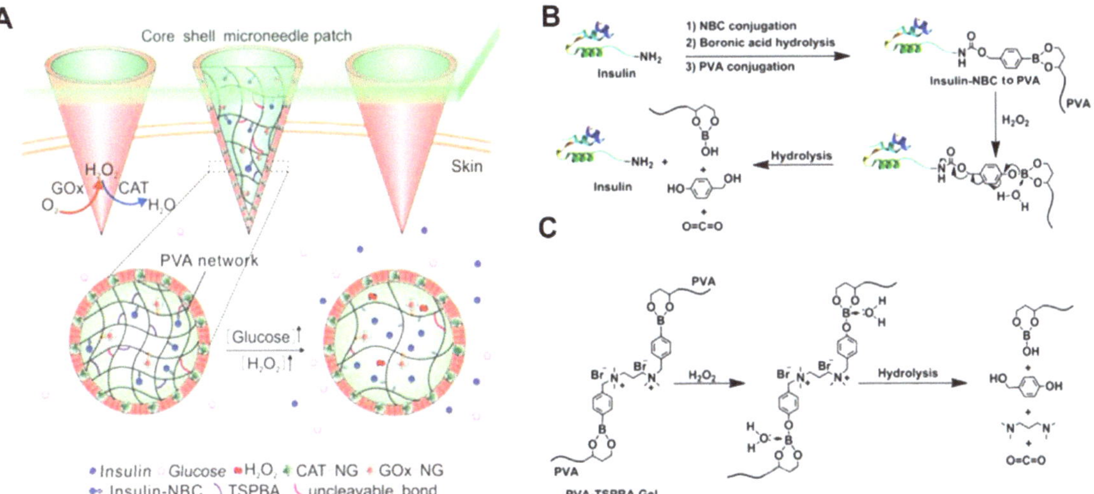

Figure 6. Schematic representation of the glucose-responsive insulin delivery system using H_2O_2-responsive PVA-TSPBA gel. (**A**) Insulin is triggered to release by a hyperglycemic state from the core matrix of the PVA-TSPBA MN patch, and the local inflammation can be greatly reduced by the catalase-embedded PVA-TSPBA shell. (**B**) Modification of insulin with NBC and H_2O_2-responsive release. (**C**) H_2O_2 responsiveness mechanism. The figure was adopted from ref. [89] with permission from the American Chemical Society.

5.3. Treating Rheumatoid Arthritis

Rheumatoid arthritis is a systemic disease involving multiple joints. Early drug treatment is mostly oral administration, but long-term oral anti-rheumatoid arthritis medication often brings about serious side effects, while local injection drug treatment methods not only require specialist handling but may also pose the risk of joint damage and infection. HFMs are suitable for the delivery of drugs for this disease because of their painless and minimally invasive delivery of various active molecules.

Cao et al. designed modified hyaluronic-acid-fabricated HFMs, loaded with reverse deoxythymidine and cholesterol-modified deoxythymidine, which had a protective effect against cartilage/bone erosion in mice joints [90]. Compared to dissolving microneedles, HFMs not only increased the loading of nucleic acid aptamers into the cavity of the microneedle mold but also allowed the loading of HFMs to be controlled by adjusting the concentration of the aptamer solution, avoiding waste and loss of aptamers during preparation.

6. Summary and Outlook

As a new type of TDD method, HFMs have been increasingly researched for TDD, mainly in the treatment of cardiovascular diseases, tumors and diabetes mellitus, with the greatest advantage being the controlled release of the drug and the targeted drug delivery in the original lesion. In addition, the HFMs are prepared with certain functional modifications, such as some photothermal materials (e.g., gold nanorods, Prussian blue, and indocyanine green) and photosensitizers (e.g., protoporphyrin, zinc titanocyanine, and titanium dioxide), which induce the HFMs to produce exogenous stimuli (changes in temperature, magnetic fields, and light) or endogenous stimuli (changes in pH, enzymes, and redox gradients), while combining the delivered drug molecules, antibodies, nucleic acids, etc., to achieve targeted treatment of diseases. However, HFM development still faces potential biosafety issues during prolonged use, batch industrial production and sterilization challenges, which poses a huge challenge for their clinical application and is a hot topic for future research.

Author Contributions: Conceptualization, Y.Z. and N.F.; methodology, X.H. and J.L.; software, X.H., J.L. and H.R.; validation, J.L. and M.L.; investigation; Y.Z., N.F. and Y.H.; writing—original draft preparation, X.H. and J.L.; writing—review and editing, Y.Z., N.F. and Y.H.; supervision, Y.Z., N.F., H.R. and M.L.; project administration, X.H. and J.L.; funding acquisition, Y.Z., N.F. and Y.H. All authors have read and agreed to the published version of the manuscript.

Funding: This work was supported by the National Natural Science Foundation of China (Grant number 82074031), the Program for Professor of Special Appointment (Eastern Scholar) at Shanghai Institutions of Higher Learning (Grant number TP2020054), and the Program for Shanghai High-Level Local University Innovation Team (SZY20220315).

Institutional Review Board Statement: Not applicable.

Informed Consent Statement: Not applicable.

Data Availability Statement: Not applicable.

Conflicts of Interest: The authors declare no conflict of interest.

References

1. Alkilani, A.Z.; McCrudden, M.T.; Donnelly, R.F. Transdermal Drug Delivery: Innovative Pharmaceutical Developments Based on Disruption of the Barrier Properties of the stratum corneum. *Pharmaceutics* **2015**, *7*, 438–470. [CrossRef] [PubMed]
2. Yin, Z.; Kuang, D.; Wang, S.; Zheng, Z.; Yadavalli, V.K.; Lu, S. Swellable silk fibroin microneedles for transdermal drug delivery. *Int. J. Biol. Macromol.* **2018**, *106*, 48–56. [CrossRef] [PubMed]
3. Verbaan, F.J.; Bal, S.M.; van den Berg, D.J.; Groenink, W.H.; Verpoorten, H.; Luttge, R.; Bouwstra, J.A. Assembled microneedle arrays enhance the transport of compounds varying over a large range of molecular weight across human dermatomed skin. *J. Control. Release* **2007**, *117*, 238–245. [CrossRef] [PubMed]
4. Kanikkannan, N. Iontophoresis-based transdermal delivery systems. *BioDrugs* **2002**, *16*, 339–347. [CrossRef] [PubMed]
5. Denet, A.R.; Vanbever, R.; Preat, V. Skin electroporation for transdermal and topical delivery. *Adv. Drug Deliv. Rev.* **2004**, *56*, 659–674. [CrossRef]
6. Wenande, E.; Tam, J.; Bhayana, B.; Schlosser, S.K.; Ishak, E.; Farinelli, W.A.; Chlopik, A.; Hoang, M.P.; Pinkhasov, O.R.; Caravan, P.; et al. Laser-assisted delivery of synergistic combination chemotherapy in in vivo skin. *J. Control. Release* **2018**, *275*, 242–253. [CrossRef]
7. Svenskaya, Y.I.; Genina, E.A.; Parakhonskiy, B.V.; Lengert, E.V.; Talnikova, E.E.; Terentyuk, G.S.; Utz, S.R.; Gorin, D.A.; Tuchin, V.V.; Sukhorukov, G.B. A Simple Non-Invasive Approach toward Efficient Transdermal Drug Delivery Based on Biodegradable Particulate System. *ACS Appl. Mater. Interfaces* **2019**, *11*, 17270–17282. [CrossRef]
8. Azagury, A.; Khoury, L.; Enden, G.; Kost, J. Ultrasound mediated transdermal drug delivery. *Adv. Drug Deliv. Rev.* **2014**, *72*, 127–143. [CrossRef]
9. Polat, B.E.; Hart, D.; Langer, R.; Blankschtein, D. Ultrasound-mediated transdermal drug delivery: Mechanisms, scope, and emerging trends. *J. Control. Release* **2011**, *152*, 330–348. [CrossRef]
10. Courtenay, A.J.; McCrudden, M.T.C.; McAvoy, K.J.; McCarthy, H.O.; Donnelly, R.F. Microneedle-Mediated Transdermal Delivery of Bevacizumab. *Mol. Pharm.* **2018**, *15*, 3545–3556. [CrossRef]
11. Donnelly, R.F.; McCrudden, M.T.C.; Alkilani, A.Z.; Larraneta, E.; McAlister, E.; Courtenay, A.J.; Kearney, M.C.; Singh, T.R.R.; McCarthy, H.O.; Kett, V.L.; et al. Hydrogel-Forming Microneedles Prepared from "Super Swelling" Polymers Combined with Lyophilised Wafers for Transdermal Drug Delivery. *PLoS ONE* **2014**, *9*, e111547. [CrossRef]

12. Migdadi, E.M.; Courtenay, A.J.; Tekko, I.A.; McCrudden, M.T.C.; Kearney, M.C.; McAlister, E.; McCarthy, H.O.; Donnelly, R.F. Hydrogel-forming microneedles enhance transdermal delivery of metformin hydrochloride. *J. Control. Release* **2018**, *285*, 142–151. [CrossRef]

13. Koutsonanos, D.G.; del Pilar Martin, M.; Zarnitsyn, V.G.; Sullivan, S.P.; Compans, R.W.; Prausnitz, M.R.; Skountzou, I. Transdermal influenza immunization with vaccine-coated microneedle arrays. *PLoS ONE* **2009**, *4*, e4773. [CrossRef]

14. Lee, S.J.; Lee, H.S.; Hwang, Y.H.; Kim, J.J.; Kang, K.Y.; Kim, S.J.; Kim, H.K.; Kim, J.D.; Jeong, D.H.; Paik, M.J.; et al. Enhanced anti-tumor immunotherapy by dissolving microneedle patch loaded ovalbumin. *PLoS ONE* **2019**, *14*, e0220382. [CrossRef] [PubMed]

15. Paredes, A.J.; Ramoller, I.K.; McKenna, P.E.; Abbate, M.T.A.; Volpe-Zanutto, F.; Vora, L.K.; Kilbourne-Brook, M.; Jarrahian, C.; Moffatt, K.; Zhang, C.; et al. Microarray patches: Breaking down the barriers to contraceptive care and HIV prevention for women across the globe. *Adv. Drug Deliv. Rev.* **2021**, *173*, 331–348. [CrossRef] [PubMed]

16. Volpe-Zanutto, F.; Ferreira, L.T.; Permana, A.D.; Kirkby, M.; Paredes, A.J.; Vora, L.K.; Bonfanti, A.P.; Charlie-Silva, I.; Raposo, C.; Figueiredo, M.C.; et al. Artemether and lumefantrine dissolving microneedle patches with improved pharmacokinetic performance and antimalarial efficacy in mice infected with Plasmodium yoelii. *J. Control. Release* **2021**, *333*, 298–315. [CrossRef] [PubMed]

17. Vora, L.K.; Moffatt, K.; Tekko, I.A.; Paredes, A.J.; Volpe-Zanutto, F.; Mishra, D.; Peng, K.; Raj Singh Thakur, R.; Donnelly, R.F. Microneedle array systems for long-acting drug delivery. *Eur. J. Pharm. Biopharm.* **2021**, *159*, 44–76. [CrossRef]

18. Donnelly, R.; Douroumis, D. Microneedles for drug and vaccine delivery and patient monitoring. *Drug Deliv. Transl. Res.* **2015**, *5*, 311–312. [CrossRef]

19. Zhang, Y.; Yu, J.; Kahkoska, A.R.; Wang, J.; Buse, J.B.; Gu, Z. Advances in transdermal insulin delivery. *Adv. Drug Deliv. Rev.* **2019**, *139*, 51–70. [CrossRef]

20. Rzhevskiy, A.S.; Singh, T.R.R.; Donnelly, R.F.; Anissimov, Y.G. Microneedles as the technique of drug delivery enhancement in diverse organs and tissues. *J. Control. Release* **2018**, *270*, 184–202. [CrossRef]

21. Huang, Y.; Park, Y.S.; Moon, C.; David, A.E.; Chung, H.S.; Yang, V.C. Synthetic skin-permeable proteins enabling needleless immunization. *Angew. Chem. Int. Ed. Engl.* **2010**, *49*, 2724–2727. [CrossRef]

22. Ye, Y.; Yu, J.; Wen, D.; Kahkoska, A.R.; Gu, Z. Polymeric microneedles for transdermal protein delivery. *Adv. Drug Deliv. Rev.* **2018**, *127*, 106–118. [CrossRef]

23. McAlister, E.; Dutton, B.; Vora, L.K.; Zhao, L.; Ripolin, A.; Zahari, D.; Quinn, H.L.; Tekko, I.A.; Courtenay, A.J.; Kelly, S.A.; et al. Directly Compressed Tablets: A Novel Drug-Containing Reservoir Combined with Hydrogel-Forming Microneedle Arrays for Transdermal Drug Delivery. *Adv. Healthc. Mater.* **2021**, *10*, e2001256. [CrossRef]

24. Li, M.; Vora, L.K.; Peng, K.; Donnelly, R.F. Trilayer microneedle array assisted transdermal and intradermal delivery of dexamethasone. *Int. J. Pharm.* **2022**, *612*, 121295. [CrossRef]

25. Yadav, P.R.; Dobson, L.J.; Pattanayek, S.K.; Das, D.B. Swellable microneedles based transdermal drug delivery: Mathematical model development and numerical experiments. *Chem. Eng. Sci.* **2022**, *247*, 117005. [CrossRef]

26. Donnelly, R.F.; Singh, T.R.; Garland, M.J.; Migalska, K.; Majithiya, R.; McCrudden, C.M.; Kole, P.L.; Mahmood, T.M.; McCarthy, H.O.; Woolfson, A.D. Hydrogel-Forming Microneedle Arrays for Enhanced Transdermal Drug Delivery. *Adv. Funct. Mater.* **2012**, *22*, 4879–4890. [CrossRef]

27. Lutton, R.E.; Larraneta, E.; Kearney, M.C.; Boyd, P.; Woolfson, A.D.; Donnelly, R.F. A novel scalable manufacturing process for the production of hydrogel-forming microneedle arrays. *Int. J. Pharm.* **2015**, *494*, 417–429. [CrossRef]

28. Ita, K. Transdermal Delivery of Drugs with Microneedles-Potential and Challenges. *Pharmaceutics* **2015**, *7*, 90–105. [CrossRef]

29. Donnelly, R.F.; Mooney, K.; Mccrudden, M.T.C.; Vicente-Perez, E.M.; Belaid, L.; Gonzalez-Vazquez, P.; Mcelnay, J.C.; Woolfson, A.D. Hydrogel-Forming Microneedles Increase in Volume During Swelling in Skin, but Skin Barrier Function Recovery is Unaffected. *J. Pharm. Sci.* **2014**, *103*, 1478–1486. [CrossRef]

30. Martin, C.J.; Allender, C.J.; Brain, K.R.; Morrissey, A.; Birchall, J.C. Low temperature fabrication of biodegradable sugar glass microneedles for transdermal drug delivery applications. *J. Control. Release* **2012**, *158*, 93–101. [CrossRef]

31. Zhang, Y.S.; Khademhosseini, A. Advances in engineering hydrogels. *Science* **2017**, *356*, eaaf3627. [CrossRef] [PubMed]

32. Rajoli, R.K.R.; Flexner, C.; Chiong, J.; Owen, A.; Donnelly, R.F.; Larraneta, E.; Siccardi, M. Modelling the intradermal delivery of microneedle array patches for long-acting antiretrovirals using PBPK. *Eur. J. Pharm. Biopharm.* **2019**, *144*, 101–109. [CrossRef] [PubMed]

33. Eltayib, E.; Brady, A.J.; Caffarel-Salvador, E.; Gonzalez-Vazquez, P.; Zaid Alkilani, A.; McCarthy, H.O.; McElnay, J.C.; Donnelly, R.F. Hydrogel-forming microneedle arrays: Potential for use in minimally-invasive lithium monitoring. *Eur. J. Pharm. Biopharm.* **2016**, *102*, 123–131. [CrossRef]

34. Dimatteo, R.; Darling, N.J.; Segura, T. In situ forming injectable hydrogels for drug delivery and wound repair. *Adv. Drug Deliv. Rev.* **2018**, *127*, 167–184. [CrossRef] [PubMed]

35. Singh, T.R.R.; McCarron, P.A.; Woolfson, A.D.; Donnelly, R.F. Investigation of swelling and network parameters of poly(ethylene glycol)-crosslinked poly(methyl vinyl ether-co-maleic acid) hydrogels. *Eur. Polym. J.* **2009**, *45*, 1239–1249. [CrossRef]

36. Tsou, T.L.; Tang, S.T.; Huang, Y.C.; Wu, J.R.; Young, J.J.; Wang, H.J. Poly(2-hydroxyethyl methacrylate) wound dressing containing ciprofloxacin and its drug release studies. *J. Mater. Sci. Mater. Med.* **2005**, *16*, 95–100. [CrossRef]

37. Chew, S.W.T.; Shah, A.H.; Zheng, M.; Chang, H.; Wiraja, C.; Steele, T.W.J.; Xu, C. A self-adhesive microneedle patch with drug loading capability through swelling effect. *Bioeng. Transl. Med.* **2020**, *5*, e10157. [CrossRef]
38. Tsioris, K.; Raja, W.K.; Pritchard, E.M.; Panilaitis, B.; Kaplan, D.L.; Omenetto, F.G. Fabrication of Silk Microneedles for Controlled-Release Drug Delivery. *Adv. Funct. Mater.* **2012**, *22*, 330–335. [CrossRef]
39. Hardy, J.G.; Larraneta, E.; Donnelly, R.F.; McGoldrick, N.; Migalska, K.; McCrudden, M.T.; Irwin, N.J.; Donnelly, L.; McCoy, C.P. Hydrogel-Forming Microneedle Arrays Made from Light-Responsive Materials for On-Demand Transdermal Drug Delivery. *Mol. Pharm.* **2016**, *13*, 907–914. [CrossRef]
40. Elim, D.; Fitri, A.M.N.; Mahfud, M.A.S.; Afika, N.; Sultan, N.A.F.; Hijrah; Asri, R.M.; Permana, A.D. Hydrogel forming microneedle-mediated transdermal delivery of sildenafil citrate from polyethylene glycol reservoir: An ex vivo proof of concept study. *Colloids Surf. B Biointerfaces* **2023**, *222*, 113018. [CrossRef]
41. Yu, M.; Lu, Z.; Shi, Y.; Du, Y.T.; Chen, X.G.; Kong, M. Systematic comparisons of dissolving and swelling hyaluronic acid microneedles in transdermal drug delivery. *Int. J. Biol. Macromol.* **2021**, *191*, 783–791. [CrossRef]
42. Sabbagh, F.; Kim, B.S. Ex Vivo Transdermal Delivery of Nicotinamide Mononucleotide Using Polyvinyl Alcohol Microneedles. *Polymers* **2023**, *15*, 2031. [CrossRef]
43. Ruan, S.; Zhang, Y.; Feng, N. Microneedle-mediated transdermal nanodelivery systems: A review. *Biomater. Sci.* **2021**, *9*, 8065–8089. [CrossRef]
44. Sabbagh, F.; Kim, B.S. Recent advances in polymeric transdermal drug delivery systems. *J. Control. Release* **2022**, *341*, 132–146. [CrossRef]
45. Romanyuk, A.V.; Zvezdin, V.N.; Samant, P.; Grenader, M.I.; Zemlyanova, M.; Prausnitz, M.R. Collection of analytes from microneedle patches. *Anal. Chem.* **2014**, *86*, 10520–10523. [CrossRef]
46. Chang, H.; Zheng, M.; Yu, X.; Than, A.; Seeni, R.Z.; Kang, R.; Tian, J.; Khanh, D.P.; Liu, L.; Chen, P.; et al. A Swellable Microneedle Patch to Rapidly Extract Skin Interstitial Fluid for Timely Metabolic Analysis. *Adv. Mater.* **2017**, *29*, 1702243. [CrossRef]
47. Caffarel-Salvador, E.; Brady, A.J.; Eltayib, E.; Meng, T.; Alonso-Vicente, A.; Gonzalez-Vazquez, P.; Torrisi, B.M.; Vicente-Perez, E.M.; Mooney, K.; Jones, D.S.; et al. Hydrogel-Forming Microneedle Arrays Allow Detection of Drugs and Glucose In Vivo: Potential for Use in Diagnosis and Therapeutic Drug Monitoring. *PLoS ONE* **2015**, *10*, e0145644. [CrossRef]
48. Kearney, M.C.; Caffarel-Salvador, E.; Fallows, S.J.; McCarthy, H.O.; Donnelly, R.F. Microneedle-mediated delivery of donepezil: Potential for improved treatment options in Alzheimer's disease. *Eur. J. Pharm. Biopharm.* **2016**, *103*, 43–50. [CrossRef]
49. He, R.; Niu, Y.; Li, Z.; Li, A.; Yang, H.; Xu, F.; Li, F. A Hydrogel Microneedle Patch for Point-of-Care Testing Based on Skin Interstitial Fluid. *Adv. Healthc. Mater.* **2020**, *9*, e1901201. [CrossRef]
50. Yang, S.; Feng, Y.; Zhang, L.; Chen, N.; Yuan, W.; Jin, T. A scalable fabrication process of polymer microneedles. *Int. J. Nanomed.* **2012**, *7*, 1415–1422. [CrossRef]
51. Kamoun, E.A.; Chen, X.; Eldin, M.S.M.; Kenawy, E.R.S. Crosslinked poly(vinyl alcohol) hydrogels for wound dressing applications: A review of remarkably blended polymers. *Arab. J. Chem.* **2015**, *8*, 1–14. [CrossRef]
52. Yang, S.Y.; O'Cearbhaill, E.D.; Sisk, G.C.; Park, K.M.; Cho, W.K.; Villiger, M.; Bouma, B.E.; Pomahac, B.; Karp, J.M. A bio-inspired swellable microneedle adhesive for mechanical interlocking with tissue. *Nat. Commun.* **2013**, *4*, 1702. [CrossRef] [PubMed]
53. Sabbagh, F.; Kim, B.S. Microneedles for transdermal drug delivery using clay-based composites. *Expert Opin. Drug Deliv.* **2022**, *19*, 1099–1113. [CrossRef]
54. Guo, Z.; Liu, H.; Shi, Z.; Lin, L.; Li, Y.; Wang, M.; Pan, G.; Lei, Y.; Xue, L. Responsive hydrogel-based microneedle dressing for diabetic wound healing. *J. Mater. Chem. B* **2022**, *10*, 3501–3511. [CrossRef] [PubMed]
55. Arsalan, A.; Raya, D.; Hala, D.; Haytam, K.; Aiman, A.A. Preparation and characterization of flexible furosemide-loaded biodegradable microneedles for intradermal drug delivery. *Biomater. Sci.* **2022**, *10*, 6486–6499. [CrossRef]
56. Yang, L.; Yang, Y.; Chen, H.; Mei, L.; Zeng, X. Polymeric microneedle-mediated sustained release systems: Design strategies and promising applications for drug delivery. *Asian J. Pharm. Sci.* **2022**, *17*, 70–86. [CrossRef]
57. Courtenay, A.J.; McAlister, E.; McCrudden, M.T.C.; Vora, L.; Steiner, L.; Levin, G.; Levy-Nissenbaum, E.; Shterman, N.; Kearney, M.C.; McCarthy, H.O.; et al. Hydrogel-forming microneedle arrays as a therapeutic option for transdermal esketamine delivery. *J. Control. Release* **2020**, *322*, 177–186. [CrossRef]
58. Yu, K.Y.; Yu, X.M.; Cao, S.S.; Wang, Y.X.; Zhai, Y.H.; Yang, F.D.; Yang, X.Y.; Lu, Y.; Wu, C.B.; Xu, Y.H. Layered dissolving microneedles as a need-based delivery system to simultaneously alleviate skin and joint lesions in psoriatic arthritis. *Acta Pharm. Sin. B* **2021**, *11*, 505–519. [CrossRef]
59. Suksaeree, J.; Charoenchai, L.; Madaka, F.; Monton, C.; Sakunpak, A.; Charoonratana, T.; Pichayakorn, W. Zingiber cassumunar blended patches for skin application: Formulation, physicochemical properties, and in vitro studies. *Asian J. Pharm. Sci.* **2015**, *10*, 341–349. [CrossRef]
60. Aung, N.N.; Ngawhirunpat, T.; Rojanarata, T.; Patrojanasophon, P.; Pamornpathomkul, B.; Opanasopit, P. Fabrication, characterization and comparison of alpha-arbutin loaded dissolving and hydrogel forming microneedles. *Int. J. Pharm.* **2020**, *586*, 119508. [CrossRef]
61. Bal, S.M.; Caussin, J.; Pavel, S.; Bouwstra, J.A. In vivo assessment of safety of microneedle arrays in human skin. *Eur. J. Pharm. Sci.* **2008**, *35*, 193–202. [CrossRef]
62. Loizidou, E.Z.; Inoue, N.T.; Ashton-Barnett, J.; Barrow, D.A.; Allender, C.J. Evaluation of geometrical effects of microneedles on skin penetration by CT scan and finite element analysis. *Eur. J. Pharm. Biopharm.* **2016**, *107*, 1–6. [CrossRef]

63. Vora, L.K.; Courtenay, A.J.; Tekko, I.A.; Larraneta, E.; Donnelly, R.F. Pullulan-based dissolving microneedle arrays for enhanced transdermal delivery of small and large biomolecules. *Int. J. Biol. Macromol.* **2020**, *146*, 290–298. [CrossRef]
64. He, M.C.; Chen, B.Z.; Ashfaq, M.; Guo, X.D. Assessment of mechanical stability of rapidly separating microneedles for transdermal drug delivery. *Drug Deliv. Transl. Res.* **2018**, *8*, 1034–1042. [CrossRef]
65. Chen, B.Z.; Ashfaq, M.; Zhang, X.P.; Zhang, J.N.; Guo, X.D. In vitro and in vivo assessment of polymer microneedles for controlled transdermal drug delivery. *J. Drug Target.* **2018**, *26*, 720–729. [CrossRef]
66. Lee, I.C.; He, J.S.; Tsai, M.T.; Lin, K.C. Fabrication of a novel partially dissolving polymer microneedle patch for transdermal drug delivery. *J. Mater. Chem. B* **2015**, *3*, 276–285. [CrossRef]
67. Donnelly, R.F.; Morrow, D.I.; McCrudden, M.T.; Alkilani, A.Z.; Vicente-Perez, E.M.; O'Mahony, C.; Gonzalez-Vazquez, P.; McCarron, P.A.; Woolfson, A.D. Hydrogel-forming and dissolving microneedles for enhanced delivery of photosensitizers and precursors. *Photochem. Photobiol.* **2014**, *90*, 641–647. [CrossRef]
68. Donadei, A.; Kraan, H.; Ophorst, O.; Flynn, O.; O'Mahony, C.; Soema, P.C.; Moore, A.C. Skin delivery of trivalent Sabin inactivated poliovirus vaccine using dissolvable microneedle patches induces neutralizing antibodies. *J. Control. Release* **2019**, *311–312*, 96–103. [CrossRef]
69. Kalluri, H.; Banga, A.K. Formation and closure of microchannels in skin following microporation. *Pharm. Res.* **2011**, *28*, 82–94. [CrossRef]
70. Roy, R.K.; Thakur, M.; Dixit, V.K. Hair growth promoting activity of Eclipta alba in male albino rats. *Arch. Dermatol. Res.* **2008**, *300*, 357–364. [CrossRef]
71. Aljuffali, I.A.; Sung, C.T.; Shen, F.M.; Huang, C.T.; Fang, J.Y. Squarticles as a Lipid Nanocarrier for Delivering Diphencyprone and Minoxidil to Hair Follicles and Human Dermal Papilla Cells. *Aaps J.* **2014**, *16*, 140–150. [CrossRef] [PubMed]
72. Wang, W.; Chen, L.; Huang, X.; Shao, A. Preparation and Characterization of Minoxidil Loaded Nanostructured Lipid Carriers. *AAPS PharmSciTech* **2017**, *18*, 509–516. [CrossRef] [PubMed]
73. Yazdani-Arazi, S.N.; Ghanbarzadeh, S.; Adibkia, K.; Kouhsoltani, M.; Hamishehkar, H. Histological evaluation of follicular delivery of arginine via nanostructured lipid carriers: A novel potential approach for the treatment of alopecia. *Artif. Cells Nanomed. Biotechnol.* **2017**, *45*, 1379–1387. [CrossRef]
74. Zhang, L.W.; Yu, W.W.; Colvin, V.L.; Monteiro-Riviere, N.A. Biological interactions of quantum dot nanoparticles in skin and in human epidermal keratinocytes. *Toxicol. Appl. Pharmacol.* **2008**, *228*, 200–211. [CrossRef]
75. Tang, L.; Zhang, C.; Song, G.; Jin, X.; Xu, Z. In vivo skin penetration and metabolic path of quantum dots. *Sci. China Life Sci.* **2013**, *56*, 181–188. [CrossRef] [PubMed]
76. Filipe, P.; Silva, J.N.; Silva, R.; de Castro, J.L.C.; Gomes, M.M.; Alves, L.C.; Santus, R.; Pinheiro, T. Stratum Corneum Is an Effective Barrier to TiO2 and ZnO Nanoparticle Percutaneous Absorption. *Skin Pharmacol. Phys.* **2009**, *22*, 266–275. [CrossRef]
77. Gonzalez-Vazquez, P.; Larraneta, E.; McCrudden, M.T.C.; Jarrahian, C.; Rein-Weston, A.; Quintanar-Solares, M.; Zehrung, D.; McCarthy, H.; Courtenay, A.J.; Donnelly, R.F. Transdermal delivery of gentamicin using dissolving microneedle arrays for potential treatment of neonatal sepsis. *J. Control. Release* **2017**, *265*, 30–40. [CrossRef]
78. Shende, P.; Sardesai, M.; Gaud, R.S. Micro to nanoneedles: A trend of modernized transepidermal drug delivery system. *Artif. Cells Nanomed. Biotechnol.* **2018**, *46*, 19–25. [CrossRef]
79. Coulman, S.A.; Birchall, J.C.; Alex, A.; Pearton, M.; Hofer, B.; O'Mahony, C.; Drexler, W.; Povazay, B. In Vivo, In Situ Imaging of Microneedle Insertion into the Skin of Human Volunteers Using Optical Coherence Tomography. *Pharm. Res. Dordr.* **2011**, *28*, 66–81. [CrossRef]
80. Arya, J.; Henry, S.; Kalluri, H.; McAllister, D.V.; Pewin, W.P.; Prausnitz, M.R. Tolerability, usability and acceptability of dissolving microneedle patch administration in human subjects. *Biomaterials* **2017**, *128*, 1–7. [CrossRef]
81. *ISO 10993-1*; FDA Guidance. Use of International Standards. Biological Evaluation of Medical Devices—Part 1: Evaluation and Testing within a Risk Management Process. FDA: Silver Spring, MD, USA, 2020.
82. Ananda, P.W.R.; Elim, D.; Zaman, H.S.; Muslimin, W.; Tunggeng, M.G.R.; Permana, A.D. Combination of transdermal patches and solid microneedles for improved transdermal delivery of primaquine. *Int. J. Pharm.* **2021**, *609*, 121204. [CrossRef]
83. Vicente-Perez, E.M.; Larraneta, E.; McCrudden, M.T.C.; Kissenpfennig, A.; Hegarty, S.; McCarthy, H.O.; Donnelly, R.F. Repeat application of microneedles does not alter skin appearance or barrier function and causes no measurable disturbance of serum biomarkers of infection, inflammation or immunity in mice in vivo. *Eur. J. Pharm. Biopharm.* **2017**, *117*, 400–407. [CrossRef]
84. Al-Kasasbeh, R.; Brady, A.J.; Courtenay, A.J.; Larraneta, E.; McCrudden, M.T.C.; O'Kane, D.; Liggett, S.; Donnelly, R.F. Evaluation of the clinical impact of repeat application of hydrogel-forming microneedle array patches. *Drug Deliv. Transl. Res.* **2020**, *10*, 690–705. [CrossRef]
85. Chu, L.Y.; Ye, L.; Dong, K.; Compans, R.W.; Yang, C.; Prausnitz, M.R. Enhanced Stability of Inactivated Influenza Vaccine Encapsulated in Dissolving Microneedle Patches. *Pharm. Res.* **2016**, *33*, 868–878. [CrossRef]
86. Chen, M.; Quan, G.; Wen, T.; Yang, P.; Qin, W.; Mai, H.; Sun, Y.; Lu, C.; Pan, X.; Wu, C. Cold to Hot: Binary Cooperative Microneedle Array-Amplified Photoimmunotherapy for Eliciting Antitumor Immunity and the Abscopal Effect. *ACS Appl. Mater. Interfaces* **2020**, *12*, 32259–32269. [CrossRef]
87. Huang, S.; Liu, H.; Huang, S.; Fu, T.; Xue, W.; Guo, R. Dextran methacrylate hydrogel microneedles loaded with doxorubicin and trametinib for continuous transdermal administration of melanoma. *Carbohydr. Polym.* **2020**, *246*, 116650. [CrossRef]

88. Chen, S.; Matsumoto, H.; Moro-Oka, Y.; Tanaka, M.; Miyahara, Y.; Suganami, T.; Matsumoto, A. Smart Microneedle Fabricated with Silk Fibroin Combined Semi-interpenetrating Network Hydrogel for Glucose-Responsive Insulin Delivery. *ACS Biomater. Sci. Eng.* **2019**, *5*, 5781–5789. [CrossRef]

89. Wang, J.; Ye, Y.; Yu, J.; Kahkoska, A.R.; Zhang, X.; Wang, C.; Sun, W.; Corder, R.D.; Chen, Z.; Khan, S.A.; et al. Core-Shell Microneedle Gel for Self-Regulated Insulin Delivery. *ACS Nano* **2018**, *12*, 2466–2473. [CrossRef]

90. Cao, J.; Su, J.; An, M.; Yang, Y.; Zhang, Y.; Zuo, J.; Zhang, N.; Zhao, Y. Novel DEK-Targeting Aptamer Delivered by a Hydrogel Microneedle Attenuates Collagen-Induced Arthritis. *Mol. Pharm.* **2021**, *18*, 305–316. [CrossRef]

MDPI AG
Grosspeteranlage 5
4052 Basel
Switzerland
Tel.: +41 61 683 77 34
www.mdpi.com

Biomedicines Editorial Office
E-mail: biomedicines@mdpi.com
www.mdpi.com/journal/biomedicines

www.ingramcontent.com/pod-product-compliance
Lightning Source LLC
LaVergne TN
LVHW071935160726
843515LV00011B/2615